Contaminated Land

Reclamation and Treatment

NATO • Challenges of Modern Society

A series of edited volumes comprising multifaceted studies of contemporary problems facing our society, assembled in cooperation with NATO Committee on the Challenges of Modern Society.

Contaminated Land

Reclamation and Treatment

Edited by

Michael A. Smith

Building Research Establishment
Garston, United Kingdom

SPRINGER SCIENCE+BUSINESS MEDIA, LLC

Library of Congress Cataloging in Publication Data

Main entry under title:

Contaminated land.

(NATO challenges of modern society; v. 8)
"Published in cooperation with NATO Committee on the Challenges of Modern Society.
Includes bibliographies and index.
1. Soil pollution—Congresses. 2. Reclamation of land—Congresses. 3. Waste disposal sites—Congresses. 4. Factory and trade waste—Congresses. 5. Water, Underground—Purification—Congresses. I. Smith, Michael A., 1943-   . II. North Atlantic Treaty Organization. Committee on the Challenges of Modern Society. III. Series.
TD878.C66   1985   333.73   85-3562

ISBN 978-1-4757-0757-1   ISBN 978-1-4757-0755-7 (eBook)
DOI 10.1007/978-1-4757-0755-7

A report on the NATO CCMS Pilot Study on Contaminated Land
initiated with the United Kingdom in November 1981

© 1985 Springer Science+Business Media New York
Originally published by Plenum Press New York in 1985

The Council of the North Atlantic Treaty Organisation (NATO) established the 'Committee on the Challenges of Modern Society' (CCMS) in 1969. The CCMS was charged with developing meaningful environmental and social programmes that complement other international programmes, and with showing leadership, first, in solution of existing problems and, second, in development of long-range goals for environmental protection in the NATO sphere of influence and in other countries as well.

The Pilot Study on Contaminated Land was initiated by the CCMS in November 1981 with the United Kingdom (Department of the Environment) as the pilot nation. Contaminated land was defined for the purposes of the Pilot Study as 'land that contains substances that when present in sufficient quantities or concentrations, are likely to cause harm directly or indirectly to man, to the environment, or on occasions, to other targets (see main text for fuller description). The inaugural meeting of the study group was attended by representatives of six countries: Canada, Denmark, Federal Republic of Germany, Netherlands, United Kingdom and the United States of America. France subsequently joined the Study Group.

The inaugural meeting decided that the Pilot Study should concentrate on the problems of long-term remedial action for contaminated sites and the study developed into a series of Projects, on aspects of this subject and on a number of related topics each led by a Project Leader assigned by one of the participant countries. Thus the Study, and this is reflected in this Report, did not attempt to cover all aspects of the problem of contaminated land, but rather to highlight a number of the important ones. The Study Group presented its conclusions and recommendations in a Summary Report to the Committee. This is reproduced as Appendix J.

The Pilot Study Director wishes to acknowledge the contribution made by all those participating in the work of the Pilot Study Group but particularly those who acted as Project Leaders. The names and addresses of the Study Group, including Project Leaders, are given in Appendix A. The programme of the Study Group is given Appendix C.

The various Chapters were prepared by the respective authors after review of draft reports by the other participants at meetings of the Study Group and by correspondence. The Study Group met together as a 'group of experts' and views expressed in the report do not necessarily represent the official views of all or any of the participant countries.

Good use was made of the CCMS Fellowship Scheme to further the Study and a number of Fellows contributed directly to the preparation of this report. The names and addresses of CCMS Fellows participating in the Pilot Study are given in Appendix B and the Projects they undertook are described briefly in Appendix E.

The Editor wishes to express his warmest thanks to those who reviewed and co-ordinated the development of contributions from their respective countries.

Contaminated sites can present complex and challenging problems. Solutions to such problems require a multi-disciplinary approach during investigation, assessment, and any remedial action. The problems are not only technical in nature, attention must also be given to the social and financial aspects.

The term 'contaminated land' covers a spectrum of different sites (from small sites with easily identified surface deposits of solid waste to large areas containing deposits of unidentified chemical wastes which are already causing serious damage to the environment). The problems are compounded if a site has already been built upon for housing.

Once it has been recognised that a site may be contaminated, proving the existence of contamination is not usually difficult. Deciding whether the contamination matters is usually more difficult since our knowledge and experience is deficient. The potential for immediate injury to workers is easy to recognise and tackle but the assessment of risks to the long-term health of someone living on the site is more difficult. In the present state of knowledge, a considerable amount of judgement has to be exercised in deciding what response is required on a particular site. This can lead to controversy particularly where justifiable social and political pressures lead to demands for 'ultimate' or 'permanent' solutions, which may not be available at all or only at inordinate cost.

Time will always be a difficult factor to handle. On the one hand, potential hazards to health may take decades to manifest themselves and environmental consequences take time to become apparent; on the other hand remedial actions have to last 'permanently' ie at least as long as the planned use of the land. The same kind of problems arise in other areas. Thus most NATO countries have recently adopted policies to limit the environmental consequences of the disposal of domestic refuse and hazardous wastes to land. Usually the legislation implies that sites will remain secure for a few decades or degradative processes will have removed the problems during a similar time period. The present approach to radioactive waste shows it is possible to operate conceptually with very much longer time scales. Proposals for safe disposal are based on a very careful construction of geologically safe storage facilities. New landfill waste disposal sites are to an increasing extent also making greater use of favourable environments and being constructed to a higher standard. But, contaminated sites occur where they occur and local circumstances will constrain what can be done about the time factors.

Some treatments of contaminated land can result in an ultimate solution to the problems; the contamination is removed or rendered harmless. Other treatments leave the contaminants in the ground. These 'stabilisation' or 'containment' options rely on engineering solutions and on engineers for their execution. They should be treated like any other engineering operation. Thus critical points in the design should be identified and proper design and safety factors calculated from an assessment of the risks.

Monitoring is a normal activity in significant civil and structural engineering such as bridges, dams and roads. Indeed people would feel less safe if this were not so. A similar approach is often appropriate for contaminated sites. Monitoring may vary from a simple walk-over and visual inspection to the installation of permanent instrumentation. It should be presented as a way of managing the residual risks arising from any uncertainties in current knowledge or practices. Closely related to monitoring is the possible need for routine maintenance. For obvious reasons when considering treatment options, there is a tendency to choose those that do not require a commitment to regular routine maintenance. However, for example, a covering system that includes vegetation is dynamic and the loss of vegetation and consequent erosion of the soil cover may lead to an expensive failure of a system as a whole. Whilst maintenance may be a necessary part of a remedial treatment, it is important that the system is not so dependent on regular maintenance that its omission could result in rapid overall failure.

Research on remedial actions presents difficulties because of the lack of long-term experience. It is also difficult to simulate field conditions in the laboratory and to devise realistic testing regimes. In this respect too there are parallels with civil and structural engineering. With research into the long-term behaviour of construction materials it is accepted that even an accelerated laboratory procedure may take several years and the results must then be correlated with observations in the field. A similar philosophy and programme of long-term research is needed in order to develop more technically sound and cost effective remedial measures.

Observations on contaminated land reclaimed many years ago can provide useful evidence although such investigations suffer from a lack of baseline data. Much more can be learnt from the proper evaluation of current and future remedial actions where adequate baseline data is more likely to exist and where instruments can be installed as necessary at the outset. The distinction between evaluation and monitoring is important: evaluation is closely related to research and may require elaborate instrumentation and frequent inspections: monitoring is, as pointed out above, best considered as a component part of remedial action. Again there is an analogy with civil engineering practice, many advances in geotechnical science have come from the study of real life structures such as dams instrumented at the time of construction.

Since knowledge is limited and laboratory studies difficult, advances in the more novel areas of treatment are most likely if some projects are considered in part or in whole as experimental or demonstration projects requiring proper evaluation. This implies acceptance by the authorities concerned that the project may at least partially fail and that further remedial action may be required. The consequences of failure must be considered and the necessary response worked out. Those with responsibility for particular projects, will be reluctant to try experimental measures, but the rewards of lower costs and improved systems in the future require that such reluctance be overcome.

M A Smith
Pilot Study Director
Building Research Establishment
November, 1984

CONTENTS

CHAPTER 1: BACKGROUND

M A Smith

CHAPTER 2: THE LONG-TERM EFFECTIVENESS OF REMEDIAL MEASURES

K Stief

CHAPTER 3:   ON-SITE PROCESSING OF CONTAMINATED SOIL

W H   Rulkens, J W   Asink, and W J Th van Gemert

CHAPTER 4:   IN-SITU TREATMENT

D E Sanning

CHAPTER 5: COVERING SYSTEMS

               G D R Parry and R M Bell

CHAPTER 6:   MANAGEMENT AND TREATMENT OF GROUNDWATER:
             AN INTRODUCTION

             K A   Childs

CHAPTER 7:   IN-GROUND BARRIERS AND HYDRAULIC MEASURES

             K A   Childs

CHAPTER 8:   TREATMENT OF CONTAMINATED GROUNDWATER

             K A Childs

CHAPTER 9:   MATHEMATICAL MODELLING OF POLLUTANT TRANSPORT
             BY GROUNDWATER AT CONTAMINATED SITES

             K A   Childs

CHAPTER 10:   TOXIC AND FLAMMABLE GASES

             S C   James, R N   Kinman and D L   Nutini

CHAPTER 11:   RAPID ON-SITE METHODS OF CHEMICAL ANALYSIS

R E Montgomery, D P Remeta and M Gruenfeld

CHAPTER 12:   FORMER IRON AND STEELMAKING PLANTS

              D L Barry

CONTENTS

CHAPTER 13:   OVERVIEW, CONCLUSIONS AND RECOMMENDATIONS

M A   Smith

BACKGROUND

M A SMITH

Building Research Establishment

United Kingdom

1   INTRODUCTION

'Contaminated land' was defined for the purposes of the CCMS
Pilot Study as 'land that contains substances that, when present in
sufficient quantities or concentrations, are likely to cause harm,
directly or indirectly, to man, to the environment, or on occasions
to other targets'. The emphasis is on the presence of potentially
harmful contaminants rather than on past use. This definition
embraces both old industrial sites that have become contaminated
owing to their former usage, and hazardous waste 'problem' sites
(OECD) or 'uncontrolled' hazardous waste sites (US Environmental
Protection Agency). It also implies that a 'problem' is only defined
to exist following a proper site investigation and after the
evaluation of all available information specific to the site and
taking into account the intended use.

It is generally agreed that there are three main stages to
dealing with contaminated sites:

  (i)   identification
 (ii)   assessment of hazard
(iii)   remedial action

All participants in the Pilot Study identified remedial measures
as the subject about which, despite active research programmes in
progress in a number of countries, much remained to be learnt,
especially about long-term effectiveness. It was considered that
benefits could be achieved by learning from each other's experiences.
Site identification and assessment, whilst of general interest, were
not considered appropriate subjects for the Pilot Study.

Remedial actions can be divided into two broad types:

(i)   elimination or control of an immediate hazard to people or the environment;

(ii) provision of a permanent, or at least long-term, solution to the problem.

The participants in the Pilot Study decided that they could most usefully direct their attentions towards the latter. Such 'permanent' or 'ultimate' solutions have to fulfil very demanding requirements related to the nature of the contamination, the geology and hydrogeology of the site and the use to which the land is, or will be, put. In order to understand these requirements it is first necessary to understand something of the origin of contamination and of the hazards that it can present.

2   SOURCES OF CONTAMINATION

Contamination can be broadly defined as the existence in the ground of hazardous substances at concentrations above 'normal' background levels. Substances giving rise to concern include:

Certain elements and their compounds (eg Pb, As, Cd, Hg etc)
Organic chemicals
Oils and tars
Toxic, explosive and asphixiant gases
Combustible materials
Radioactive materials
Biologically active materials
Asbestos and other hazardous minerals

The 'concentration' actually present can range from a few mg/kg of a trace element to thousands of drums of unidentified organic chemicals stacked on the land surface.

Contamination of the ground itself commonly arises from disposal of wastes, in one form or another, to properly designated landfills, to uncontrolled dumps and tips or to areas within the boundaries of industrial sites. It may also arise from accidental spillages and leakages during plant operation and from the transportation and stockpiling of raw materials, wastes, and finished products. Widespread contamination can arise from the deposition of air and water borne emissions. Land can also become contaminated from overdosage with metal-contaminated sewage sludge and other similar organic wastes.

Contaminated sites, particularly those of industrial origin, frequently present other problems. They are often 'filled' or 'made' ground and consequently badly compacted, they often contain massive foundations and underground pipework, tanks and other structures, and there may be abandoned and derelict, unsafe and contaminated buildings still standing.

Any former industrial site may be contaminated but some industries have a high probability of producing contaminated sites. These include:

    Mining and extractive industries
    Smelting and refining, steel works, etc
    Scrap yards
    Gas works
    Waste disposal
    Wood preservation
    Tanning and associated trades
    Asbestos
    Pesticide manufacture or use
    Railway yards
    Chemical and allied products
    Explosives and munitions
    Metal treatment and finishing
    Paints
    Sewage works and farms
    Oil storage
    Oil production
    Dockland areas
    Acid/alkali plants
    Pharmaceutical, perfumes/cosmetics/toiletries

Although generally, and for the purposes of the Pilot Study, contaminated land should be viewed in terms of problems arising from the presence of particular substances rather than from previous land use, there is nevertheless a link between past land use and the nature of contamination that is likely to be found. Thus in the United Kingdom some guidance has been provided about the problems likely to be encountered on former coal-carbonisation sites[1,2] (gas works and coke works), sewage farms[3], scrap-yards[4] and sites used for the disposal of domestic refuse[5].

3   THE NATURE OF THE PROBLEM

Whether the contamination found on a particular site will be a 'problem' will depend upon many site specific factors and how they influence the degree of risk to a set of 'targets' arising from the 'hazards' presented by the contaminants that are present. A number of schemes to rate sites in order of priority for action have been drawn up in the participant countries based upon more or less elaborate risk/hazard/target assessments.

Targets potentially at risk from contaminated sites include:

(i)   Workers engaged in investigation, remedial or construction activities.
(ii)   Eventual residents (housing) or users (of schools, factories, recreational areas, etc) including particularly sensitive groups in the population such as small children.
(iii) A wider population owing to pollution of aquifers and water courses, and wind blown pollution.
(iv)   Animals, plants, aquatic life, etc.
(v)   Building structures, services and materials, eg by corrosion, fire or explosion.
(vi)   Investment if site closure or evacuation is required, or there is major damage to buildings.

Hazards/exposure combinations include the following:

(A) Physical hazards

    (a) explosion and fire
    (b) subsidence
    (c) corrosion of structures
    (d) effects on mechanical properties of soil

(B) Toxic hazards
  (i)   by inhalation

    (a) dust
    (b) toxic gases
    (c) asphyxiant gases

  (ii)   by ingestion

    (a) from fingers, etc (particularly children)
    (b) contamination of food

  (iii) by direct ingestion

    (a) uptake of toxicants by edible plants
    (b) contamination of water

  (iv)   by contact

    (a) skin irritation, etc
    (b) absorption through skin
    (c) retardation or death of plants, etc

The hazards to man may be further categorised as short term, such as those posed to workers by the presence of highly toxic or corrosive substances or flammable gases, or may be long term, such as might arise from the presence in garden soils of carcinogens, radioactive materials, or of elements, such as cadmium, that are taken up by food plants and may accumulate in the body to an undesirable extent over very many years.

The experience of dealing with contaminated land has differed in each of the participant countries leading to a difference of emphasis on particular risks that might arise or are actually occurring.

The emphasis in the United Kingdom is on the restoration of contaminated land so that it can be re-used for some beneficial use such as housing, public open space, agriculture or industry. This need arises from the two-fold pressures of (i) large areas of former industrial land in already built-up areas which become vacant owing to the decline of traditional industries, such as steelmaking and docks, and which need to be regenerated to ensure the social and economic health of the cities affected and (ii) the need to protect scarce resources of agricultural land from unnecessary development for industry or housing. Permanent loss of the land would rarely be an acceptable solution to a problem site in the UK.

The other extreme can be represented by the concern in the United States for remedial action on very many 'uncontrolled' hazardous waste sites that are already causing, or are seen as likely to cause, overt environmental damage. There is a strong emphasis on protection and restoration of groundwater quality. Land tends not to be a scarce resource and the emphasis thus tends to be on 'containment' or 'isolation' of problems, and possible after use is rarely a primary factor to be taken into account at the present time.

Between these extremes there have been differing experiences. Thus in the Netherlands an active search has been made for hazardous waste 'problem' sites whilst at the same time research is in progress on rehabilitating former industrial land for housing. Germany also has had to deal with contamination arising from a variety of sources other than waste disposal sites, including gas works sites and areas around smelters which have become contaminated with toxic metals. In Canada, whilst the situation is not dissimilar to that in the United States, since land is not a particularly scarce resource, there have nevertheless been 'problems' with former industrial sites, eg coking plants. In general, large deposits such as mine tailings are sufficiently remote from major habitation areas, and industry is relatively young and segregated from such areas, so that problems of conversion of land from industrial to residential use have not arisen.

The existence of problems arising from similar industrial legacies of the participant countries was given specific recognition by one of the CCMS Fellows contributing to the Project in choosing to examine the problems posed for redevelopment by former steelworks sites (see Chapter 12). The steel industries in a number of countries have undergone major reductions in output with consequential closure of plants leaving very large areas of land vacant and derelict. The coking plant areas present similar problems to coal carbonisation plants; other parts of sites are frequently buried in great depths of slags and other wastes. An understanding of the contamination of these sites is essential to their re-use and to the rehabilitation of the associated communities.

4   OPTIONS FOR REMEDIAL ACTION

Actions taken in response to the identification of a contamination problem may in the first instance be designed to provide a
temporary amelioration of an immediate hazard or they may be designed
to control pollution of the environment on a longer-term basis eg to
limit the amount of pollutant reaching a water course rather than to
totally prevent the pollutant escaping from the site. However, in
all but a minority of cases, the aim must be to find some permanent
or 'ultimate' solution that will either eliminate the identified risk
or at least reduce it to an 'acceptable' level.

The most direct and obvious solution in many cases appears to be
to remove the offending materials, deposits and contaminated ground,
from the site for disposal or treatment elsewhere. Unfortunately
this is not always practicable, nor is it always desirable, nor is it
necessarily a permanent solution in the broadest sense.

Excavation and transport of contaminated material may be
hazardous, present other environmental hazards (eg nuisance of
traffic movements) and is costly. It may not be possible to find a
suitable disposal site within an economic distance either because of
its toxicity or its bulk. Further, redeposition may simply move the
problem elsewhere to be rediscovered by later generations. Transportation of material off site for treatment by incineration or
chemical means can present similar problems. There are also likely
to be community objections to being the recipients of somebody else's
waste.

Thus there are strong environmental, economic and social
reasons for dealing with contamination where it is found. This is
the main theme of the CCMS Pilot Study on Contaminated Land although
some attention was paid to problems of a somewhat different
character.

Most of the options available for treatment of contaminated land
consist of, or rely on, actions of an engineering type. In
considering options it is therefore important to learn the essential
lessons that can be drawn from a consideration of the generality of
civil and structural engineering works. These are:

(i)   A recognition that the properties of materials and components
       used in a structure may deteriorate with time to such an
       extent that the structure has a foreseeable and limited life.
(ii)   Consequent upon this, recognition of a need to monitor
       performance and to carry out maintenance to ensure continued
       performance to an acceptable standard.
(iii) A need to design for natural and man-made interventions
       of a catastrophic type, eg 1 in 100 year flood level.

The study was restricted to a consideration of 'chemical' contaminants present in the ground as solids, liquids or gases and did not concern itself with biological, radioactive or combustible contaminants.

5   STRUCTURE OF THE STUDY

The study evolved into a series of seven projects each led by a separate country (see Table 1.1) with a supporting exchange of information on case studies, and research in progress to aid in their development. A number of other topics were identified as of interest (see Appendix D) but were not pursued owing to lack of resources.

As already stated above, the Study was very much concerned with dealing with contamination where it is found. Three main methods of treatment were identified and the results are presented in Chapters 3 to 7:

  (i)   On-site treatment – defined as being methods of decontami-
          nating or otherwise reducing the environmental impact of the
          bulk of contaminated materials on a site by: excavation;
          treatment to detoxify, neutralise, stabilise or fix; and
          usually redeposition on site.
  (ii)   In-situ treatment – defined as the treatment without
          excavation of the bulk material on a contaminated site.
  (iii) Macro-encapsulation – in which a site is contained within
          a 'box' usually consisting of a cover, a peripheral barrier
          and sometimes a base (obviously very difficult to install).

The liquid phase within and on contaminated sites is of great importance. It is the usual means by which the contamination moves out into the wider environment and the treatment option should not be adopted without first considering how to deal with the liquid phase. This may be in the form of a neat undiluted contaminant, a strongly contaminated aqueous phase permeating or leaching from the soil or, at the other extreme, a very lightly contaminated groundwater remote from the site. For example:

  (i)   excavation may be required below groundwater level –
          groundwater must then be pumped, treated and disposed of;
  (ii)   in-situ treatment by leaching with water or other solvent will
          produce a contaminated waste stream.

Methods for the treatment of heavily contaminated leachates, surface waters, etc, are well established and continuously being improved upon. Such methods were therefore not included in the Pilot Study. However, the study looked at the liquid phase in three respects:

Table 1.1 List of Projects Together With Brief Descriptions

A   In-situ treatment of contaminated sites

Methods of treating without excavation the bulk material on a contaminated
site by detoxifying, neutralising, degrading, immobilising or otherwise
rendering harmless contaminants where they are found.

B   On-site processing of contaminated soils

Methods of decontaminating or otherwise reducing the potential
environmental impact of the bulk of contaminated material on a site by:
excavation; treatment to detoxify, neutralise, stabilise or fixate; and
usually redeposition on-site.

C   Covering and barrier systems

Systems designed to prevent the migration of contaminants vertically or
laterally, or to prevent ingress of surface or groundwater into
contaminated sites.

D   Control and treatment of groundwater

Primarily concerned with those operations designed to control or treat the
liquid phase on contaminated sites including design of cut-off systems
mathematical modelling, and groundwater treatment.

F   Rapid on-site methods of chemical analysis

Methods of chemical analysis allowing determinations to be made on 'soil',
water and air samples on-site so as to speed and reduce the costs of site
investigation.

G   Long-term effectivesness of remedial measures

Overall problem of the design of long-term effective remedial measures.
Collection of information on examples of remedial and restoration actions
that can be demonstrated to have worked for a number of years and methods
for the evaluation of sites for long-term effectiveness of remedial
measures.

H   Toxic and flammable gases

Concerned with; volatile organic emissions from contaminated sites,
production, migration and control of gases from land disposal sites
including typical landfill gases such as methane and carbon dioxide.

M   Contamination and specific industries*

Collection and dissemination of information on the contamination characteristics of specific industries or processes. Experience has shown that
links can be made between previous uses and the nature of contamination.

---

* Contribution from CCMS Fellow - not covered by Pilot Study
  Group.

(i)   The use of hydraulic measures as an adjunct to other treatment options.

(ii)   The treatment of contaminated groundwater involving extraction, treatment and re-injection or by in-situ means. Such treatment will often be required in association with other remedial measures particularly those involving containment of a site.

(iii) The modelling of groundwater movement. An understanding of the probable consequences of containment and other hydrological control measures is essential to their design.

These three aspects are reviewed in Chapters 7, 8 and 9 respectively.

Chapters 3 to 8 review the options available for their respective forms of treatment. A different approach was adopted for the project on toxic and flammable gases (see Chapter 10). It was thought necessary to review the problems presented by such gases and the modes of generation and dispersal before considering the methods available for dealing with them.

Underlying the Chapters (3-10) on the practical problems of reclamation of contaminated sites is the question of what is meant by 'permanent' solutions and indeed whether they are possible. Most of the on-site treatment options result in destruction or at least removal of the contaminants from the site in question (although there may be a residue for disposal elsewhere) and would seem therefore to be truly permanent. However many of the other treatment options rely on technology that is as yet unproved in this application, although they may be well established engineering practices for other purposes. The question of what can be achieved, and what the aim should be, in looking for long-term effectiveness is explored in depth in Chapter 2. One of the themes emerging from this and the subsequent Chapters is the need for evaluation of remedial measures as carried out, and how such an evaluation should be carried out is also discussed in Chapter 2.

It is important to recognise that the term 'effectiveness' can be used to mean different things during the stages that any remedial measure will go through. It can be applied to the performance of a component part of a remedial system (eg cut-off wall) or to the system as a whole (eg cut-off wall plus groundwater pumping). A distinction can be made between theoretical effectiveness and installed effectiveness. Long-term effectiveness can be looked at as effectiveness on an arbitrary scale at a point in time or as the ability of the system to continue to perform to an acceptable standard over a prolonged period of time ('performance' may be a better term for this latter concept). These concepts are explored in more depth in Chapter 2.

One project differed substantially in subject matter from the other, and that was concerned with rapid methods of analysis for use in the investigation of contaminated land. A proper evaluation of any contaminated site requires a thorough investigation to determine what contaminants are present, how much of them there is, and where they are in order that the risks to human health and safety and to the environment or other targets can be properly assessed, and in order that appropriate remedial actions can be identified. Taking samples of deposited materials for subsequent analysis in a laboratory tends to be both time consuming and expensive. A need is seen for rapid methods of on-site analysis that, even if they do not meet the rigorous evidential standards required in some countries, would nevertheless enable a preliminary identification and quantification of contamination to be made on-site. This problem is addressed in Chapter 11.

6   CONTRIBUTIONS BY CCMS FELLOWS

The successful completion of the Pilot Study was greatly assisted by the opportunity presented by the CCMS Fellowship scheme to increase the number of active participants in the study. Eleven Fellowships were awarded for work in connection with the Study; six in 1982 and five in 1983. The projects undertaken by Fellows are listed in Appendix E. In some cases the Fellows contributed directly to the preparation of the reports on the major projects (see Chapters 3 and 5). In other cases they assisted their national representatives to collate and present information available in their own countries relevant to the various projects. Some on the other hand, or in addition, produced contributions of their own for inclusion in this final report of the Pilot Study (mention has already been made of the work on steel industry sites which is described in Chapter 12) or for publication independently. In addition each of the CCMS Fellows attended one or more of the study group meetings and joined in the discussions of draft reports and of the overall problem.

7   OUTCOME OF THE PILOT STUDY

The major practical output from the Pilot Study is of course this report and the information it contains, particularly on treatment options and on the approach to the identification and selection of such options that should be adopted. Each project leader was asked to include in their reports 'conclusions and recommendations' and these are brought together in Chapter 13 together with some overall conclusions and recommendations for action agreed by the study group as a whole.

8   NATIONAL EXPERIENCES

As part of the general exchange of information on the problems
of contaminated land, various participant countries made written and
verbal presentations on the situation in their respective countries
and there were usually a number of such presentations and individual
case studies at the study group meetings. Much of the written
information provided was reproduced in the various progress reports
of the study.

9   REFERENCES

1   D C Wilson and C Stevens, 'Problems arising from the Redevelop-
      ment of Gasworks and Similar Sites', UKAEA Harwell Laboratory
      Report AERE-R-10366, HMSO, London (1982), available from Central
      Directorate on Environmental Pollution, Department of the
      Environment, London.
2   Interdepartmental Committee on the Redevelopment of Contaminated
      Land (ICRCL). 'Notes on the redevelopment of gas works sites',
      (4th edition), Department of the Environment, Central
      Directorate on Environmental Pollution, London (1983), ICRCL
      19/79.
3   Interdepartmental Committee on the Redevelopment of Contaminated
      Land (ICRCL), (2nd edition), Notes on Sewage Works and Farms',
      Department of the Environment, Central Directorate on
      Environmental Pollution, London (1983), ICRCL 23/79.
4   Interdepartmental Committee on the Redevelopment of Contaminated
      Land (ICRCL), 'Notes on Scrapyards and Similar Sites', 2nd
      edition, DOE, CDEP, London (1980), ICRCL 42/80.
5a   Interdepartmental Committee on the Redevelopment of Contaminated
      Land (ICRCL), 'Notes on the redevelopment of landfill sites,'
      4th edition, Department of the Environment, Central Directorate
      on Environmental Pollution, London (1983), ICRCL 17/78.
5b   M A Smith, Redevelopment of contaminated land: landfill sites,
      in: 'Proc Symposium Engineering Behaviour of Industrial and
      Urban Fills, Birmingham 1979', Midlands Geotechnical Society,
      Birmingham, (1979), pp B49-B70.

LONG-TERM EFFECTIVENESS OF REMEDIAL MEASURES

K Stief

Umweltbundesamt, Federal Republic of Germany

1   INTRODUCTION

Where remedial measures need to be taken to protect people from harm and to protect the environment, it is reasonable to seek to devise remedial actions that give a high degree of protection (have a high degree of 'effectiveness') at the lowest possible cost. Except for a few isolated cases, not only must there be a high degree of effectiveness at the time the remedial measures are carried out but they must remain effective in the long-term or be reparable or renewable in planned time spans. This is obviously a difficult task, particularly, if there is not a clear understanding of what 'effectiveness' means or for what length of time the measures must be designed to be effective.

In the majority of reclamation or remedial schemes, the measures to be employed have not been demonstrated to be effective in practice; instead they must be based on professional judgement of what is likely to be effective. It is often necessary to predict behaviour over a long time period (which might range from 20 years to an indeterminate time span) on the basis of very little practical experience.

Discussion of 'long-term effectiveness' or performance must necessarily be in both philosophical and technical terms. The philosophical aspect includes the questions: what does 'effective' mean and what does 'long-term' mean? The technical aspect, in addition to the selection of appropriate remedial measures having the required effectiveness and anticipated long-term performance, also includes the design of the monitoring system needed to obtain data to enable an evaluation of effectiveness both in the short term and in the long term.

2   SCOPE

This chapter is concerned with the general philosophical and
technical aspects; with the questions that must be asked about any
remedial measure in order to assess its likely immediate and long-
term effectiveness, and the general approach required for an
evaluation of long-term performance or effectiveness.

The CCMS Pilot Study on Contaminated Land has included a
detailed review of several different types of remedial action that
are seen as alternatives to total excavation of contaminants:

-    on-site processing of contaminated soil
-    in-situ treatment of contaminants
-    macroencapsulation or containment of contaminated areas by
     using:
        covering systems
        vertical barrier systems
        underground (horizontal) lining systems
-    hydraulic measures, including groundwater treatment either
     in situ or after extraction.

These remedial measures, dealt with in detail in other chapters,
will in general terms:

-    render contaminants harmless,
-    prevent release of contaminants, or
-    decrease release of contaminants.

On-site processing is in a special category. Whilst the aim in
relation to the treated soil will usually be to render the
contaminants harmless, consideration must also be given to how to
deal with any contamination that has moved outside of the area to be
excavated. Thus, some additional steps, having different objectives,
may also need to be taken.

For an evaluation of effectiveness (both short term and long
term) it is important to know:

-    to what extent will the measure be effective at the time of
     installation?
-    how long after installation will the full effect be realised?

-    will effectiveness decrease with time?

This last question is the fundamental one in relation to long-
term effectiveness and it has no general answer. It can only be
answered in relation to a specific remedial action designed to take
account of conditions present at the contaminated site and the design
objectives that have been set. The most important consideration in

setting the design objectives is the use to which the land is to be put and whether this use can be guaranteed or controlled in the future.

If the remedial action does not involve the removal or rendering harmless of the contaminants, the site may need to be monitored over possibly long periods of time. It should be noted that for the purposes of the Pilot Study a distinction has been made between monitoring performance and an evaluation of performance. The former may say a measure has failed: the latter aims to say why.

3   LONG-TERM EFFECTIVENESS FROM A PHILOSOPHICAL
      POINT OF VIEW

Most of man's activities in structural or civil engineering are not permanent. Bridges and dams for example have a design life, and a real life that may be extended by maintenance or improvement works. Continued monitoring of their condition is frequently considered essential and is often required by legislation.

Similarly all remedial measures that do not result in contaminants being removed or destroyed are likely to need some attention to ensure continued effectiveness so as to prevent renewed hazard to people or the environment. Once this possible lack of permanency is accepted, then the main philosophical problem regarding the design of remedial measures has been overcome.

The next step is normal engineering practice. A judgement must be made as to how long the remedial measure in question will remain (adequately) effective. In order to do this reliable information is required based on practical experience or derived from research. Available data must be carefully checked for reliability and relevance to the case in hand. Except in rare cases the aim will be to design remedial measures that will remain effective for a very long time period; 100 years would not be an unreasonable design life. However, so limited is our knowledge that this may be frequently unrealistic. It may be more realistic to design with confidence for a comparatively short life (eg 10 years) than to make unrealistic claims for 100 years. This raises two important issues:

  (i)   the need to monitor the treatment to match the confidence
        in the design;
  (ii) the possibility of remedying any future weakness or
        failure of the treatment.

When discussing remedial action on contaminated land it is necessary to keep in mind the following:

(a) will the contaminant be rendered harmless with time?

(b) is it possible to control the use of the site, both the land
     surface and the ground beneath the site (eg groundwater
     extraction), once remedial action has been completed? and
(c) what environmental impacts may arise elsewhere from movement of
     contaminants from the site, eg contaminated groundwater or
     contaminated soil taken to a treatment plant or disposal site?

     Remedial measures for contaminated land have to be affordable at
a community, regional, state or national level. The priority
attached to treatment of a particular site and to particular hazards
it may present, will differ in the various levels of society. What
may be of highest priority from a community's point of view, may be
of low priority from the national government's point of view. When
public money has to be spent it is important to achieve a long-term
solution and not simply to apply a cosmetic treatment with little
long-term effect. Also, financial constraints may lead to
unrealistic claims being made about long-term effectiveness. It is
better to be honest about what is practical in technical and monetary
terms. It is consequently stressed again that remedial measures may
require long-term maintenance and monitoring and good record keeping,
whenever contamination is not fully removed or destroyed in-situ.

     It is useful for the purposes of discussion to categorise
remedial actions into:

 (i) those which decrease the rate of contaminant release and
(ii) those which prevent contaminant release

Contaminants can be:

(a) non-degradable, or
(b) degradable.

     The following combinations are possible: i-a, i-b, ii-a, ii-b.
To realise remedial actions that totally prevent release of non-
degradable contaminants (ii-a) is the hardest task and will
subsequently need the most effort to monitor, maintain, and possibly
renew the measures. Depending upon how the design objectives are set
it could be concluded that it would be impossible to achieve such a
treatment successfully and that in practical terms only remedial
actions to decrease contaminant release (albeit initially to very low
levels) were possible.

     The life-time that is required of a remedial measure, and the
associated monitoring and maintenance, will depend on the relationship
of the amount of contaminants available to the rate of release. This
life-time may be calculated. Similarly if the contaminants present
are rendered harmless by degradation in time or by in-situ treatment.
The life-time required of any renewal measures may be shorter than
that required of the initial measures and they may be simpler.

The aim should be to develop highly durable methods and
materials of construction, eg for a cut-off wall or bottom lining,
but always bearing in mind the danger of poor installation standards,
or damage during installation or subsequently. The frequent
difficulty of persuading people that it is possible to predict
performance from laboratory simulations should be noted.

To assess the initial and the long-term effectiveness of
remedial actions, specific parameters have to be monitored. These
can be:

(a)   to determine the behaviour (properties) of the system of
      remedial measures
(b)   to determine the environmental conditions in the area influenced
      by the contaminated site (land)
(c)   to do both.

In every case it is necessary to ask:

'What parameters should be measured, how often and by what
means?'

The answer will depend on whether the system to be investigated
is a dynamic one or a static one (system here means contaminated land
and the associated remedial measures). The conservative approach is
to assume that the remedial measure is itself a dynamic part of the
system, with worsening properties, and that the contamination is
static, ie of constant potential environmental hazard. The latter
assumption is probably true for inorganic or persistent contaminants
but may be too pessimistic for degradable organic pollutants.

It seems advantageous, when monitoring a dynamic system, for the
initial parameter coverage to be sufficiently broad to define the
problem. Once these initial data have been collected, the parameters
monitored should be reduced as much as possible to reduce costs.
Periodically, the coverage should be broadened again to test the
hypothesis that the limited coverage provides an adequate check on
the situation.

When sites are being considered for new development (whether as
housing or amenity or for commercial buildings) it is important that
the users or purchasers of the land are made aware of the original
problem and the confidence in the treatment applied.

4   LONG-TERM EFFECTIVENESS FROM A TECHNICAL POINT OF VIEW

4.1   General

In the majority of completed reclamation or remedial schemes, measures have been employed which, on the basis of available knowledge and professional judgement, are taken as effective, but which have not been demonstrated to be so.

It will continue to be necessary in future years to propose remedial measures and predict behaviour over a long time period, on the basis of very little practical experience. Thus information on already restored sites that subsequent monitoring has shown to have been successful to date, will be of great value.

In order to obtain better quality information on schemes that have been carried out and in order that the information should be available as soon as possible, suitable monitoring systems should be installed at the time of restoration. These need to be cost effective, durable and robust so that they can be a continuing source of data for evaluating the success of the various remedial actions.

Requirements for long-term durability of materials, for long-term effectiveness of remedial measures and for monitoring installations to measure overall long-term performance, have to be discussed in relation to the different types of remedial measures that may be used alone or in combination with others. These are:

-       on-site processing of contaminated materials
-       in-situ treatment of contamination
-       covering and capping the contaminated land
-       in-ground macroencapsulation (barrier systems)
-       hydraulic measures.

When discussing both the initial and the long-term effectiveness it is important to keep in mind that the measures chosen may be affected by anticipated and unanticipated external factors. Factors which can be anticipated and therefore designed for include: water pressure on cut-off walls, chemical reactions between cut-off wall material and identified contaminants, microbial attack and damage to liners or covers by burrowing animals. Whether of course such impacts will be planned for depends on the foresight of the people designing the remedial action. Unanticipated factors may include subsequent uncon-trolled excavation on a remedial site, or exceptional rainfall not designed for (allowance for normal, say once in 100 year storms or otherwise as deemed necessary should be part of the design brief) or chemical reactions due to unidentified contaminants.

It is very important to develop and use remedial measures that need very little maintenance, that do not need extensive control, and that are robust in the sense that their effective-ness is not unduly dependent upon the quality of installation or the avoidance of subsequent accidental damage.

     For each type of remedial measure, as listed above and discussed in subsequent Chapters, it is possible to recommend a set of questions that can be posed about the measure. These questions can be asked in general terms or in relation to a specific site but the answers to general questioning must be reviewed on a site-specific basis for each proposed practical application. Some questions and important points relating to the selection of remedial actions in general are listed in Table 2.1. Similar lists for each of the treatment options dealt with in subsequent chapters are listed in Tables 2.2 to 2.6.

4.2   On-site Processing of Contaminated Materials

     On-site processing of contaminated materials is defined in this CCMS Pilot Study to encompass: methods of decontaminating or other-wise reducing the environmental impact of the bulk of contaminated material on a site by: excavation; treatment to detoxify, neutralise, stabilise or fixate; and usually redeposition on-site.

     On-site processing of contaminated materials in general presents the fewest questions concerning long-term effectiveness in relation to the land that is to be treated (see Table 2.2). The aim of most of the processes described in Chapter 3 is either to destroy the contaminants or to separate them from the soil in which they are found. The treated soil is then returned to the ground. Clearly the objective will always be for this treatment to be 100% effective, and proper controls can be installed to ensure that this is indeed the case. There may, however, be some residual contamination left in the treated soil but this will have been defined before the process is adopted as being at a level that will not be a 'problem'. There will also in most cases be either the separated contaminants or some other wastes to be disposed of elsewhere. That these may cause an adverse environmental impact elsewhere, does not detract from the high degree of total effectiveness of the process in relation to the site. What will detract from the overall effectiveness (but not of the process as operated) is that some residual contaminantion may be left in the ground. This may either have to be designated as 'acceptable' or dealt with by supplementary measures such as those discussed in other Chapters.

     If the on-site process is fully effective at the time of treat-ment in destroying or separating the contaminants, the question of long-term effectiveness does not arise.

     However, there are on-site processes in which the contaminants are not destroyed or separated but rather stabilised (for example, a soluble form may be changed into a less soluble form or the processes for stabilisation and/or solidification of hazardous wastes may be applied and the treated material returned to the ground) and here the

Table 2.1   Crucial Points and Questions Regarding Long-term
            Effectiveness of Remedial Measures in General

Crucial points:

Very little practical experience with long-term effective measures or materials.

Only few or poor monitoring devices installed at the time of restoration.

Maintenance and monitoring have to be paid for over long periods of time.

Accidental damage, eg of in-ground barriers, may decrease effectiveness of the remedial measure.

At contaminated sites usually only a restricted area can be treated or encapsulated. Contaminants in other (surrounding) areas are left as they are.

Effectiveness of measures is critically related to material suitability and quality of application.

Questions:

Is it probable that contaminants left in soil in the untreated part of the site will migrate into the environment or affect the remedied area?

If so under what external influences:

(a)   independent of the remedial action?
(b)   caused by the remedial action?

-     will the impact of those contaminants be sudden or continuous?
-     what is the probable area of influence?
-     how can the migration or resulting effects be monitored?

Are the tasks to be fulfilled by the remedial actions to be monitored?

-     at what rate may contaminants migrate into water, soil or air from the treated area?
-     what is the total amount of contaminants that may be released from the treated site?
-     for what time span must the remedial measure be effective?
-     how can the effectiveness be measured?

Is it possible to measure the performance of the remedial measure?

-     if so by what parameters?

What will influence the performance?

-     the contaminants to be treated?
-     external impacts?

How can those impacts be measured?

How can remedial measures be maintained or repaired?

What restrictions for use of the remedied site are necessary to ensure effectiveness in the long-time?

-     how will control be ensured?
-     how long will it be necessary to exercise control?

Will excavation and subsequently disposal of contaminated materials be more difficult owing to the remedial measure, if this becomes ineffective?

---

question of long-term effectiveness will arise –in respect of the
immobilisation of the contami–nants. It is probable that some
authorities, will not be willing to approve such processes that,
whilst acceptably effective at the time of application (they would
otherwise not be used), carry some implication of reducing
effectiveness. The factors likely to influence the effectiveness of
stabilisation/solidification processes at the time of application and
later are mentioned briefly in Chapter 3. Assuming that the
stabilisation/solidification has been properly carried out, the
basic questions concern resistance to external environmental factors
such as infiltration of acid precipitation or groundwater, freeze–
thaw action and vegetation growth. If these are in any way critical
then steps should be taken to control them, such as installation of a
liner in the hole in which the material is to be re–deposited and
imposition of suitable cover. The initial and long–term effectiveness of these supplementary measures will then govern the long–term
effectiveness of the overall remedial action scheme.

The evaluation of behaviour once treated material is redeposited should centre on:

(i)   the possible entry of contaminants into the surrounding
      ground
(ii) the physical and chemical condition of the re–deposited
      material.

It should be possible to monitor assumptions made on these
points through permanent sampling wells installed when the material
is redeposited and by taking cores. Changes in monitored parameters
should give an early warning of any adverse changes occurring. It
should be clear at the outset of any studies what the response is to
be to adverse results.

The physical and chemical condition in–situ after redeposition
can only be examined by taking cores. This can only be done at
infrequent intervals and carries with it some risk of damage to the
remedial action scheme (eg damage to liner if drilling goes too deep,
allowing increased infiltration, escape of vapours). Such sampling
could be carried out in response to indications of changes given by
periodic monitoring or as part of research into the performance of
this type of remedial action. The sampling must be properly
representative and the results of analysis reproducible, because
time–dependent changes, eg of contaminant concentration or
solubility, are to be monitored. The US EPA ULP-leaching test is
suitable for leaching cores. Although this test will not directly
simulate real in–situ behaviour of the treated contaminants, it
should provide reproducible leaching behaviour under the test
conditions.

Table 2.2   Crucial Points and Questions Concerning
             On-site Treatment Methodologies

Crucial points:

When designing the remedial measure the contaminant concentration to
be left in the ground has to be decided.

The treated soil will usually contain residual contaminants.

Additives will be used in treating the contaminated soils, which may
be hazardous for water or corrosive.

After redeposition of the treated soil or material it may be polluted
by contaminated groundwater unless the groundwater table is lowered
or the groundwater is treated.

Questions:

Will effects from redeposited materials cause or enhance mobilisation
and migration of contaminants outside the treated area?

Is it likely that contaminants will be released in the long-term from
the treated (eg by detoxification, neutralisation, immobilisation
possibly using additives) material. What contaminants? What amounts?
Under what external influences? In what timespan?

How could immobilisation be measured in-situ? How could deterioration be identified?

Which external influences should be monitored to get timely indications for remobilisation of contaminants? At what values is the
remedial measure to be judged ineffective?

Will contaminants left in ground or groundwater outside the treated
area affect the treated soil in the long-term, eg by mobilisation of
immobilised contaminants?

How could further decrease of immobilisation be prevented? Which
supplemental remedial measures should be taken?

It may be possible to install lysimeters in the excavated pit
before the treated soil is redeposited, to measure in-situ leached
contaminants although there may be difficulty in selecting a
representative location.

4.3 In-situ Treatment

In-situ treatment of contaminants is defined in this CCMS Pilot
Study to encompass: methods of treating, without excavation, the
bulk material on a contaminated site by detoxifying, neutralising,
degrading, immobilising or otherwise rendering harmless contaminants
where they are found.

There are major questions (see Table 2.3) concerning the in-
situ treatment processes described in Chapter 4 regarding both the
effectiveness with which they can be applied initially and their
subsequent behaviour when the contaminants are not destroyed,
rendered permanently harmless, or separated.

The difficulties surrounding effectiveness of application
are discussed in Chapter 4. The most serious difficulty is to ensure
intimate and uniform contact between the treatment agent and the
contamination. Most of the processes described involve the injection
of fluids into the ground and there are numerous factors that govern
how easily and well this can be done.

The in-situ processes can be broadly divided into those that are
intended to provide a 'permanent' solution and those which render the
contaminants less 'available' or otherwise stabilise the situation.
However, as explained in Chapter 4 there is not necessarily a clear
line between the two types of process since the stability of
particular chemical species and a judgement of the probability of
stable environmental conditions are involved.

The treatment of contaminated material in place by fusion/
vitrification is intended to produce a permanent solution. The final
product should have the physical and chemical stability of an igneous
or metamorphic rock, usually of similar composition to a granite
(this obviously depends on the composition of the soil). Such
products are considered likely to be acceptable for the long-term
disposal of radioactive wastes, and although 'storage' conditions are
different, it is reasonable to suppose that, provided the
vitrification process is properly carried out,the treatment will
remain fully effective for even geological time spans. However
checks on the composition, stability, resistance to leaching, etc,
would prudently be required in each case of application.

When the in-situ treatment involves removal of contaminants from
the ground, effectiveness can be checked by monitoring and sampling

Table 2.3   Crucial Points and Questions Concerning
            In-situ Treatment Methdologies

_____

Crucial points:

Contaminants are distributed heterogeneously vertically and
horizontally.

A variety of contaminants is usually present.

Even very costly investigations will not result in complete knowledge
of the distribution of contaminants in soil.

To treat contamination in-situ, it often is necessary to inject
agents into the soil. To get the reactions wanted, agents will be
used in excess and this may cause water pollution.

Questions:

Will mobilisation and migration of contaminants outside the in-situ
treated area be enhanced by eg injection agents, reaction products or
microbial activities?

Are the immobilised contaminants likely to become mobilised by
environmental conditions or specific use of land?   What contaminants?
At what rate?

How should immobilisation or decrease of immobilisation be measured?
At what threshold values have the remedial measures to be considered
ineffective?

What environmental conditions should be monitored to get timely
indications of potential mobilisation of contaminants?

What can be done to stop decreasing effectiveness of the in-situ
treatment?

_____

at the time of application and treatment can be continued until the
designed level of decontamination is achieved. Provided any
supplementary remedial measures also work the question of long-term
effectiveness does not arise.

   Processes designed to destroy contaminants in the ground by
chemical or microbial action can be viewed in a similar manner. It
should be possible to check effectiveness at the time of application
to see that design objectives are met. These will include dealing
with any contaminating by-products of the treatment process and such

considerations as to whether selectively bred microbes can be safely left to die out (from starvation) or must in turn be destroyed less they become an environmental problem in their own right (a microbe that destroys an organic contaminant may also be capable of attacking an organic construction or liner material). The question of long-term effectiveness should not arise; the problem would be residual contamination and the side effects of the remedial process.

Many of the potential in-situ treatment processes do however promote questions concerning their potential long-term effectiveness, and this can be separated from the effectiveness of initial application. The general concept is that the contaminant is rendered less available to the environment by chemical reaction or change in the physical state of the ground. Most chemical reactions are reversible and most materials used to bring about physical changes are susceptible to chemical and/or physical deterioration.

A simple case of chemical action of potentially limited life is the application of lime to metal-contaminated soil at the surface. This is a very simple form of in-situ treatment that should, by raising the pH, reduce the uptake of most metals by plants. However, lime is lost from soil each year under the influence of rain which is invariably acid owing to the presence of dissolved carbon dioxide. The rate of loss depends on the rainfall, texture of the soil and amounts of calcium in the soil and in the UK varies from 125 to 1000 kg per hectare annually. The lime loss in drainage is even greater when rain water contains acids, beside carbonic acid, such as those likely to be found in urban or other settings where the atmosphere is polluted with sulphur dioxide. The maximum effect on soil pH occurs about 4 years after liming, and thereafter the pH decreases slowly. Maintenance applications are likely to be required at intervals of about 3 to 7 years depending on soil type. In other words the contaminants are held in a metastable state and long-term effectiveness (almost a misnomer) is utterly dependent on regular maintenance and monitoring of soil conditions. Preferably, not only should the soil be monitored to see that the pH remains within the design range, but also the vegetation itself should be the subject of periodic examination.

Examples of potential physical and chemical instabilities are provided by grout materials (this applies equally when such materials are used to form vertical or horizontal barriers as when used to treat a mass of material in situ). Cementitious grouts are susceptible to attack by many chemicals including acids and sulphates. If exposed close to the surface they may suffer from freeze-thaw cycles. Components can be selectively leached from gelling grouts by moving groundwater or percolating rainwater. Both short-term and long-term properties are highly dependent upon the raw materials used (eg cement type) and the proportions used. The latter is dependent upon the degree of control during treatment but may be inadvertently changed during injection into the ground (for example permeation is highly dependent on porosity and components with larger grain sizes may be selectively stripped out or settle). Little is known about how contaminants are likely to influence properties at the point of application and even less about long-term performance. Reliance will need to be placed on laboratory experimentation, a not altogether satisfactory procedure.

The crucial question of chemical stability raises numerous supplementary questions about the factors that will influence stability and the consequences of a chemical change. A gradual change in which a harmful chemical is slowly dispersed to the environment say over 100 years may be acceptable but a sudden release, such as might occur from flooding with seawater, could have catastrophic consequences.

Environmental factors that may have an effect on chemical stability include; changes in pH, microbial action, infiltration of other chemicals such as fertilisers, flooding with seawater, changes in groundwater levels, changes in redox potentials.

Two examples of probable effects of land use following in-situ treatment can be mentioned. First a change in agricultural practice involving perhaps application of heavy doses of fertiliser - particularly nitrogenous. Second, use of the site for industrial purposes may raise a potential for (further) spills of chemicals, broken drains, etc.

Supplementary remedial actions may be critical to the overall effectiveness of a remedial scheme. For example it may be necessary to prevent infiltration of acid rain by means of an imposed covering system. Depending upon how essential it is to keep the rain out there might be a requirement for a physical and/or chemical barrier, both of which may have limited lives. The time and rate at which infiltration occurs may then become the critical factors.

The immediate effectiveness of application of 'stabilisation' measures can probably be adequately judged by soil and water samples taken from within and around the site treated. Long-term performance is clearly more difficult to assess.

The effectiveness of surface applied agents such as lime can be comparatively easily monitored through soil and vegetation samples. To monitor and evaluate the in-ground treatments that are the main subject of Chapter 4 is much more difficult. The three monitoring points already discussed in relation to on-site processing are available:

(i) fluids within the site

(ii)   fluids outside the site
(iii) 'soil' samples from within the site,

and the same constraints apply. In addition, however, critical
environmental factors may need to be monitored to give an early
warning of possible problems. For example, whether a surface water
or in-ground drainage system is working properly, the composition of
water that may be entering the site, and any deterioration of
supplementary measures.

4.4 Cover Systems for Contaminated Sites

Cover systems for contaminated sites have been defined in this
study to encompass: systems designed to prevent the migration of
contaminants vertically upwards, or to prevent ingress of surface
water into contaminated sites. They are discussed in Chapter 5 in
terms of the many, and sometimes conflicting functions they must
perform. With very few exceptions, for example trafficability during
installation, these functions are associated with natural processes
that are time dependent and that generally become increasingly
adverse with time (problems arising from the generation of methane
are among the exceptions since the quantity of methane will fall with
time). This is not to say that the effectiveness necessarily and
inherently decreases with time but rather that covering systems may
not be fully tested until many years after installation. For example,
a tree root will generally grow ever larger. In addition covering
systems require the use of either natural soils or man-made materials,
vegetation layers and engineering constructions such as drainage.
Soils are complex dynamic systems, man-made materials generally
deteriorate with time, vegetation changes both in size and nature
without any intervention by man, and drains can become blocked.
Crucial points and questions regarding covering systems are listed in
Table 2.4.

The methods available for the monitoring and evaluation of
performance of covering systems depend upon the nature of the system
and the function that is to be examined. They range from
comparatively simple to very difficult, both at the time of
installation and in subsequent years. The time dependent changes
that may affect performance of the system are major complicating
factors in making an evaluation of performance. Most covering
systems will be multi-functional and no single monitoring measure
will provide information on performance in relation to all functions.
Each function may need to be looked at separately although it may be
sufficient to concentrate on one or two key functions.

Very few of the functions of a covering system can be tested
for effectiveness at the time of installation. Exceptions are
possibly those involving a fluid flow. Thus a gas collection system

Table 2.4   Crucial Points and Questions Regarding
            Covering Systems

Crucial points:

Contaminants will not themselves be changed or rendered harmless by
the remedial measure.

Cover systems in general will only decrease infiltration or
precipitation, not prevent infiltration altogether.

Cover systems in general will only decrease contaminant migration,
not prevent migration altogether.

Cover systems could be affected by critical environmental conditions,
eg plant roots, rodents, settlement.

Cover systems from mineral materials may be eroded by wind or water.

Cover systems with membranes as sealing layer may be damaged
following erosion of the protective layer.

Some plastic membranes may be affected by chemical and microbiological attack.

Questions:

Will the contaminants be rendered harmless during the expected life-
time of the cover system?

Do environmental impacts occur as soon as the cover system becomes
ineffective?

What are the environmental conditions causing the decrease of
effectiveness of the cover system?

At what rate will the changes occur and does the rate matter?

At what point in time will the most critical environmental conditions
occur?

How could the decreasing performance of the cover system be measured?

Can performance of the cover system be increased by maintenance and
repair?

How does the performance of the cover system depend on the
performance of various parts of the system?

How could monitoring data support decision making on repair or
renewal of the remedial measure?

How susceptible is the system to the use to which the land is put or
to mans unwitting intervention? How can this influence the answers
to the above questions?

---

can be demonstrated to be collecting gas and the site can be
monitored for gas to see if it is absent from where it should be
absent. Similarly a water collecting and drainage system is amenable
to assessment through flow measurements and chemical analysis but it
is unlikely to be possible to detect any hole in a membrane that is
allowing water to enter the contaminated ground until some time
has elapsed. Functions such as those to sustain vegetation,
prevent erosion, prevent root penetration and prevent upward
movement of contaminants; are time-dependent and cannot be
evaluated until some years have elapsed. In addition, it may be
difficult to do so without damaging the functioning of the system
or with any degree of confidence about the representative nature
of the test area.

The simplest and most essential monitoring and evaluation
measure is the systematic and planned site inspection involving
mainly visual observation but with some use of simple instrumentation, eg for gas detection. Among factors that can be noted
are: extent, nature and quality of vegetation; signs of animal
activity; signs of erosion or settlement; whether surface
drainage is satisfactory; any signs of leachate; the condition of
any buildings or services on the site. Such surveys are
inexpensive and easy to do and should be made in different
seasons. Proper records and photographs should be kept.

When the functions of the cover include prevention of upward
movement of contaminants and/or uptake by plants, soil and
vegetation samples can be taken. If an indicator is required
samples can preferably be taken from plants (and parts thereof)
likely to show the most marked response. Since the processes
involved are likely to be slow and any changes in soil or plant
small, it is essential that proper and adequate baseline data are
collected and that reference samples from early samplings are
preserved for subsequent re-analysis to allow for changing
analytical methods.

Certain physical characteristics of the system may be
capable of long-term, even automatic, monitoring. For example,
instrumentation for the long-term monitoring of geotechnical
measurements of earth dams and embankments (eg settlement,
groundwater levels, pore water pressures) is well established and
others such as soil suction should be amenable to instrumental
monitoring. Such monitoring should reveal any long-term trends
or sudden changes.

Monitoring of fluid from within and around the site from
fixed points should be possible and may provide useful
information. However physical sampling of the ground beneath the
cover may be extremely difficult without damaging the system, eg
punching a hole through a membrane or clay sealing layer.

If it is accepted that some of the materials used in the
system are susceptible to deterioration, then it becomes very
difficult to see how their state can be monitored without severe risk
of damage to the functioning of the system and attempts to look at
such things as root penetration may be equally disruptive. For
membranes something can be done using double layers with a detection
system for the offending agent, gas or liquid, between the two
layers. If the condition of materials, including soils is to be
studied in place then it is probably preferable to designate a study
area at the time of installation and carry out such extra work as may
be necessary to make subsequent excavation safe. Indirect evidence
can be achieved by establishing at the time of installation test
specimens or systems under controlled and accelerated conditions.
Thus early warning of a material failure may be obtained.

The fact that changes are likely to be gradual rather than
sudden is an advantage. But it equally means that it may be
necessary to continue apparently unproductive monitoring over very
long periods of time and it may be difficult to judge when conditions
have deteriorated to a degree that further remedial action is
required. As a simple example, if plants die the need for action is
obvious but if cadmium levels in food crops increase marginally the
situation is far from simple. It may be necessary to sample crops in
several successive years to be sure that the effect is real and even
then a judgement must be made as to whether it matters on health
grounds and also whether it is the first indication of serious
failure to come.

4.5   In-Ground Macroencapsulation

In-ground macroencapsulation has been defined to encompass:
systems designed to prevent migration of contaminants vertically or
laterally, or to prevent ingress of groundwater into contaminated
sites.

The factors governing the effectiveness of in-ground barrier
systems (vertical and horizontal) are discussed in Chapter 7 and
questions concerning their effectiveness are listed in Table 2.5.
Where the vertical barrier is designed to stop further leaching of
contaminants by groundwater flow it should be fairly easy to
determine effectiveness at the time of installation using well
established procedures such as groundwater monitoring wells. But
this approach serves mainly to check the overall improvement of
groundwater quality. It does not determine the effectiveness of
every square metre of the vertical barrier although this is what is
required to guarantee proper construction and to enable repairs of
leaks shortly after installation due to bad construction, or later on
due to erosion or chemical reactions between barrier materials and
contaminants. When the barriers are supplementary to some other

treatment, eg in-situ, the situation may be more difficult since the polluting agent(eg an injected chemical) and the driving force of a continuous supply of liquid, will not exist after the barrier is installed. Once treatment begins, a monitoring well outside the barrier will indicate whether any unwanted pollutant is escaping, but by then of course it is too late. The certainty of a high quality installation of a barrier thus may be the critical factor in deciding whether to employ an in-situ treatment method particularly when this involves mobilisation of the contaminants or use of substances hazardous to water.

Long-term performance would appear to be largely a materials problem. Whether the barrier consists of a bentonite cut-off wall, a plastics membrane or an injected chemical or cementitious grout, the questions essentially are similar to those for other remedial treatments:

(i)   Will there be any inherent and unavoidable changes in material properties?

(ii)   To what extent are such changes affected by external factors including interaction with contaminants?

(iii) At what rate will the changes occur and does the rate matter?

(iv)   To what extent is the system sensitive to physical damage, eg insertion of ground piles, accidental breaches of cut-off wall, root penetration?

Such systems are essentially (as discussed in Chapter 7) concerned with the control of movement of fluids (gas barriers are mentioned in Chapter 10). As such, water samples taken from appropriate locations provide an apparently obvious means of monitoring performance (but see below). Depending on the purpose of the barrier these may need to be taken on just one or both sides of the barrier. An example of the latter is when the aim is a controlled net flow of water into the macroencapsulated area in order to determine whether any unwanted reversal of flow is occurring. In many cases, monitoring of groundwater levels will also be needed.

The taking of water samples from inside or outside the barrier is not as straightforward as at first it might appear. Even to find the proper parameter to analyse for and to determine appropriate locations for sampling will be difficult. The large numbers of samples to be analysed may also be a problem. The main difficulty, however, is that: it may not be possible to be sure whether the analysed contaminant comes from inside the encapsulated area or was already present on the outside. Therefore, it will be difficult to make a decision to repair a section of a cut-off wall based on the detection of a particular parameter in a water sample taken outside of a vertical barrier. A combination of water and material analysis seems to be necessary, combined with constructional solutions that allow for the location of leaks or weaknesses in the barrier.

Table 2.5   Crucial Points and Questions Regarding
            In-Ground Macroencapsulation

Crucial points:

Contaminants will not be rendered harmless by the encapsulation itself.

Macroencapsulation systems will only decrease ingress of groundwater
and migration of contaminants, not prevent them.

Macroencapsulation systems could be affected by external factors,
eg drillings.

Materials used for macroencapsulation may interact with contaminants,
and are likely to deteriorate.

During construction of macroencapsulation systems, substances harmful
to water may be used in excess and pollute groundwater.

Questions:

Will the contaminants be rendered harmless during the expected life-
time of the macroencapsulation system?

Does contaminant migration start again as soon as the macroencapsulation system becomes ineffective?

What external environmental conditions will cause the decrease of
effectiveness of the macroencapsulation system? Will there be any
inherent and unavoidable changes in material properties?

To what extent could interaction of materials and contaminants be
expected? At what rate will the changes occur? Does the rate
matter?

What parameters to be measured may indicate decreasing effectiveness
of the macroencapsulation system? How and how frequently should the
parameters be measured?

By what maintenance or repair could decreasing effectiveness be
stopped?

Are there specific parts of the macroencapsulation system that will
deteriorate first and most severely?

How could monitoring data support decision-making on repair or
renewal of the remedial measure?

A determination of how the barrier itself is behaving is much more difficult and, as for covering systems any physical sampling system presents a risk of damage to the barrier. Controlled and accelerated testing of the materials may be helpful.

4.6 Hydraulic Measures

Hydraulic measures as remedial measures at contaminated sites are defined to encompass: operations designed to control or treat the liquid phase on contaminated, sites ie groundwater modification systems and in-situ or on- or off-site treatment of contaminated groundwater.

Hydraulic measures, other than barrier systems as already discussed in 4.5 consist esentially of two components:

  (i)   control of groundwater flow by pumping and/or infiltration
        of water
 (ii)   treatment of groundwater either in-situ or outside the
        aquifer in a water treatment facility.

The purpose of controlling groundwater flow is to cut-off the source of contamination from the surrounding groundwater and/or to stop migration of contaminants towards a target to be protected, eg drinking water supply well. These measures are only effective as long as the pumps are working.

Whether contaminants are to be rendered harmless in-situ in the aquifer, or to be removed from the aquifer, control of the ground-water flow will be essential. (These measures are likely to be more technically effective and more cost-effective if a groundwater flow simulation model or a contaminant migration simulation model is employed in their design - see Chapter 9.)

In-situ treatment of contaminants in groundwater aquifers presents, in general, difficulties similar to those in-situ treatment of contaminated soils (see Tables 2.3 and 2.6). Frequently, however the distribution of contaminants in the aquifer is more homogeneous so that the treatment may be easier. If the contaminants are chemically or microbially destroyed the question of long-term effectiveness does not arise. If, however, the contaminants are immobilised or reversibly rendered harmless, then there is a need for long-term monitoring and appropriate measures to maintain the performance, eg control of land use and groundwater extraction.

In general, when groundwater is treated outside the aquifer and treated water subsequently infiltrated into the ground, the question of long-term effectiveness does not arise provided the measure is carried out until the objectives of the remedial measures are met

Table 2.6   Crucial Points and Questions Regarding
            Hydraulic Measures

Crucial points:

The effectiveness of hydraulic measures depends on pumping and an
infiltration of groundwater.

Question:

Can it be guaranteed that pumping will be continued as long as
contaminant migration is likely to occur?

(the aim will invariably be to reduce contaminants to a level at
which the water can be returned to the environment). However the
effect of treated and infiltrated water in the aquifer must be
considered since there would be undesirable secondary effects, (eg
heavy metals in the aquifer may mobilise owing to a decrease to pH of
groundwater during treatment).

When hydraulic measures are planned for contaminated sites care
should be taken to avoid penetration of the upper layer into deeper
groundwater layer owing to poor design, implacement or use of
extraction or monitoring wells and to avoid an extension of the contaminated groundwater plume because of changed groundwater flows.
Proper hydrogeological investigation and possibly use of groundwater
models may help to prevent this (see Chapter 9).

5   RESEARCH

An evaluation of existing knowledge reveals obvious subjects
on which information is missing with which to assess long-term
effectiveness. Research can help to fill some of these gaps in
knowledge and some research requirements have been identified in the
Chapters on remedial measures that follow. The essential need for
long-term test programmes lasting 10, 20 or more years is emphasised.

6   REGISTER OF EFFECTIVE LONG-TERM TREATMENT AND
     MONITORING STUDIES

As part of the supporting information exchange to the Pilot
Study an attempt was made to establish a Register of Effective Long-
term Treatments and Monitoring Studies'. Very little information was
submitted by the participant countries, reflecting the lack of sites
that fulfil the criteria and the lack of evaluation or monitoring
studies. Notwithstanding this lack of hard information this is still

considered a valid and useful exercise to be carried out at an international level if an appropriate mechanism to operate it can be established.

It is strongly recommended that a record should always be made of remedial actions that are carried out, preferably in a central (national/state) register. The information to be put into the register must of necessity be brief and generally indicate where more detailed information (ie the full records) can be obtained.

It is hoped that this report will serve to make people more conscious of the need to keep appropriate records and to carry out proper monitoring and evaluation exercises.

7   CONCLUSIONS AND RECOMMENDATIONS

(i)   Remedial actions are seen as falling into two broad groups:

    (a)   those that remove contaminants or render them harmless,

    (b)   those that stabilise or contain contaminants.

(ii)   The question of long-term effectiveness does not arise for the first group, if the measures are carried out properly.

(iii)   In general, however, such are the practical problems in carrying out the treatment, (eg to render harmless all contaminants in-situ in inhomogeneous ground, using substances harmful to water, or subject to deterioration) that the question of long-term effectiveness must be asked.

(iv)   All other remedial measures offer only a solution of limited or uncertain duration unless other mechanisms, such as microbial degradation, reduce the level of contamination.

(v)   In theory some in-situ treatments may be effectively permanent, for example if contaminants can be converted to a form that is insoluble under any foreseeable environmental conditions. In practice, however, it will be difficult to ensure that all contamination is treated.

(vi)   Macroencapsulation systems, including covering systems, will be particularly vulnerable to a loss of effective-ness with time.

(vii)   Monitoring of parameters that will:

    (a)   describe the behaviour of the remedial measure,

(b) describe conditions within and outside the contaminated
      area,
(c) describe behaviour of materials used for encapsulation,

is required with any remedial measure because it will never
be possible to demonstrate long-term effectiveness in
practice in advance of the treatment being carried out.

(viii) Establishment of an adequate set of baseline data before
        and immediately after treatment, will be helpful in
        evaluating the remedial actions.

(ix) Evaluation of the remedial actions, and the reduced impacts
       of remedied contaminated sites, may lead to reduced
       monitoring requirements, as long as the environmental
       conditions influencing the success of the remedial action
       stay constant.

(x)   From the point of view of long-term effectiveness of remedial
       measures it seems to be reasonable, first, to reduce the
       environmental impacts of contaminated sites by macro-
       encapsulation including covering with assistance of hydraulic
       measures or excavation and temporary storage of the contami-
       nated material and second, to treat the contaminants in-situ
       or on-site to render them harmless.

(xi) Following this concept it becomes obvious that available
       technologies for macroencapsulation should be used as
       temporary solutions, which need to be effective as long
       as contaminants inside the encapsulated area pose
       environmental problems.

(xii) Research is necessary to:

        (a) develop methods to identify decreasing effectiveness of
              remedial measures in time, eg leak detection,
              mobilisation of immobilised contaminants.
        (b) develop methods of treatment to render contaminants
              harmless reliably, preferable by in-situ treatment.

(xiii) It is strongly recommended that a record should always be
        made of remedial actions, and that a brief summary of the
        most important information should be put into a central
        register. This record will form part of a most important
        information base, of help in the distribution of knowledge
        and experience, on a national or inter-national level.

ON-SITE PROCESSING OF CONTAMINATED SOIL

W H Rulkens, J W Assink and W J Th van Gemert

TNO, The Netherlands

1   INTRODUCTION

On-site processing of contaminated soil typically involves three process steps carried out at the contaminated site; excavation, treatment and redeposition of the soil.

In this Chapter the types of treatment methods considered suitable for the on-site treatment of contaminated soil are discussed.   Excavation redeposition and eventual transport are not dealt with. At the moment only a few of these on-site treatment processes are really operational. Most of them have still to be developed to the stage of full-scale operation and there is still a long way to go before most of these processes can be evaluated and used in practical situations. The on-site treatment methods discussed are also suitable for treatment of contaminated soils at a central location elsewhere. This can be favourable for economic or practical reasons when only small amounts of soil have to be treated or for reasons of legislation or avoidance of local nuisances at the contaminated site.

The following know-how is relevant to a discussion of on-site treatment processes that may be developed for use under practical conditions:

-    Unit operations used in the chemical process- and oil-industry, for example extraction, evaporation, stripping, distillation, drying, dewatering, suspension and neutralisation.[1,2]

-    Typical processes used for treating hazardous wastes,[3] such

as leaching, incineration, heat treatment, land farming of
sludges of the oil-industry and stabilisation (immobilisation,
solidification) methods.

- Processes used for the purification of municipal and industrial
  waste water streams[4,5] for example biological treatment,
  coagulation, flocculation, settling, flotation, neutralisation,
  activated-carbon adsorption, ion exchange and sludge
  dewatering.

- Typical processes applied in the excavation, transportation and
  handling of soil and processing of ores.

- Soil properties[6] (eg physical, chemical, geohydrological)

- Typical properties of the contaminants often present in
  contaminated soil.

The following aspects must be considered before an actual on-
site treatment can be applied:

- Psycho-social aspects concerning the people living near the
  site during and after the treatment.

- The impact of the treatment method upon environment and
  human beings.

- The long-term effects of the treated site.

- Whether it is permissible to return the treated soil to the
  excavated site despite some residual contaminants.

- Reuse of the site.

- Types of waste streams resulting from the treatment and the way
  of disposing of these waste streams.

- Relevant legislation.

1.1 Scope

For each type of process discussed the underlying principle of
the process is given first, followed by a brief description of the
complete process scheme based on this principle. Furthermore the
area of application is discussed. Finally the most important limitations of the process are mentioned. The essential characteristics of
all on-site treatment processes discussed are summarised in Table 3.5.
Some information is also given about the estimated costs of the
treatment methods. After discussion of the various types of

processes, some recent developments and applications are discussed.
Finally the main conclusions are given and a few recommendations are
made. The following general remarks should be noted:

-   The emphasis is on purification by removing or destroying the
    contaminants present in the soil, for these techniques give an
    ultimate solution for contaminated sites in contrast to
    stabilisation methods.

-   The type of soil contamination where the contaminants are
    present in closed vessels or drums and the treatment of former
    waste disposal sites are not considered. In the first case
    remedial action consists primarily of a cautious excavation and
    handling of drums and disposal of the drums as hazardous waste.
    Most former waste disposal sites are difficult to deal with
    because of the inhomogeneous characteristics of the material,
    both size, shape and type of particles.

-   Whether or not a contaminated site needs remedial action, and
    the quality standards a clean soil has to satisfy are not
    discussed.

-   The degree of removal or immobilisation of contaminants is
    strongly dependent on factors like type of soil, type of
    contaminants and intensity of treatment method. Tests are
    always needed to assess the practical effectiveness of the
    treatment method chosen and therefore the efficiencies of the
    various methods are not discussed. Nor is the problem of waste
    streams resulting from the several treatment methods discussed,
    although in some cases these waste streams are substantial and
    limit the applicability of the method.

1.2   On-site Treatment Methods Available

    The following on-site treatment methods are available for the
removal or destruction of contaminants:

-   Extraction (leaching)

-   Mechanical separation (sieving, classification, etc)

-   Flotation

-   Thermal treatment

-   Thermal treatment

-   Stripping

-     Chemical treatment

-     Microbiological treatment

-     Ion exchange

-     Electrophysical and electrochemical methods

An additional group of treatment methods can be distinguished in which the leachability of contaminants in the bulk of the soil is reduced, the so-called stabilization methods.

These various methods, with the exception of ion exchange, stripping techniques like vacuum stripping and air stripping, and electrophysical methods (electro-osmosis, electrophoresis, electro-dialysis), which will probably have only limited applicability for cleaning up soils, are considered in detail below.

2   EXTRACTION (LEACHING)

2.1   Introduction

The contaminated soil is mixed with an extracting agent to transfer the contaminants from the soil particles to the extracting agent. With this mixing step it is possible to remove not only contaminants that are soluble in the extracting agent (in general an aqueous solution, sometimes an organic solvent), but also contaminants that are in fact insoluble in the extracting agent but form stable colloidal suspensions in it. After extraction the purified soil particles are separated from the extracting agent. The contaminants are subsequently removed from the latter.

2.2   Process Scheme and Description

The extraction process for the treatment of contaminated soil is given in Figure 3.1. The excavated soil is subjected to a preliminary treatment which may include passing it over a wide mesh screen to remove large objects and reduce the size of clods. It is then transported to the extractor where soil and extracting agent are mixed intensively. After extraction, soil and extracting agent are separated. The soil is subjected to a post-treatment to remove remaining extracting agent and contaminants. This can be done by washing, neutralisation, evaporation, etc., dependent on the type of contaminants, type of extracting agent and type of soil. The extracting agent leaving the extraction apparatus generally contains the finest soil particles (eg < 80 um) which may be removed by a second separation step. The fine-particle fraction thus obtained can also be subjected to a post-treatment in order to remove (adsorbed) contaminants. The extracting agent leaving the second separation step is treated to remove the contaminants. These contaminants are generally separated as a concentrated sludge which has to be disposed of by incineration or by dumping at a suitable place. Other options for disposal may be chemical treatment, electrochemical detoxification or further concentration of contaminants.

There are several modifications of the extraction process, dependent on the type of soil to be treated and the type of contaminants to be removed. The following types of equipment can be considered:

Pre-treatment:

-   Crusher

-   Sieve

-   Mechanical scrubber (mutual rubbing or scrubbing of soil particles in a slurry)

Extraction and separation of soil and extracting agent:

-   Mixer/settlers in line

-   Screw extractor

-   Fluidised bed

-   Mixers and hydrocyclones in line

-   Settlers

-   Sieve belt

-   Vacuum belt filter

-   Hydrocyclones

-   Rotating sieves

Post-treatment of soil:

-   Vacuum belt filter with a washing installation

-   Screen or dewatering screen with a washing installation

-   Neutralisation eg by spraying chemicals

-   Heating (for the evaporation of organic extracting agents)

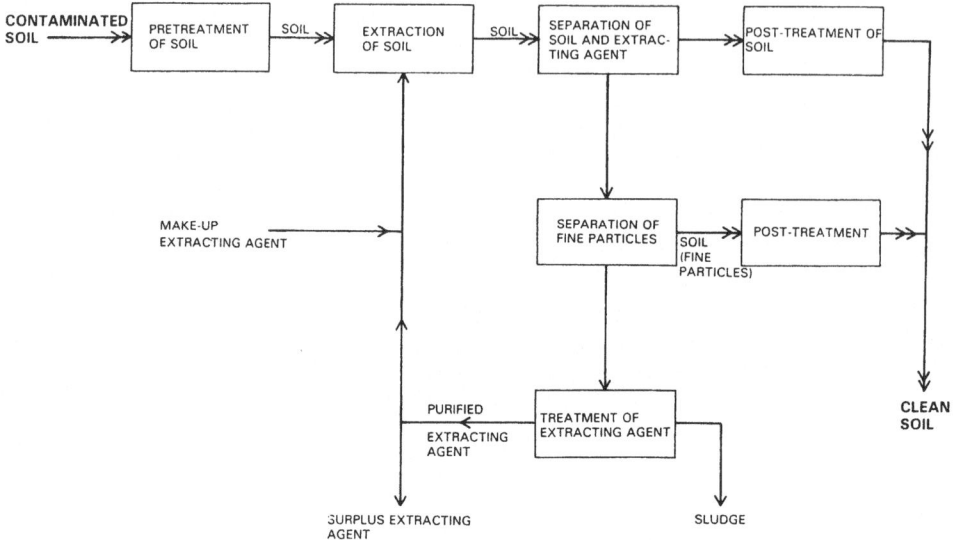

Figure 3.1 Extraction of Contaminated Soil

Separation of fine particles:

−   Tiltable plate separators

−   Sedimentators

−   Hydrocyclones

−   Centrifuges

−   Flotation units

The following methods can be considered for the treatment of extracting agent:

(when water or water with additives is used as extracting agent)

−   Coagulation/flocculation followed by settling and dewatering of the sludge

−   Ion exchange

−   Activated carbon adsorption

−   Reverse osmosis

−   Ultrafiltration and microfiltration

-    Chemical conversion (ozone, hydrogen peroxide)

-    Electrochemical conversion

-    Stripping

-    Precipitation

-    Flotation

(when an organic solvent is used as extracting agent)

-    Evaporation, stripping

-    Distillation

-    Secondary extraction

-    Electrochemical conversion

2.3 Potential Applications

Heavy metals. The following extracting agents can be
considered for the removal of (heavy) metals from contaminated
soil by formation of true solutions:

-    HCl   (Cd, Cu, Zn, Ni, Cr, As, Sb, Pb)

-    H_2SO_4 (Cd, Cu, Zn, Ni, Cr, As, Sb)

-    HNO_3 (Cd, Cu, Zn, Ni, Cr, As, Sb, Pb)

-    NaOH   (Pb, Zn, metallo-organic compounds)

The removal of heavy metals is largely dependent on the form
of the metals present in soil (dissolved ions, organic and inorganic
complexes, organically bound, physically or chemically absorbed,
occluded, precipitated, particles). It should be noted that certain
chemicals such as SO_4^{2-} and NO_3^- containing electrolytes may
be considered hazardous to the environment and extracting agents such
as H_2SO_4 and HNO_3 must therefore be removed from the treated
soil before redeposition. Furthermore, (partial) recycling of the
extracting agents must be adopted in order to cut costs and minimise
emissions to the sewer system.

It is also possible to make use of complex-forming agents, alone
or in combination with acid or alkali. Combinations of alkali
extraction and acid extraction may be necessary.

Cyanides. For the removal of cyanides or cyanide complexes aqueous NaOH can be used. However, iron cyanide complexes strongly adhere to small soil particles (clay, silt). These particles cannot be cleaned with aqueous NaOH, but can be washed out by classification (see 3.2.3).

Hydrocarbons and halogenated hydrocarbons. Extraction of hydrocarbons can be achieved either by using an aqueous solution of HCl, NaOH, Na_2CO_3, detergents, or organic solvents. Aqueous NaOH- or Na_2CO_3-solutions have dispersing properties and are able to dissolve humus-like substances in the soil. This makes it possible to remove contaminants from the soil that are insoluble in water and preferentially adhere to humus. As regards the use of organic solvents it should be noted that the soil to be treated generally contains a small amount of water. This means that extracting agents are preferred which are miscible with water, eg ethanol, isopropanol and acetone, although in all cases the applicability of these agents should be tested.

2.4 Limitations

The extraction process is favourable for application if the soil to be treated consists only of sand particles, but unfortunately this is seldom the case. Usually organic, humus-like substances and clay particles are also present. These can result in the following difficulties:

- Some compounds of heavy metals (but also organic contaminants) are preferentially adsorbed or absorbed by humus-like substances. Extraction with acids can then be difficult. For the removal of organics, extraction with alkali is often more effective than acid extraction. However, in general most of the humus-like substances dissolve in alkali resulting in large amounts of sludge being formed when the extracting agent is purified. It is extremely difficult in practice to separate the contaminated compounds from the humus-like substances.

- The amount of extracting agent needed can increase markedly if large amounts of humus-like substances are present.

- Clay particles cannot be separated effectively from aqueous extracting agents by means of simple separation techniques such as settling (see also classification in 3.2.3).

- Clay particles have strong adsorption properties with respect to most heavy metals. These particles can not be easily cleaned by extraction, but may be washed out by classification (see 3.2.3).

– If clay particles are not removed and purified in a post-
  treatment process, they concentrate in the sludge resulting from
  treatment of the extracting agent. The result can be a
  substantial amount of contaminated sludge which cannot easily be
  disposed of. In general when normal settling and decantation
  methods are applied, most of the particles smaller than 30–60 um
  are washed from the soil.

Organic extracting agents have to be removed completely from
the treated soil before redeposition; this can be achieved by
evaporation and washing for example. Mainly because of the
highly restrictive environmental requirements it is likely that
organic solvents will only be applicable as extracting agents on
a limited scale. The high cost of these extracting agents is
another limiting factor.

2.5 General Remarks

A counter-current flow of extracting agent and contaminated soil
is a prerequisite of an efficiently operating extraction process.

2.6 Special Safety Aspects

When specific contaminants and extracting agents come into
contact with one another, the possibility exists of hazardous
compounds being formed. Safety measures have to be taken especially
when these compounds are gases. An illustrative example is HCN which
can be formed when cyanides come into contact with acids.

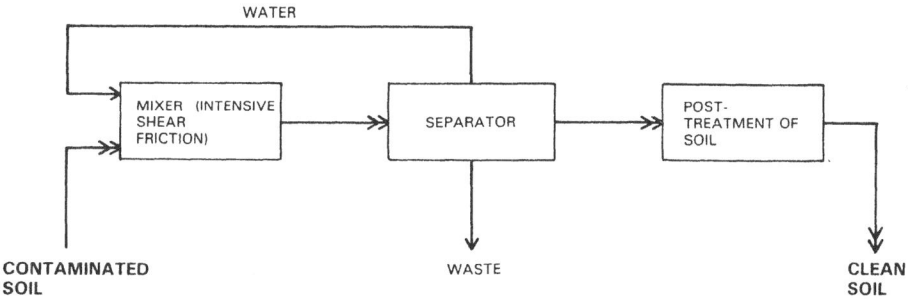

Figure 3.2   Separation Based on Differences in Specific Gravity

3   MECHANICAL AND PHYSICAL SEPARATION

3.1   Introduction

Treatment techniques for contaminated soil that are based on a
mechanical or physical separation between the original soil particles
and the phase consisting of solid or liquid contaminants, can be
grouped as follows:

—   Separation based on differences in specific gravity.

—   Separation based on differences in particle size.

—   Separation based on differences in settling velocity in a
     suitable liquid (or gas).

—   Separation based on different magnetic properties.

—   Separation by flotation based on differences in surface
     properties of the particles to be separated (discussed in
     Section 4).

3.2   Process Description and Potential Applications

3.2.1   Separation Based on Differences in Specific Gravity
The contaminated soil and a suitable liquid (in most cases water) are
intensively mixed (see Figure 3.2). Mechanical means such as water

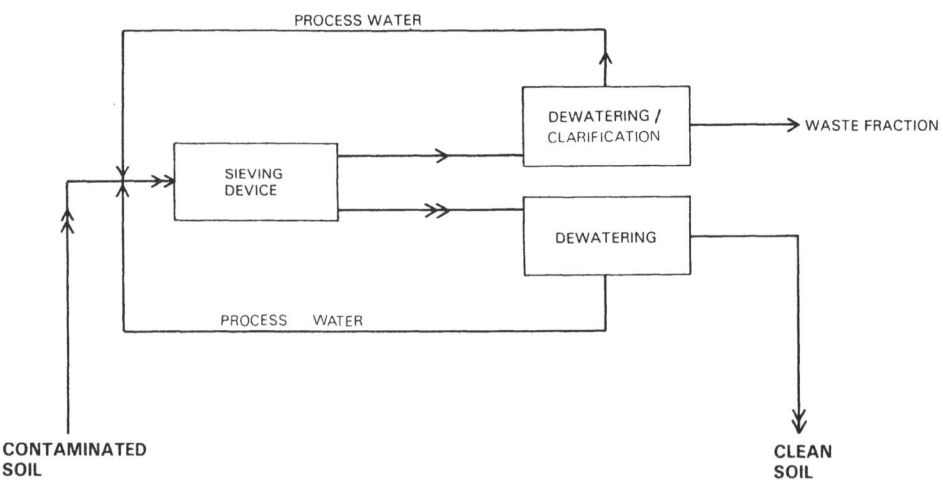

Figure 3.3   Separation by Wet Sieving

jet, water air jet and scrubbing devices are often necessary. Afer
mixing the soil phase is separated from the liquid phase. The latter
contains the contaminants, usually as a floating layer, but in some
cases part of the contaminants may be present as an emulsion or
suspension.

The contaminants are separated from this liquid phase as
a sludge or liquid. The purified liquid can be reused. The
separator can be a two-stage separator, for example the first
stage for separating the cleaned soil from the contaminated
liquid and the second stage for the separation of liquid and
contaminants.

This process is particularly applicable to the treatment of
sandy soil contaminated with oil or other substances with lower
density than water and not soluble in water.

3.2.2 Separation Based on Differences in Particle Size.
The process consists of a wet (or dry) sieving (see Figure 3.3).
With wet sieving post-treatment of the soil is required for
dewatering. A wide variety of sieving devices and dewatering screens
are commercially available.

Dry sieving can be applied down to about 100 μm when the soil
is sufficiently dry, otherwise larger mesh sizes must be used. Wet
sieving is possible down to about 30 μm.

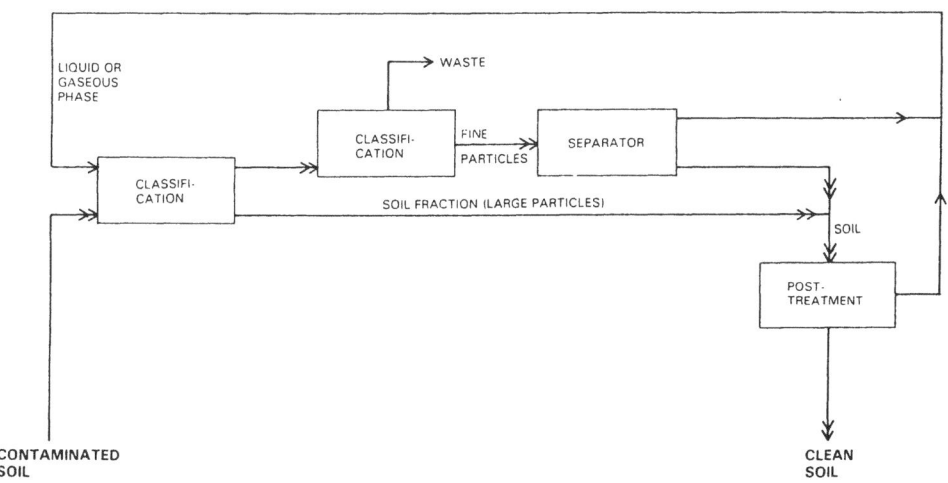

Figure 3.4   Separation by Classification (Two Stage)

Fields of application may be soils contaminated with hazardous particles that differ in size from the other soil particles (for example fluorescention powders and clods of crude oil on beaches). The contaminated particles must have a particle-size distribution within a relatively narrow range in order to ensure a sufficiently small waste stream or mass stream for secondary treatment.

3.2.3 Separation Based on Differences in Settling Velocity.
The process (see Figure 3.4) consists of a wet or dry classification step where the soil particles and the contaminants are separated. The liquid or gas flow carrying the contaminated particles is treated in a second process step in which the contaminants are removed.

For efficient and correct classification, the contaminated soil should be pretreated to ensure that all soil particles can be classified individually. In dry classification the soil must first be dried in such a way that no agglomerates are formed. Pretreatment for wet classification may be carried out in a high performance mixer to form a slurry of the soil in water (a so-called scrubber).

The kind of apparatus to be used for the classification process itself depends on the range of particle sizes to be separated and on the phase (liquid or gas) used for classification.

Types of apparatus for liquid-based classification (water):

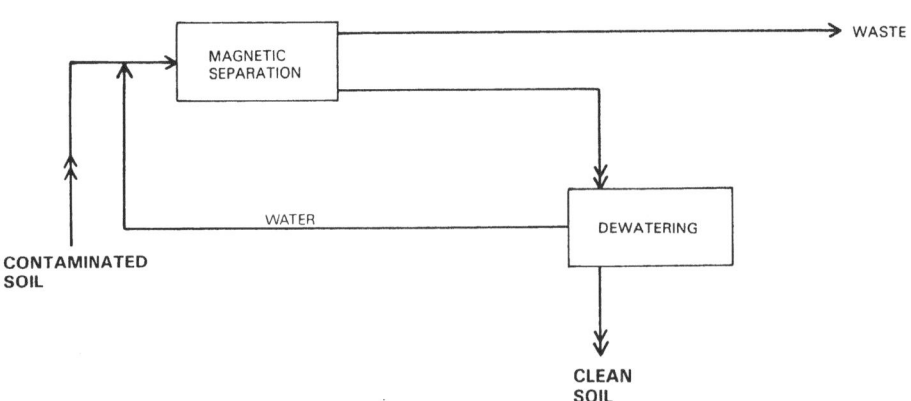

Figure 3.5 Treatment of Soil Contaminated with Magnetic
         Hazardous Particles

-    Upflow column or jet sizer (countercurrent flow)

-    Screw classifier

-    Rake classifier

Types of apparatus for gas-based classification (air):

-    Zigzag air classifier (countercurrent flows)

-    Cross-flow classifiers.

The classifying technique is applicable to sandy or slightly loamy soils with the contaminants present as particles or adsorbed to certain soil particles (eg clays) having settling velocities different from the settling velocity of the clean soil particles.

It should be noted that wet classification has the same main characteristics as extraction, with settling and decantation as the principal separation step for soil and extracting agent. This makes it possible to develop a treatment installation in which extraction and classification are combined (see 12.1.c and Appendix F).

3.2.4 Separation Based on Different Magnetic Properties. The contaminated soil is slurried with water and then passed through a magnetic field (see Figure 3.5). A technique that can be used is High Gradient Magnetic Separation (HGMS).

The applicability of this technique is probably limited. The most important group of contaminated soils to be treated by this technique are those containing iron or magnetite particles doped with hazardous materials (heavy metals).

3.3 Limitations

The most important limitation in the separation treatment processes is the presence of clay particles (difficult to separate from the water phase). In addition the presence of humic-acid like substances can hinder the settling of particles. It is likely that in practice it will often be difficult or even impossible to achieve a good separation between 'clean' soil and contaminated wastes.

4 FLOTATION

Flotation is a process for separating certain types of particles from suspensions. Flotation processes are widely used, particularly in ore processing and waste water treatment. They have recently been applied for the recovery of oil from oil sands or tar sands. In the

case of soil treatment the process can be applied for the selective
removal of contaminants from a suspension of contaminated soil in
water. A prerequisite, however, is that the contaminants are present
as distinct liquid or solid particles.

The flotation process consists basically of three steps:

-    Treatment with a suitable chemical

-    Flotation

-    Removal of concentrated contaminants and post-treatment.

In the first step suitable flotation agents are added to a
suspension of contaminated soil in water and usually mixed
intensively. These agents should adhere selectively to the
contaminated particles and their function is to give hydrophobic
properties to the surface of the contaminants.

The second step consists of the formation of small air bubbles
in the soil suspension. These air bubbles adhere to the hydrophobic
particles of contaminants and transport them to the surface of the
suspension where they are removed as a foam. The foam can be treated
in several ways, most of which will result in a concentrated sludge
of the contaminants. The contaminanted sludge contains a small
amount of the treated soil, strongly depending on the clay and silt
content of the soil, the process conditions and the type of flotation
agents used. Sometimes it is possible to reduce the amount of sludge
by removing the clay particles from the foam in a secondary flotation
step. The process scheme of the complete flotation process is shown
in Figure 3.6.

As already mentioned, selective flotation is applicable where
the contaminants are present in the soil as distinct particles or
when the contaminants are in the form of a lipophilic liquid
(potentially emulsion forming). If this condition is satisfied all
kinds of contaminants like halogenated organics, oil and metal
compounds (as particles) are treatable. There is also another mode
of flotation – ion-flotation – for removing inorganic ions from a
soil–water suspension. The flotation agents used exist as ions in an
aqueous solution and react in a suspension of soil with the
contaminating ions to form components with hydrophobic properties.
These can be lifted out from the suspension by means of air bubbles.
Ion-flotation is applicable where dissolved metals (eg Cu, Co, Cd, Ni
etc) are present in the soil suspension. However, although,
separation of dissolved copper from ore by ion flotation has been
investigated, the principle has not yet reached the stage where it
can be applied in practice, even in ore mining. It is therefore
unlikely that this method will be of practical use in soil cleaning
in the near future.

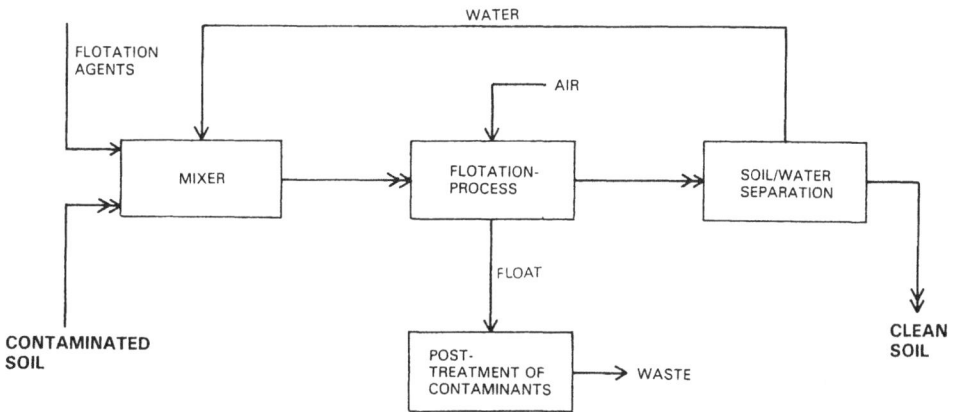

Figure 3.6   Treatment of Soil by Flotation

The main difficulty of the flotation and ion-flotation processes is to obtain such a selectivity that only the contaminants are removed and not the (small) soil particles themselves. Otherwise a large amount of residual sludge will result. Another drawback of flotation is its sensitivity to changes in soil composition; therefore thorough mixing of all the soil to be treated prior to flotation is essential.

5   THERMAL TREATMENT

5.1 Introduction

There are two principle modes of thermal treatment of soil, ie

(a) Removal of the contaminants by evaporation and subsequent treatment of the resulting gas stream; two subgroups are distinguished:

   (i)   thermal treatment by direct heat transfer (convection or radiation) from heated air (or another gas) or an open flame to the soil:

   (ii) thermal treatment by indirect heat transfer (conduction) to the soil.

(b) Destruction of the contaminants by heating (direct or indirect heat transfer) the soil to an appropriate temperature.

The gas leaving the heat treatment appliance can be treated in three different ways, ie

-    Incineration at high temperatures in an after burner.

-    Thermal treatment at moderate temperatures using appropriate catalysts.

-    Treatment at low temperatures by wet scrubbing of the gas and purification of the washing liquid.

The different principles of operation are discussed below.

5.2   Thermal Treatment by Direct Heat Transfer (convection or radiation) to Remove Contaminants by Evaporation

5.2.1   Process Scheme and Description.   In Figure 3.7 a scheme is shown in which the soil is heated by direct contact with hot (combustion) gases. The three alternatives of post-treatment of exhaust gases are included.

Treatment of soil at high temperatures in order to remove the contaminants by evaporation can be effected in a kiln, most suitably a rotary kiln. In the rotary kiln gas temperatures of up to $900^{\circ}C$ and above are possible. Volatile contaminants and water originally present in the soil will evaporate when adequate contact between the soil particles and the heat source is effected. Soil temperatures lower than $600^{\circ}C$ will often be sufficient to remove the contaminants, the required temperature depending on the boiling temperature or vapour pressure of the contaminants concerned. The air or gas supplied to the rotary kiln in Figure 3.7 is raised to a high temperature in an external heater.

Because of its relatively low heat content, a large amount of gas is necessary to provide the energy for raising the temperature of the soil to about $600^{\circ}C$. Furthermore a rather long residence time is needed because of the small heat transfer coefficient between gas and soil. Preliminary calculations show that volumes of up to approximately 1000-2000 m^3 (STP) gas per tonne of treated soil are needed for a final soil temperature of $400^{\circ}C$. When heat is supplied by an open flame burner in the kiln the volume of gas necessary to heat 1 tonne of soil to $400^{\circ}C$ will be smaller, ie approximately 700-1500 m^3 (STP) per tonne, because of more direct heat transfer (higher temperature gradients and radiation) and smaller heat losses to the surroundings. The amount of gas is mainly dependent on the water content of the soil, the maximum temperature of the gas and the option of recycling latent heat from exhaust gases.

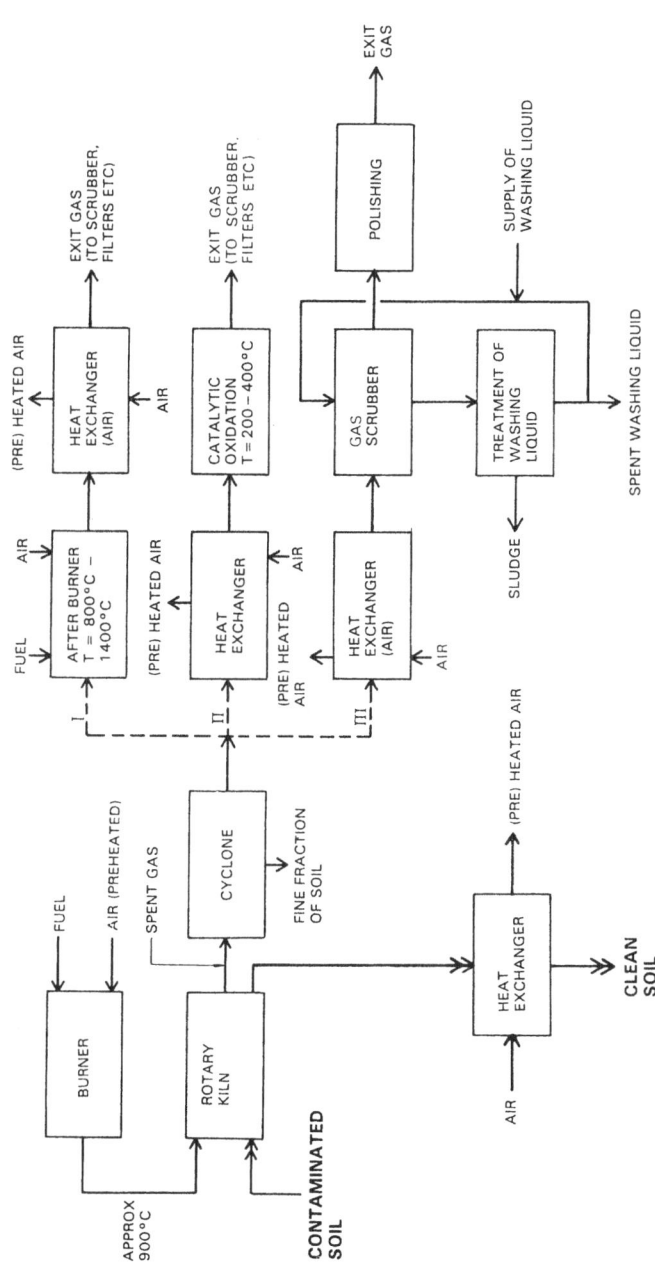

Figure 3.7   Thermal Treatment of Soil by Direct Heat Transfer and Evaporation of Contaminants

The gas leaving the rotary kiln will pass through a cyclone for separation of the fine mineral particles entrained by the exhaust gas flow. During passage through the cyclone the gas should have a sufficiently high temperature to avoid condensation of contaminants on the fine mineral particles to be separated.

Three possibilities exist for further treatment of the gas:

-   Incineration at high temperatures (I)

    The gas leaving the cyclone is burned or heated (eg in an after burner - see Appendix F) to decompose the contaminants to harmless products. In some cases (eg dioxines, PCBs) temperatures up to $1400^{\circ}C$ and residence times of several seconds are necessary.

-   Thermal treatment at moderate temperatures, using catalysts (heterogeneous catalysis) (II)

    Catalytic oxidation is an alternative to incineration in an after-burner. In principle it is possible to oxidise catalytically several organic substances at temperatures between 200 and $400^{\circ}C$, which would otherwise only be decomposed or destroyed at temperatures of $800^{\circ}C$ and higher. In general, for heterogeneous catalysis, metals like Ni, Zn/Cu, Fe/Cu, Al/Cu, Pt and Pd/C are used.

-   Treatment at low temperatures (III)

    The gas from the cyclone is cooled in an air condenser and then scrubbed with a suitable liquid. It is possible that very fine particles which have passed through the cyclone will be retained in the liquid phase. The liquid used must be such that the condensed contaminants dissolve in it, or are retained as a colloidal suspension. The spent liquid may be partly recycled or go on for further treatment. When the liquid is an organic solvent this treatment step may consist of distillation. Whenever an aqueous solution is used, normal physical-chemical treatment methods can be used, like coagulation, precipitation and adsorption. As a rule, scrubbing will be done in spray towers although other devices like packed-bed scrubbers can be used. The scrubbed air can be polished, eg by activated carbon adsorption, and then discharged.

It is possible to combine two or three methods of gas treatment and sometimes this may be essential. For example in the case of burning evaporated contaminants that contain halogenated compounds, a supplementary scrubber is needed to remove the resulting products (HCl, HBr) from the gas stream; a scrubber will also remove any dust particles (see 5.2.2).

5.2.2 Potential Applications

Removal of contaminants by evaporation

Evaporation of contaminants by thermal treatment of the soil is restricted to those cases where the contaminants have relatively high volatilities at the applied temperatures, eg hydrocarbons and halogenated hydrocarbons. Moreover, the method can be applied for the sublimation of certain metals or metal compounds. For results of practical experiences with this method see Appendix F.

Post-treatment of gas

Incineration of the spent gas leaving the cyclone is particularly favourable when hydrocarbons are involved which completely oxidise or decompose at the applied temperatures. A partial oxidation will not be acceptable because it may interfere with environmental requirements. In this connection the risk of dioxin formation during the combustion of certain pesticides can be mentioned. Incineration of halogenated hydrocarbons is a possibility only when emission of the evolved halogenides to the environment is permissible. If not, the halogenides have to be adsorbed by a washing liquid (water) before emission of the spent gas is permitted. In the case of heavy metals, post-treatment of the gas by means of an after-burner has always to be followed by a process that removes the heavy metals from the gas phase.

Catalytic oxidation of the contaminants present in the spent air is applicable to those cases where the contaminants are oxidisible to harmless oxidation products at the applied conditions. In the case of halogenated hydrocarbons catalytic oxidation has to be combined with gas-washing. Some catalytic oxidation processes are well established for industrial waste gases, eg carbon monoxide, methane, aldehydes, anthracenes, nitrogen dioxide, nitric oxide, formaldehyde, naphthaquinones, phenols, oil vapours, solvents.

Treatment of the spent gases at low temperatures in a scrubber may be the most generally applicable mode of post-treatment. With this method all remaining contaminants, such as heavy metals and halogens, can be removed from the gas phase, if the proper equipment is used and the appropriate process conditions are applied.

5.2.3 Limitations.
The most important limitations of processes involving evaporation of the contaminants followed by treatment of the resulting gas stream are:

—   Relatively large amounts of gas are necessary to supply the heat required for evaporation in the rotary kiln. Consequently, the installation required for treatment of the spent gases has large dimensions (heat exchanger, after-inceratory scrubber, catalytic oxidation equipment). This means high investment costs.

- The process requires a large amount of energy because usually a lot of water has to be evaporated and the soil itself has a large heat content at elevated temperatures. This makes it attractive to recover the latent heat from the exhaust gases to preheat the soil.

- Clay-rich soil easily forms clods or nodules in rotary kilns. Adding pebbles to the soil can reduce the amount and size of clods formed. The pebbles can be recovered by sieving.

- The practical use of catalytic oxidation is still at an exploratory stage.

- Incineration should occur under such process conditions (temperature, residence time) that formation of toxic components (eg dioxines) is avoided. Adequate safety control measures should be taken to avoid calamities (caused by the emission of unexpected contaminants). The risk of formation of toxic components may be lowered by homogenising the soil prior to treatment.

5.3 Thermal Treatment by Indirect Heat Transfer (Conduction) to Remove Contaminants by Evaporation

In Figure 3.8 a scheme is shown of this mode of evaporation of the contaminants. Indirect heating is realised by the use of heat transfer pipes in a rotary kiln. Only a small gas stream is necessary for the transport of the evaporated contaminants out of the kiln, eg 300 m^3 (STP) per tonne soil and its post-treatment is similar to that discussed in 5.2. However, owing to the gas stream being smaller the dimensions of the post-treatment installation are markedly smaller, resulting in lower investment and process costs for the post treatment installation.

The primary difficulty of this treatment process is that heat transfer from a hot gas to soil by an indirect method requires a very large heat exchange area. It is estimated that 10-20 m^2 is required for heating 1 t/h of soil up to 400°C. This is mainly due to the low heat transfer coefficients that can be obtained both on the gas side and on the soil side of the heat exchanging surfaces. The heat exchanging area can be somewhat reduced when (superheated) steam is used as a heat supplying medium instead of a gas, but the use of steam is limited to cases where temperatures of around 100°C or up to approximately 160°C (high pressure steam) are sufficient or preferred. The investment costs are rather high owing to the

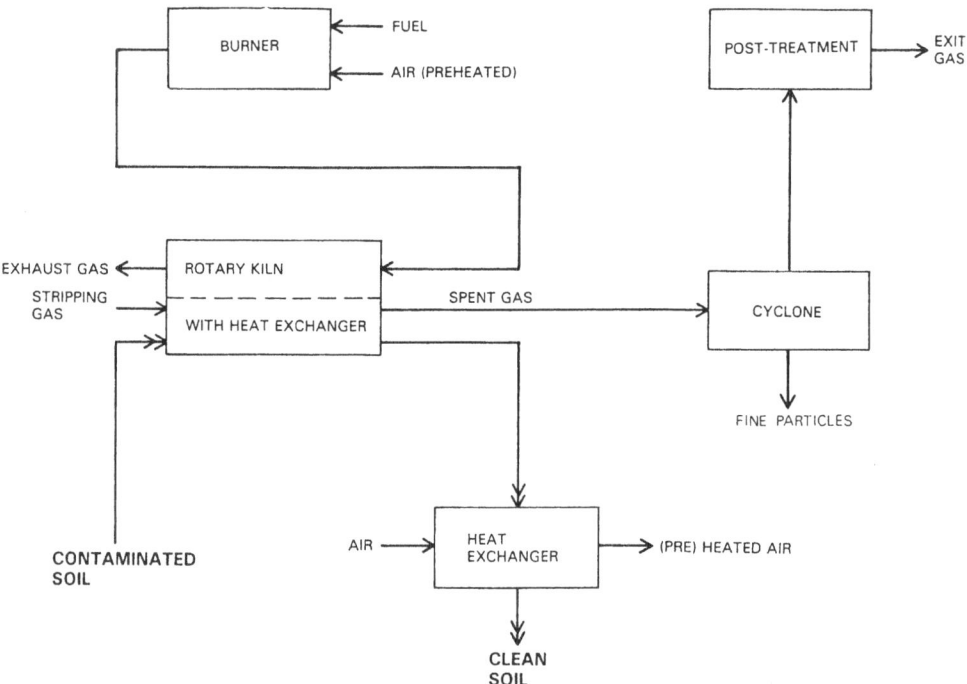

Figure 3.8   Thermal treatment by indirect heat transfer and
              evaporation of contaminants

complex construction of the heat exchanger kiln.   For other
difficulties that can be encountered, reference should be made to
5.2.

5.4   Destruction of the Contaminants by Heating or Incinerating
         the Soil

     In the thermal treatment methods mentioned in 5.2 and 5.3 the
volatile contaminants are evaporated and subsequently destroyed or
removed from the gas phase.   Another possibility using thermal
treatment is direct incineration at a temperature sufficiently high
to destroy the contaminants.   This can be achieved in a rotary kiln
but other options are methods already applied for incineration of
hazardous wastes, such as multiple hearth and fluidised bed

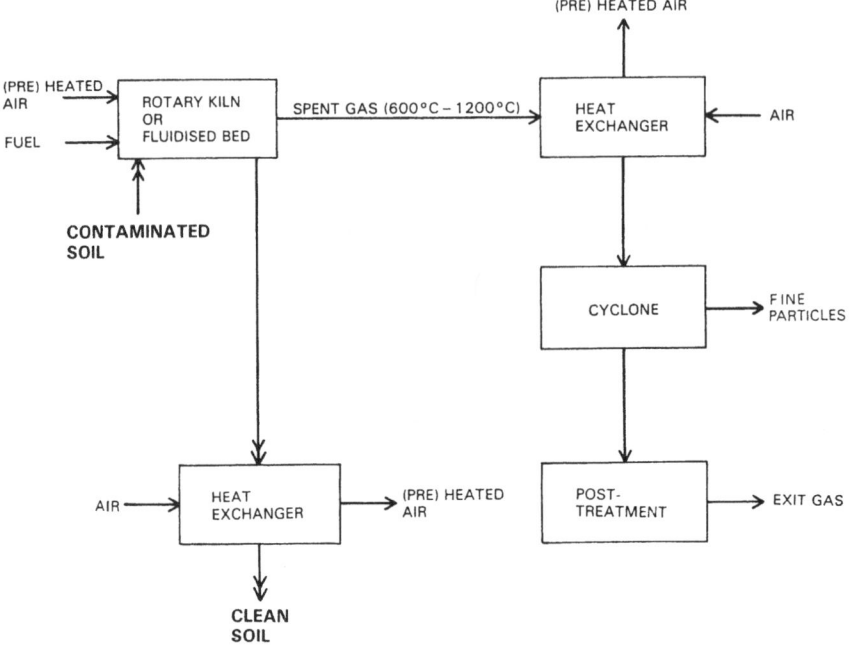

Figure 3.9   Thermal Treatment by Incineration

incinerators; possibilities like molten salt and plasma reactors are
still in the development stage and are expected to be expensive.
Figure 3.9 gives a scheme of the process. Post-treatment of the
spent gases can be carried out as already discussed in 5.2.

Destruction of contaminants present in the soil by incineration
or (direct) heating can be applied to organic contaminants with a low
thermal stability, of both high and low volatility (even for example
tar-like substances), and to cyanides.

In general incineration processes are carried out at gas
temperatures between 800 and 1200°C. The final temperature of the
soil will often be a few hundred degrees lower ie 600-1000°C.
Vitrification or sintering of the mineral part of the soil can occur
at high temperatures. To prevent this it is essential that the
temperature of the soil is as low as possible and that the soil
particles are in continuous motion. This can be achieved in a rotary
kiln or fluidised bed provided that the soil has fluidising
properties when it is in a dry state.

For a brief discussion of the various process steps which can be
distinguished in this type of treatment process and the typical
problems encountered, reference should be made to 5.2 and 5.3. The
process described is not as easy to control as the other (2-stage)
processes described, especially when the final temperature is below
1000°C (there is risk of incomplete incineration of contaminants,
eg PCB's).

5.5 Special Safety Aspects

Most thermal methods include a step in which the contaminants
are present in the gas phase. In general these compounds are of an
organic origin and are sometimes potentially explosive. Measures
should be taken to minimise explosion hazards, eg by controlling the
concentration of oxygen and hydro-carbons present in the kiln (and
afterburner).

6   STEAM-STRIPPING

6.1 Introduction

In steam-stripping volatile components are removed from a solid
or liquid phase by contacting it with steam. The process could be
applied for decontaminating soil if the contaminants which may be
either water soluble or water insoluble, are relatively volatile.

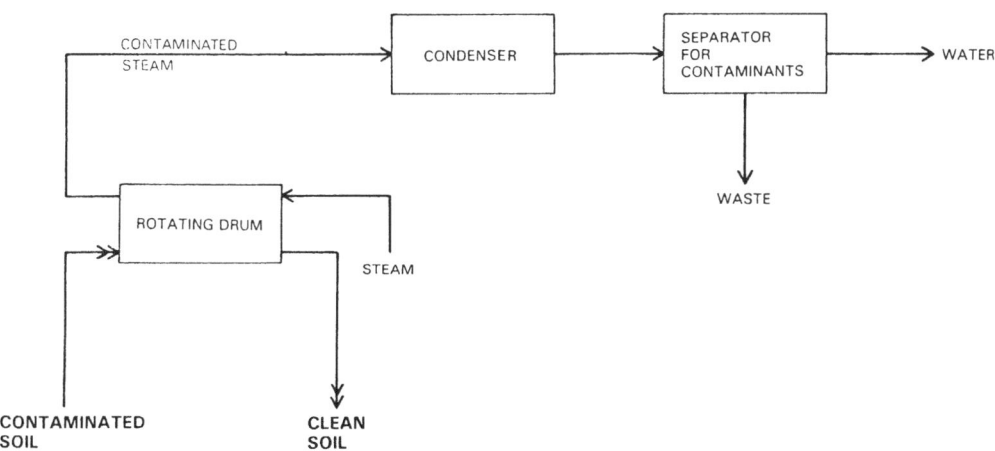

Figure 3.10 Treatment of Soil by Stream-stripping

Water insoluble contaminants are easier to remove however owing in part to the formation of low boiling azeotropes. To obtain an efficient treatment process it is important that the volatility or vapour pressure of the contaminants is high enough at the azeotropic temperature or the temperature of condensing steam and an intensive contact between gas and solid is achieved.

Another form of steam distillation is the generation of steam from the soil itself, wetted with some extra water if necessary. This can be done by direct or indirect heating, in a rotating drum, for example. The steam generated from the soil 'strips' the volatile contaminants and can be removed with the gas phase. This process partly resembles the evaporation technique described in 5.3, though the mechanism of removal of contaminants from the soil is different and the final process temperatures are only approximately 100°C (atmospheric conditions) or up to approximately 160°C (high pressure conditions).

An advantage of steam-stripping is the small gas stream and consequently the small (and therefore relatively inexpensive) apparatus required for treating the condensed gases from the steam stripping unit.

6.2 Process Scheme and Description

The equipment used for steam distillation in waste treatment generally consists of a rotating drum in which soil and steam are preferably in counter-current flow. The soil is heated directly by the (superheated) steam. Preliminary calculations show that it requires about 3 wt % of water to condense to heat the cold soil to 100°C. The evaporated contaminants are carried off with the steam flow. Steam and gaseous contaminants are both condensed in a condenser. The contaminants are then separated from the condensate. Several treatment methods are available, dependent on the type of contaminants. Activated-carbon, adsorption, simple decantation of the contaminants which are water-immiscible, separation by oil filters, oil skimmers, cyclones, centrifuges and flotation can be mentioned in this context. The process is presented schematically in Figure 3.10.

6.3 Potential Applications

The types of volatile contaminants that can be removed from contaminanted soil include:

-       Water-immiscible hydrocarbons, for example, petrol, kerosene, turpentine, benzene, toluene, xylene.

- Water–immiscible halogenated hydrocarbons, for example
  perchloro ethylene, trichloro ethylene, methylene, trichloro
  benzene, dichloro benzene.

- Water–soluble hydrocarbons, for example methanol, ethanol,
  iso–propanol, phenol.

- Volatile inorganic components, for example H_2S and NH_3.

Practical applications to soil cleaning have not yet been fully
developed. Research work should be focussed on the following problem
areas:

- Contaminated soil mostly contains some water. It is therefore
  expected that the soil, in particular soil that is rich in clay,
  has poor free-flowing properties. This can be a problem in
  effecting good contact between soil particles and steam flow.

- When no energy is recovered the total energy need of the process
  can be high. Recovery of energy is possible if vapour
  recompression is applied.

The best prospects for steam–stripping are in cleaning soils
contaminated with volatile hydrocarbons (eg halogenated solvents like
1,1,1-trichloroethane) which can cause difficulties when incinerated.

6.4 Limitations

One of the main limitations that is expected is how to remove
the final traces of contaminants from the soil in the three phase
system soil–water–steam. At low concentrations contaminants are in
general preferably adsorbed by the soil particles which are covered
with a water film. Consequently, a long residence time may be
required to achieve the desired low concentrations of contaminants.

7   CHEMICAL TREATMENT METHODS AND ELECTROLYSIS

7.1 Introduction

Chemical treatment methods and electrolysis comprise a large
number of techniques with the common feature of a chemical or redox
reaction (directly or indirectly) involving the contaminants and
resulting in harmless products. The methods discussed here do not
include stabilisation techniques ('immobilisation', fixation and
solidification), as these are discussed separately in Section 9.
For in-situ chemical treatment techniques reference should be made
to Chapter 4.

Table 3.1   Important Chemical Treatment Methods

Treatment method	Notes	Direct applicability to contaminated soil
electrolysis	1	low
chlorinolysis	2	low
neutralisation		high
hydrolysis	3	moderate
chemical oxidation		moderate
chemical reduction		moderate
ozonation	4	low
photolysis	5	low

Notes:

(1)   Electrolysis refers to the reactions of oxidation or reduction taking place at the surface of conductive electrodes immersed (in a suspension of soil) in an electrolyte, under the influence of an applied EMF.

(2)   Chlorinolysis refers to the reactions between hydrocarbons and chlorine at temperatures around $500^{\circ}C$ and pressures of about 200 atm forming carbon tetrachloride.

(3)   Hydrolysis generally refers to double decomposition reactions with water of the type: $XY + H_2O = HY + XOH$. The reactions are usually carried out at elevated temperatures and pressures, often with acid, alkali or enzyme catalysts.

(4)   Ozone gas is a powerful oxidising agent that cannot be shipped or stored, so it must be generated on site prior to use. The ozone produced is led through a suspension of soil in water or a packed soil bed.

(5)   In photolysis, chemical bonds are broken under the influence of UV or visible light. Reactor geometry must be such that adequate penetration of the light onto the contaminated soil occurs.

Table 3.2 Potential Applications of Chemical Methods and Electrolysis

Contaminant	Treatment method	Chemicals that can be used
Cyanide	Oxidation to CNO^-	$NaClO$, $Ca(ClO)_2$, $O_3(+ UV)$
	Hydrolysis to CO_2 and NH_3 or N_2	Cl_2, alkali
Cyanide complexes	Oxidation	$O_3 + UV$
Heavy metals	Oxidation to change leachability (eg Cr^{3+} to Cr^{6+}) or to enhance precipitation	$NaClO$, $KMnO_4$, O_3, H_2O_2
	Reduction (eg Cr^{6+} to Cr^{3+})	SO_2, SO_3^-, Fe^{2+}, Al
	Electrolysis	
Halogenated hydrocarbons	Hydrolysis	Aqueous acids, alkali
	Oxidation	ClO_2, O_3, H_2O_2, $KMnO_4$
	Electrolysis	
Organics, general	Oxidation (eg phenolics, aldehydes)	H_2O_2, O_3 (+ UV)
	Hydrolysis	Acids, alkali
	Electrolysis	

     Two major groups of chemical treatment methods can be
distinguished. First, the treatment of the soil without slurrying in
a liquid phase and second, the treatment of soil in suspension (in a
suitable liquid) with a suitable chemical. In addition to a direct
treatment of contaminated soil, chemical treatment methods can also
be applied in combination with other techniques, eg thermal treatment
and extraction. For instance when extraction is used, chemical
treatment can be of importance for the detoxification of the used and
polluted extracting agent.

A large number of chemical treatment methods for contaminated soil are available. Table 3.1 lists the most important chemical treatment methods including electrolysis for hazardous industrial wastes and the presumed applicability to contaminated soils. Table 3.2 lists potential treatment methods for specific contaminants.

Not all techniques are considered in the following paragraphs. Chlorinolysis and photolysis are not discussed any further in view of their limited applicability.

Some general characteristics of the chemical treatment methods of soil are:

-    intimate contact between soil and chemicals is essential;

-    contact times are frequently long;

-    chemicals are dosed in excess of the contaminants to ensure complete detoxification.

7.2 Process Schemes and Description

Only some general process schemes are given, because of the large number of different methods and the lack of experience in applying these techniques to soil. Nor is attention given to chemical methods in the post-treatments of solutions and gases other than those arising from chemical treatment.

The first group of processes are those which consist simply of mixing contaminated, relatively dry soil with a chemical (Figure 3.11). As noted before, the chemicals will be added in excess, and therefore this method can be used only when the chemical itself is harmless to the environment or quickly loses its chemical stability. It is desirable that the soil be relatively dry after the treatment

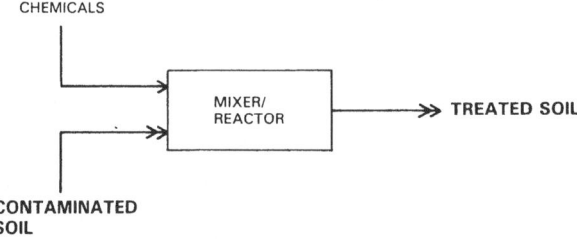

Figure 3.11 Treatment of Soil by Mixing with Suitable Chemicals

before it is re-deposited, and so the chemicals should preferably be added in a solid form or as liquid in a low volume ratio to the soil. An important example of this kind of treatment method is pH-adjustment of soil. Several kinds of apparatus can be used for the mixing step, for instance rotating mixers and screw conveyors.

A second group of processes are those in which a separation step between soil and chemicals is needed (Figure 3.12). This can be the case with chemicals in solution (eg diluted sodium hypochlorite) or gases (eg ionised air, ozone). Soil is brought into contact with a suitable chemical reagent. After a certain reaction time in the mixer-reactor the mixture the soil is separated from the reactive phase as completely as possible. In some cases, upgrading of the reactive phase may be possible, otherwise the used chemicals should be detoxified and discharged. Upgrading of the reactive phase generally results in a waste stream of reacted matter and sometimes some soil particles.

The treated soil generally needs post-treatment to remove or detoxify any hazardous material present (either the contaminants or the used chemicals). This will also result in a waste stream in most cases.

Both the first and the second groups of processes (Figures 3.11 and 3.12) include treatment methods such as neutralisation, hydrolysis, chemical oxidation and reduction, and ozonation. There is little information available on the treatment of soil by ozonation, chemical reduction and hydrolysis.

Electrolysis is a somewhat different type of process (see Figure 3.13). It is not known if it can be applied directly to a mixture of soil and a suitable liquid phase, but there have been some laboratory experiments (TNO, Netherlands). Preliminary results with respect to the destruction of organic compounds are promising, but the process has not yet been applied on a larger scale. Some experience with the removal of mobile metal ions is also available[7]. In any case one can assume that electrolysis is a suitable technique as a secondary treatment step of, for example, extraction liquids.

7.3 Potential Applications

The chemical methods considered are derived from treatment methods for hazardous wastes in general and have not yet been developed and adapted for treating contaminated soils. The list of possible applications in Table 3.2 is therefore probably not exhaustive and may be too optimistic in some cases.

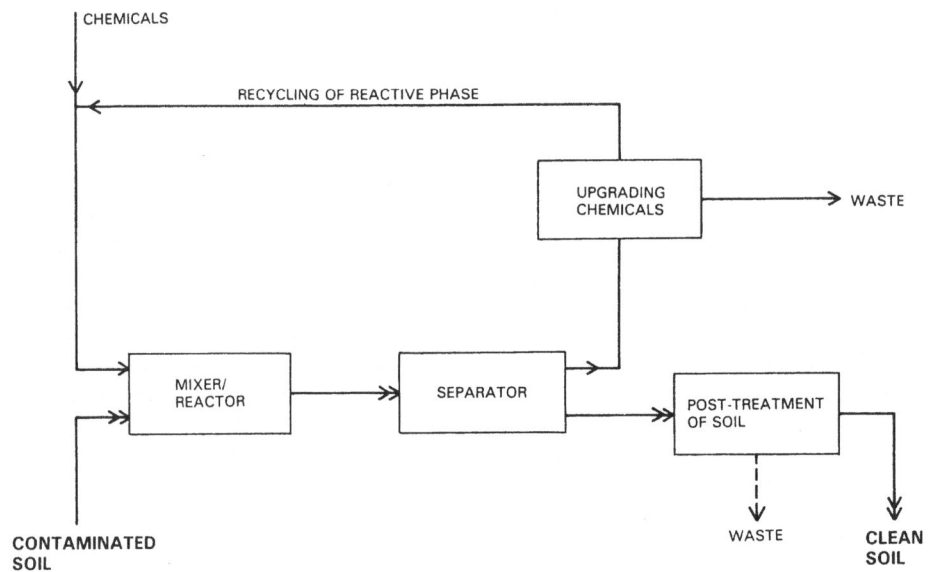

Figure 3.12 Chemical Treatment of Soil by an Excess of
            Reactive Liquids or Gases

7.4 Limitations

As mentioned before, none of the chemical methods that are
considered here have yet been developed for treating contaminated
soils. Therefore much research and development work is needed.

The following problems are expected:

-   Natural soils consist of several compounds among which there are
    many of an organic nature (humic-acid like substances).
    Chemicals added to contaminated soil will often also attack
    these natural compounds. It is possible that the products
    resulting from these reactions will not be acceptable from an
    environmental point of view.

-   A long reaction time will often be needed to ensure that all
    contaminants are completely converted. Therefore large reactors
    will be required for on-site treatment if immediate redeposition
    after mixing is not allowed.

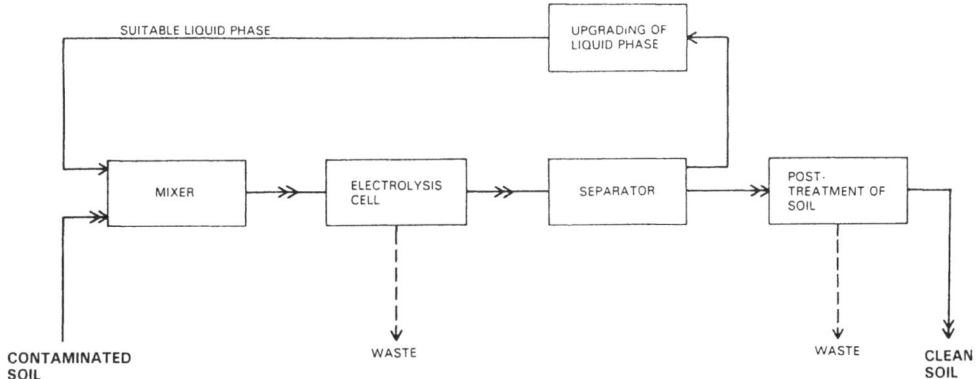

Figure 3.13   Treatment of Soil by Electrolysis

- Chemical agents are expected to be used in excess, in order to allow for competitive reactions and to ensure a complete reaction. Mostly, the chemicals used are not harmless to the environment, so an extra separation and post-cleaning step are needed. Furthermore, special safety measures are needed when these chemicals are used.

8   MICROBIOLOGICAL TREATMENT

8.1   Introduction

Many hazardous chemicals present in soil can be decomposed or converted to harmless components such as H_2O and CO_2 by microbiological methods. To effect microbiological degradation in an actual situation of soil contamination a number of conditions have to be fulfilled. These are:

- Presence of sufficient oxygen for aerobic processes and a full depletion of oxygen for anaerobic processes.

- Presence of nutrients (mainly P and N).

- Presence of sufficient water.

- Concentrations non-toxic to the micro-organisms.

-     Sufficient availability of the contaminants to the micro-
      organisms.

-     Presence of the appropriate micro-organisms.

-     Favourable temperature.

     In general, microbiological degradation processes are very slow.
This is especially the case with anaerobic degradation so for the
on-site treatment of contaminated soil the usually more rapid aerobic
microbiological processes are to be preferred.

8.2 Process Description, Applicability and Limitations

     Several modifications are possible for the microbiological
treatment of contaminated soil. A schematic presentation of the
basic process scheme is given in Figure 3.14. The soil to be treated
is mixed with nutrients (for example in a rotating drum), suitable
micro-organisms and water (if necessary) and placed in an aerated
reactor. In this reactor, which can be a fermentor, a simple
container or a suitable pit, microbiological degradation of the
contaminants takes place. Other possible methods are composting
(stockpiling) and techniques derived from land farming of sludges.
These techniques have been well known for many decades, but have so
far not found general application in treating contaminated soils.

     The method is applicable to all contaminants which, under
certain circumstances, are biodegradable, for instance:

-     cyanides (also anaerobic conversion) and nitril-compounds

-     simple aromatics, such as benzene, toluene, xylene, phenol

-     n-alkanes

-     simple cycloalkanes

-     low chlorinated C-1 and C-2 compounds (eg $C_2H_4Cl_2$)

-     certain pesticides.

     Not or hardly biodegradable are heavy hydrocarbon compounds,
polynuclear aromatics or highly substituted compounds such as PCBs.
The degradation of these compounds can take several years, even under
optimal conditions.

     A second method of microbiological treatment is bacterial
leaching of metals. It is known from studies on mine waste dumps,
containing only traces of the original metals, that the leachates may

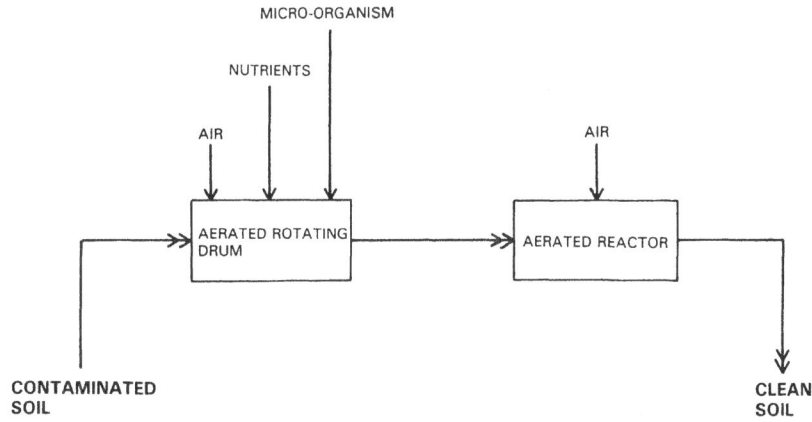

Figure 3.14   Microbiological Treatment of Soil

contain considerable amounts of these metals. Different strains of
bacteria (Thiobacillus) are capable of oxidising sulphide containing
ores to form more soluble sulphate salts, deriving the energy for
their life cycle from sulphur and iron-oxidising reactions. It is
known that pyrite mine wastes containing sulphides of copper, zinc,
nickel, cobalt, manganese, uranium or molybdenum can be leached in
this way, but they are very slow and likely to be more suitable for
in-situ treatment.

Another possibility for biological treatment is the formation
of (volatile) metallo-organic compounds. Examples are arsine
(AsH_3) and methylised mercury, both being very toxic. The volatile
can subsequently be removed from the soil by gas stripping.

A prerequisite for on-site application in a mobile installation
is that the micro-biological conversion rate of the contaminants is
fast enough. Otherwise a long residence time and consequently a
large reactor volume are necessary. Methods that make no use of
special apparatus to contain the soil during treatment, for example
stock piling, land farming and treatment in pits, are relatively
inexpensive (low investments) and are able to handle large amounts of
soil during a long period. The disadvantage of such methods is that
the conditions during the process of microbiological conversion are
not easily controllable. It is stressed that conversion rates are
strongly dependent on prevailing conditions such as temperature,
moisture content, available nutrients, etc.

It may be permissible in some cases to redeposit the soil –
mixed with micro-organisms, water and nutrients – before the
microbiological degradation has been completed. In such cases low
microbiological degradation rates are acceptable. This means that
anaerobic microbiological degradation processes can also be used.

Microbiological treatment can also be implemented as a unit
process for treating contaminated extraction agents (resulting from
both on-site and in-situ extraction). This should provide a less
heterogeneous feed stream for the micro-organisms and the conditions
of the microbiological treatment can be more easily optimised for a
high degradation efficiency than when a soil is treated directly (see
also Chapter 8 for microbiological treatment of groundwater etc).

9   STABILISATION TECHNIQUES

9.1   Introduction

Stabilisation processes are not intended (nor do they provide
the possibility) to remove the contaminants from a certain soil or
waste, but rather to eliminate physically or chemically the hazardous
nature of the contaminated soil (or waste, sludge or liquid) so that
it can be handled, stored or dumped in a safe way. Basically, the
technique aims at minimising the leachability of the contaminants in
the bulk of the soil. Other aims of stabilisation can be reducing
dust emissions and improving the bearing capacities of the soil
treated. A minimisation of the mobility or leachability can be
achieved by the following mechanisms:

(a)   a chemical reaction with the contaminants to form hardly
      soluble compounds (chemical immobilisation).

(b)   isolating the contaminants from infiltration water by adding
      hydrophobic compounds or chemicals that will form hydrophobic
      compounds.

(c)   adding chemicals able to fixate (available) water and thus
      influence the (micro-)leachability.

(d)   vitrification.

(e)   influencing pH-value and redox-potential in order to achieve
      minimal solubility of the contaminants in leachate or
      groundwater.

Mechanisms (a) and (d) will seldom occur in commercial
processes. Mechanism (c) is the most common one. Mechanism (e) is
in fact a special case of mechanism (a).

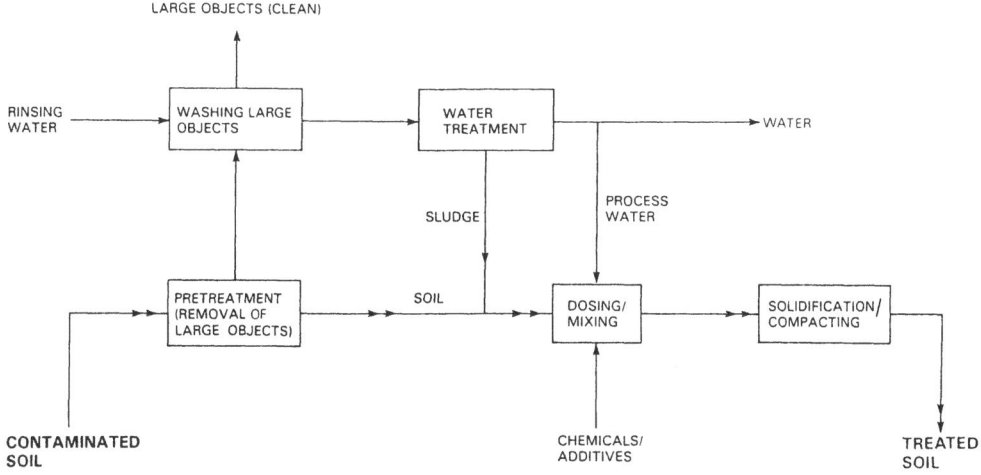

Figure 3.15 Process Scheme for Stabilisation Techniques

Stabilisation techniques are considered here as the collection of stabilisation, solidification and immobilisation techniques. For in-situ stabilisation techniques reference should be made to Chapter 4.

9.2 Process Scheme and Description

9.2.1 General. Most on-site stabilisation methods use relatively simple techniques for mixing contaminated soil and suitable chemicals or other additives[8]. There are two main options for the mixing step. In the first the contaminated soil is spread in thin layers on a suitable plot of ground and chemicals are added to each layer by spraying and/or ploughing. The next layer of soil is placed on top of the treated soil after a certain reaction time has been allowed.

The second option comprises simple mixing of soil and chemicals in a mobile installation, eg a rotating drum. This technique can be particularly suitable for those cases where emissions of hazardous gases can be expected. The treated soil (mostly a slurry) can be either transported by tanker-lorry to the final destination or pumped into a mould to solidify on-site. Figure 3.15 gives a general impression of the relevant process steps, including an option for pretreatment of the soil.

Although the treated soil could be redeposited on-site (or another suitable location) in some cases the reuse of the product could be considered, such as for road construction works (in the case of a good bearing capacity), soil stabilisation and anti-traffic noise walls.

The commercially available stabilisation techniques were in most cases originally developed for the treatment of waste sludges, but several are also suitable for the treatment of contaminated soils. Most of these stabilisation methods use one of the following compounds, or combination of compounds, as active chemicals: cement, lime, gypsum, pozzolans, bitumen, thermoplastics and/or organic polymers (resins). Several recipes use special additives such as water glass and fly ashes.

The following paragraphs describe the most important methods for stabilising contaminants. Stabilisation techniques based on treatment with organic polymers as well as vitrification techniques will not be discussed, for these techniques are specially developed for very hazardous chemical and nuclear wastes. Chapter 4 deals with an in-situ vitrification technique.

9.2.2 Methods Based on Cement Products. Cements have been used for centuries in construction works. Currently, in addition to Portland cements, a variety of cements are available made from blends of Portland cements with pozzolans or hydraulic slags (see below and 9.2.4). Stabilisation involves mixing the soil with a cement based product and water in sufficient quantities. In general, the amount of cement is not negligible compared with the amount of soil treated. The amount of soil is in consequence significantly enlarged by the treatment.

Several additives may be used to improve the physical and chemical properties of the final product, eg waterglass (sodium silicate). The exact recipes for these cement-based stabilisation processes are, in general, industrial secrets.

Significant amounts (1-5% in soil) of organic compounds like peat, humus or organic contaminants can slow or even prevent the hardening of cement. Thus cement is not suitable in such cases unless the soil is first treated to render these organic compounds harmless.

Certain metal ions react with the carbonates or hydroxides in the cement mixture and form insoluble compounds. Other compounds in the cement mixture can also probably immobilise or partly immobilise metal ions by chemical fixation although this has not been proved so far. Certain metals strongly affect the final strength and setting properties of the cement, for instance salts of sodium, manganese, lead, tin and zinc. Advantages and disadvantages of stabilisation techniques based on cement are given in Table 3.3.

9.2.3 Methods Based on Lime Products. Some natural soils contain pozzolan-like materials (see 9.2.4) and this can be of benefit for stabilisation methods based on lime products. Calcium silicate or calcium aluminate gels will be formed when calcium oxide or hydroxide is added to these soils. The gels bring about a fixation of water and therefore reduce the leachability of contaminants.

Lime products can also be used in the absence of pozzolan like materials. An important example is the use of calcium oxide for oil containing wastes. Calcium oxide will react with water contained by the waste as expressed below:

$$CaO + H_2O = Ca(OH)_2 + heat$$

This reaction produces significant heat and is responsible for enlarging the surface area of the lime in such a way that free oil will be enclosed by solid calcium hydroxide (micro encapsulation). A disadvantage is that the heat generated can cause volatile compounds to evaporate.

Commercial processes based on lime products often use additives, such as pozzolans.

The advantages and disadvantages of these stabilisation techniques are summarised in Table 3.3.

9.2.4 Methods Based on Pozzolans or Hydraulic Slags. Pozzolans are materials that contain active silicates or aluminates. Examples are volcanic ashes, fly ashes, and trass.

The stabilising effect of pozzolans is based on the formation of gels when they are mixed with water-containing wastes or soils. The ultimate strengths of the gel or final product will be reached only after a considerable time (several months to some years). The presence of sufficient lime in the soil or soil mixture is essential for stable gels. The lime may be present in the soil, pozzolan or deliberately added. If Portland cement is used as a source of lime then the pozzolan takes part in the overall hydraulic reactions of the calcium silicates and aluminates that the cement contains.

Ground granulated or pellitized blastfurnace slag also exhibits hydraulic reactions in the presence of lime or other suitable alkaline activators (including Portland cements) and could also conceivably be used in stabilization.

The final product is a solid and almost impervious mass. It
is believed that several metal ions are chemically bound to silicates
and aluminates to form insoluble compounds.

9.2.5 Methods Based on Thermoplastics. Applicable thermoplastics are polyethylene, bitumen and paraffin; paraffin is only in
the stage of development. The contaminated material must first be
dried and heated before it is mixed with the thermoplastic. After
cooling and solidification the final product can be dumped.

Suitable specialised apparatus to heat and mix the mixture is
commercially available for example apparatus for asphalt production
are available with large capacities. The temperature necessary for
stabilisation with thermoplastics depends on the thermoplastic,
contaminants and apparatus used, but varies generally between 130 and
$230^{o}C$; the potential risk of fire or explosion should be
recognised. It is advisable to set a limit for the concentration of
flammable compounds. Furthermore, appropriate measures should be
taken to avoid the emission of toxic gases.

Compounds affecting the process or the physical-chemical properties of the final product are solvents (bitumen dissolves) and strongly
oxidising salts (eg nitrates, perchlorates). See also Table 3.3.

9.2.6 Minimising Mobility by Adjustment of pH and Redox
Potential. The mobility of metal ions mainly depends on the prevailing pH and redox value. If it is assumed that no stable complexes
with organic ligands are formed, a contaminated soil can be treated
with chemicals in such a way that mobile metal ions are precipitated.
It is well known that most metal salts, and especially oxides and
hydroxides, exhibit minimal solubility at pH value 8.5 to 9.
Slightly oxidising conditions are generally preferred for immobilised
metal hydroxides. Although the mobility of some organic contaminants
(eg phenols) is influenced by pH-value, this method is especially
suitable for soils contaminated with heavy metals and containing only
small amounts of organics (ie potential ligands for metal ions).

9.2.7 Miscellaneous Techniques. Some less widely applied
techniques based on epoxy resins, polypropylene, sulphur, polyesters,
phenolic resins, copolymers (eg urea formaldehyde) and vitrification,
have been developed for hazardous chemical wastes and it is not known
whether they can be applied to contaminated soils. In addition, they
are likely to be costly and may give rise to environmental
objections.

9.2.8 Combinations of Several Active Compounds. As already
mentioned above, most commercial stabilisation processes use
combinations of several active compounds in order to improve the
quality of the final product but the exact recipes of these
commercial processes are secret. The following combinations of

substances are frequently used for stabilisation of chemical wastes:

- lime and fly ashes or hydraulic slags

- cement and bentonite

- cement and silicates or aluminates.

9.3 Potential Applications

Whenever the soil is contaminated with one specific contaminant, eg an oily contaminant, acid resin or heavy metal, it is believed that suitable stabilisation processes are available. In most cases there will be more than one potential method to be considered for application. Most contaminated soils however contain a large range of compounds. Those soils cannot be treated without prior extensive tests. These tests should result in selection of the most appropriate stabilisation agent or in a special recipe for a mixture of reactants for the contaminated soil concerned. Some firms specialise in this kind of work and can in some cases offer a satisfactory solution. In other cases however, stabilization is inadequate or insufficient, owing to the large variety of contaminants present in the soil. Some processes can even enhance the mobility of certain contaminants.[9,50]

Some stabilisation techniques are less suitable for soils that are heavily contaminated with volatile compounds. Since the processes referred to release or use heat whenever applied to the contaminated soil, there is a serious risk of evaporating these volatile compounds.

Whenever the permeability of the treated soil or waste is too high (k 10^{-8} m/s) it is advisable to use coverings and/or barriers for the redeposited material to avoid infiltration of water. Specific applications are mentioned in 9.2.2 to 9.2.6.

9.4 Limitations

The limitations specific for a certain stabilisation technique have been described briefly in Section 9.2. Table 3.3 gives a summary of these limitations or disadvantages. A more general disadvantage is the risk of releasing hazardous vapours whenever heat is used or generated during the processing of soil.

Contaminants are not removed, so the treated soil can still form a potential source of hazard. This aspect can be a deciding factor in considering the use of a stabilisation method as a remedial action. In some countries a soil treated in order to stabilise or

Table 3.3 Advantages and Disadvantages of Several Stabilization Methods

Method/Chemical	Advantages	Disadvantages
1 Cement	- inexpensive chemicals - standard procedures, well developed - dewatering is not necessary - method is reasonably tolerant to chemical variations	- large quantities of cement required - final strength is influenced by several contaminants (eg organic compounds, sulphates, sodium, manganese, lead, tin, and zinc) - significant leachability in case of acids - NH_3 can be released in case of NH_4^+
2 Lime	- very inexpensive chemical - standard procedures, well known in waste treatment - well defined chemical reactions - dewatering is not necessary - fairly universally applicable	- significant leachability in case of acids - handling of lime can be difficult - added quantity of lime is not negligible
3 Pozzolans	- relatively small quantities required - long-term stability is good - well known reactions	- slow to very slow solidification - sensitive to acids
4 Thermoplastics	- minimal leachability - complete encapsulation - low biodegradability - fairly resistant to most aqueous solutions	- specialized apparatus, skilled labour - danger of explosion or fire hazard. - air pollution must be avoided - relatively expensive - drying of soil is necessary

immobilise the contaminants will still be considered hazardous, irrespective of the quality of the stabilisation process, and therefore may not be redeposited on-site. In these cases deposition elsewhere (eg hazardous waste dump sites) may be considered provided the normal legal requirements governing such deposition are met.

As mentioned before, a stabilisation method implies an extensive testing programme before a specific method can be allowed to be used on a particular site. The presence of combinations of several compounds (contaminants as well as natural constituents) can sometimes mean that no adequate stabilisation technique can be found. The US Environmental Protection Agency found[9] that most heavy metals in five types of waste sludge failed to become, or hardly became, insoluble after treatment with some commercial stabilisation processes.

Even when an appropriate method can be found for stabilisation, there are certain risks like the uncertainty of scaling-up (including undesired and unforeseen inhomogeneities of the soil) and the potential hazardous properties of a stabilised soil.

In general there are two ways in which the treated soil can become hazardous:

-   the treatment has not been executed according to the specifications, and therefore contaminants are not sufficiently immobilised;

-   owing to weathering and ageing the stabilising effect may be substantially reduced. This aspect is discussed in Chapter 2, on long-term effectiveness.

Furthermore, contaminants do slowly leach out to some extent from any stabilised soil. In general, leachate concentrations will be very low and will not be a problem for the receiving environment, but the possibility of accumulation of the released contaminants in some part of the environment should be considered in every case.

The treated soil can have a monolithic structure or be delivered in a milled form, but seldom resembles a natural soil. It will often be impossible to redeposit the treated soil on site. Special dump sites or destinations are then necessary. When the bearing capacity is sufficient the treated soil can be used for road construction works or dikes.

The leachibility of the treated soil or waste is often increased by a few freeze/thaw cycles and therefore the treated soil must be covered by a sufficiently thick layer of soil.

10   COST OF ON-SITE TREATMENT

The costs of the on-site clean up of contaminated soils are
strongly dependent on:

-    the type of contaminants, type of soil (and the interaction
     between the two and with the ecosystem),

-    the concentrations of contaminants and the extent of contamina-
     tion (spread and depth; amount of soil to be treated),

-    site situation (buildings, pipes, stratification of soil,
     uniformity, etc),

-    the need for special measures to be taken (safety, draining, air
     pollution, etc) and of course:

-    the method of remedial action chosen, and

-    the degree of removal of contaminants required.

Table 3.4   Estimated Costs of On-site Treatment Methods

Method	Range of costs* Hfl/tonne
Thermal	
- evaporation	80-150
- incineration	150-450
Steamstripping	(80-150)
Extraction	50-200
Mechanical and physical separation	
- classification	50-150
- differences in density (oil separation)	80-150
Flotation	50-150
Chemical treatment	
- 'dry' treatment	(20-80)
- 'wet' treatment	(80-200)
Biological treatment	(30-100)
Stabilization	
- lime, pozzolans	50-150
- cement	80-200
- thermoplasts	(100-1000)

* Figures in brackets are estimates

An estimate of the costs of several on-site cleaning techniques is given in Table 3.4. This table refers to the costs of the treatment itself and does not include costs such as expropriating land, legislation, excavation of soil, engineering constructions to keep the pit open or to support nearby buildings, treatment of groundwater, disposal or treatment of certain waste streams (sludges) etc. Figures in brackets refer to costs based on rough estimates of preliminary desk studies. Other figures are slightly more definite, partly because they are based on the experience gathered with actual treatment installations; see Section 11 and Appendix F. It is stressed, however, that the costs are strongly influenced by the depreciation of a particular installation. When a large market is available, rather low costs can be achieved. A small market means high costs per unit of soil. The depreciation costs can probably be cut by developing mobile installations that are easily adjustable to the specific conditions of different contaminated sites.

The total costs of remedial action can be several times higher than those in Table 3.4 owing to the excluded costs. Therefore no proper cost estimate can be made using Table 3.4 without further assessments.

11   PRACTICAL EXPERIENCE

As mentioned before, only a minority of the processes discussed have been developed to the stage of practical application[12]. Nevertheless, recent developments show that the number of available treatment methods will probably increase in the near future. In developing on-site treatment processes for actual application the technological know-how already mentioned in the introduction can be relevant (know how concerning unit operations, hazardous waste treatment, treatment of waste water and gases, transport and handling of soil). Also the typical properties of soil and the contaminants encountered in practice are important factors in developing on-site treatment processes.

A short summary is given below of on-site techniques that have been applied in practice or which are currently (summer 1983) the subject of testing programmes on a pilot plant scale. More detailed accounts of some of these, are given in Appendix F.

(a) Evaporation by direct heating at 200-300°C and treatment of gases in an after burner at approximately 800°C. The treatment installation has been developed by Ecotechniek and resembles the process described in 5.2. More details are given in F.4. Capacity: approx 20 t/h.

(b) Incineration by direct heating at 1000°C and treatment of gases in an after burner at 1200°C. This EPA-ORD Mobile

Incineration System was specially developed for treating PCB-containing waste media (including soil). A general process description is given in 5.4 and a more detailed one in reference 12. Exhaust gases from the after burner are treated by quenching with cold water, passing them through an air-filter and gas scrubber. Capacity: approx 4 t/h.

(c) Extraction, in combination with classification. A semi-continuous process installation has been developed by HBG with the primary aim of treating soils containing cyanide complexes. It is expected however, that the method will also be applicable to many other types of contaminated soils. The first tests have been run with positive results in the summer of 1983. The installation is a slightly altered version of the proposed version described in Appendix F. The main alteration is the exclusion of a series of hydrocyclones for counter current washing of fine sand and silt. These particles are only separated from the the water stream by a hydrocyclone and are subsequently stored for disposal or separate treatment (elsewhere). Furthermore, an upflow column is chosen to achieve the separation of clean, coarse sand particles from the extracting agent. Capacity: approx 5 t/h. A new installation with a capacity of 20 t/h is planned.

(d) Extraction. A mobile, prototype, field use treatment system in which a large variety of contaminated soils can be cleaned has been developed by US Environmental Protection Agency.[14] Three major treatment devices are used in the installation:

   (i) a rotating sieve drum with a water knife stripper for breaking up soil lumps and stripping chemicals from large particles;

   (ii) a 4-stage counter current extractor in which an intensive mixing is achieved by the pneumatic action of froth flotation units (flotation is not the primary aim, but can be useful in removing easily flotable materials);

   (iii) a system for cleaning and recycling of washing fluids (essentially a granular activated carbon treatment with ancillary treatment). Capacity: approx 5 t/h.

(e) Extraction of heavy metals from dredged materials by an acid treatment. The method is developed for the decontamination of harbour sediments, but is probably also applicable for contaminated soils. This technique, referred to as the leaching method of Muller[15], comprises three steps:

   —   acid treatment (HCl or aqua regia) to extract heavy metals,

   −   separation of the solids from the solvent,

   −   removal of the heavy metals from the extracting agent by a
          hydroxide and subsequently carbonate precipitation.

   The maximal capacity is approx 30 t/h.

(f) Separation based on differences in specific gravity (or more
     generally based on the lipophilic character of the
     contaminants). The firm Kroon (Netherlands) developed a mobile
     installation in which oil-like contaminants are spouted loose
     from soil particles by passing the contaminated soil through a
     water jet curtain (water velocity: 220 m/s). Soil particles
     (clean) are removed from the resulting slurry in two
     hydrocyclones. Contaminants are removed in a water-oil
     separator. Water is recycled to the water jet, but has to be
     refreshed once a day. Capacity: 10–20 t/h.

(g) Separation based on differences in specific gravity. Robertson
     Research International Ltd (UK) has developed a mobile 'dense
     media cyclone plant' for use in reclaiming areas of mine spoil
     which contain toxic metals such as lead, cadmium and zinc. The
     value of metals separated from the spoil defrays the cost of
     reclamation. The plant has successfully processed a
     lead/zinc/barites/fluorite site and a zinc/fluorite deposit.
     Capacity: 15 t/h.

(h) Flotation of contaminants in four steps by the so-called froth
     flotation principle. The firm Mosmans has built an installation
     which has been tested in Holland for the treatment of soil
     contaminated with oil (including PCAs). It may also be suitable
     for treating soils containing heavy metals, cyanide-compounds
     and chlorinated hydrocarbons (see Section 8 and reference 16).
     The separated foam is transported to a specialised waste
     processor. Capacity: approx 10 t/h.

(i) Dry chemical treatment. Experience in 'detoxifying' and
     'immobilising' Cr^{6+} containing soils (chromite roasting
     residues) exists in Japan[17]. Remedial activities consist of
     excavating the soil, mixing it up with an excess of $FeSO_4$ to
     reduce Cr^{6+} to Cr^{3+} and redepositing it at certain
     controlled sites (encapsulation by steel sheet-piles and clay
     layers). Slightly contaminated soil was redeposited on the
     excavated site after treatment.

(j) Bacterial leaching. Experience exists in extracting large mine
     waste dumps.[18-20] Leaching of wastes of metal sulphide ores
     (eg CuS) by bacteriological activity has been demonstrated as
     practicable and feasible. Also sulphide ores of zinc, uranium,
     nickel, manganese, molybdenum and cobalt can be leached by an

autogeneous process in which bacteria play an important role.
It is not known if leaching of metal compounds other than
sulphides can be enhanced by bacterial activity. Furthermore it
is unknown if these bacterial leaching processes can be applied
to normal soils containing low concentrations of (heavy)
metals.

(k)   A hot water 'fluidisation' process for cleaning oil-
      contaminated beach sand. A pilot device capable of cleaning one
      tonne per hour of oil-contaminated beach sand was built and
      tested by the University of California, Santa Barbara, USA
      around 1970[21]. The process is a variation of the hot water
      method used in the Athabasca Tar Sands Deposits and utilises
      water at $95^{\circ}C$ in a ratio of 1.2 m^3 to 1 tonne of soil in a
      fluidised bed contactor (upflow column). Tests performed with a
      sand mixture containing 1 to 2% of a 23° API crude oil showed
      that upwards of 95% of the crude oil could be removed.
      Operation with a 14° API residual oil was less satisfactory.
      The limitations of the process are concerned primarily with the
      range or distribution of sand partcile sizes that can be
      fluidised without excessive elutriation.

(1)   Stabilisation methods. At present, a wide variety of
      stabilisation techniques is commercially available[22]. These
      techniques are especially developed for chemical wastes or waste
      sludges (eg petrochemical sludges, paint wastes, acid resins,
      incineration wastes, mining tailings, waste water treatment
      sludges). Some contractors claim their procedure is also
      suitable for stabilising solid products, eg contaminated soil.

      Although all methods described in Section 9.2 are commercially
      available, most are based on the addition of lime products.
      Probably lime has also the best prospects for application to
      soils. As far as is known stabilisation techniques have not
      been used for contaminated soil to date, with one exception; the
      treatment of oil-contaminated soil with lime base products (for
      instance, the patented DCR process: Dispersion by Chemical
      Reaction). On the other hand, there is a broad experience with
      stabilising waste sludges or hazardous sludge dump
      sites[8,23,24]. The effectiveness of the latter varies strongly
      in practice.

      Some examples of commercially available processes are: DCR-
      process (lime based), Chemfix (cement based), Seal-0-Safe
      (cement based), Petrofix (cement and pozzolan based), Calcilox
      (lime based), Poz-0-Tec (lime and pozzolan based). This list is
      far from complete and is not intended to give an exhaustive
      survey of (suitable) stabilization processes.

From the information gathered it is known that there are five more techniques available for application in practice for these have passed the 'engineering' stage. However, no installation based on these principles have so far been built, and therefore no practical experience exists at the moment. These methods are ((a) to (c) refer to developments in the Netherlands):

(a)   Evaporation by indirect heating and afterburning at tempera-tures of maximum 400°C and maximum 1000°C respectively. The heat is supplied by an internal heatexchanger in a rotary kiln. If necessary the exhaust gases from the afterburner can be treated in a gas washer (scrubber) or other filter devices. General information is given in 2.3.3 The proposed capacity is 40 t/h.

(b)   Incineration by indirect heating and final treatment of the gases in an afterburner at temperatures of maximum 850°C and maximum 1200°C respectively. The proposed installation has a rotary kiln with a capacity of 8-20 t/h, depending on moisture content and final temperature of the soil in the kiln.

(c)   Incineration by direct heating in a fluidised bed at maximum 850°C. Details are unknown.

(d)   Thermal destruction in absence of oxygen (pyrolysis) at ultra high temperatures. The T M Huber Corporation (Borger, Texas, USA) has planned to construct a 50 000 tonne/year mobile unit for treating contaminated soils and silt. It is claimed that chlorinated hydrocarbons, PCBs and dioxines can be destroyed and removed from soil to a satisfactory extent. The hazardous material falls free through the reactor where it is brought to temperatures over 2000°C in a fraction of a second and chemicals are broken down into their atomic constituents. The treated soil becomes a sand-like material.

(e)   A mobile pyrolysis plant is planned by PPT GmbH & Co (Hanover, FRG). The reactor volume is 16 m^3 and the maximum temperature and pressure will be 500°C and 5 bar. A condenser and heat-generator are integrated. The plant is intended to treat oil-contaminated soils as well as oily sludges and metal-containing sludges.

Experience in the Netherlands with some of the above mentioned processes is discussed in more detail in Appendix F which also gives some background information on the manner in which these processes have developed from the solution of actual decontamination problems. The investigations described in F.1 and F.2 were carried out with regard to sites contaminated with organic bromine compounds and complex ion cyanides using an extraction process and descriptions of the proposed treatment installations are included. In addition a

clean up operation of a former gas work site with the mobile
evaporator and afterburner of 'Ecotechniek' is described in F3.

12   ADVANTAGES AND DISADVANTAGES OF ON-SITE TREATMENT

The most important characteristics of the on-site treatment
processes potentially available for the treatment of contaminated
soil have been discussed in Sections 2 to 9 and are summarised in
Table 3.5. In this table an attempt has also been made to estimate
the possibilities of developing the listed processes to the stage of
practical application. Only a minority of the processes mentioned
have yet been fully developed, tested or applied for the on-site
treatment of contaminated soil. Further experimental investigations
and studies are necessary to evaluate the several treatment processes
in more detail.

On-site treatment techniques (after excavation) have some
specific advantages and disadvantages compared with other remedial
options:

Specific advantages of on-site techniques compared with in-situ
techniques are:

-    The process conditions can be easily controlled, ie the quality
     and effectiveness of treatment is better guaranteed.

-    The number of methods suitable for application is larger because
     on-site treatment of soil can be carried out in complete
     isolation from the environment.

-    The conditions applied can be varied over a wider range without
     endangering the environment.

-    Excavation offers the possibility of dividing the excavated soil
     into separate lots which can then each be subjected to optimal
     treatment methods.

Specific disadvantages of on-site techniques compared with in-situ
techniques are:

-    The steps of excavation, transport and (re)deposition introduce
     significant extra costs.

-    Special safety measures are often needed for employees and/or
     people living nearby when the contaminated soil is excavated and
     treated in an installation.

-    During in-situ treatment some continued use of the site may be
     possible (demolition of buildings may not be necessary) whereas
     this is not possible during on-site treatment.

Table 3.5 Qualification of some Aspects of On-site Treatment Processes

Treatment process	Stage of development* (1)	Applicability (2)	Energy requirement (3)	Balance between positive and negative factors (4)	Prospects of development (5)
(A) Extraction	0/+	M, H, HH, C	+	+	+
(B) Thermal treatment					
. Evaporation by direct contact with heated gas					
− after-burning of gases	+	H, HH	−	+/0	+
− catalytic after-burning of gases	−	H, HH	−	+/0	0/−
− wet scrubbing	−	H, HH	−/0	0	0
. Evaporation or indirect contact with heated gas					
− after-burning of gases	−	H, HH	−	+/0	+
− catalytic after-burning	−	H, HH	−	+/0	0
− wet scrubbing	−	H, HH	−/0	+	+
. Incineration					
− after-burning of gases	0/+	H, HH, C	−	0	+
− catalytic after-burning	−	H, HH, C	−	0	0/−
− wet scrubbing	−	H, HH, C, M	−	0	+/0
(C) Chemical treatment	−/0	H, HH, C, M	0/−	+/0	+/0
(D) Physical separation					
. Sieving	0	(M, H, HH, C)	+	+/0	0
. Classification	0	H, HH, (C) (M)	+	+/0	+
. Magnetic	−	M (C)	0	0	−
(E) Steam stripping	−	H, HH	0	0	0
(F) Biological treatment	0/+	H, HH (C)	+	0	+
(G) Flotation	0	H, HH, C, M	+	0	+/0
(H) Stabilization					
− lime, pozzolans	0/+	M, H, HH, C	+	0	+
− cement	0/+	C, (H, HH, M)	+	0	0
− thermoplastics	0	M, H, HH, C	−/0	−/0	−/0

* 1983

Column 1:

− : Not investigated or only on a laboratory scale
0 : Investigated on a pilot plant scale
+ : Applied on-site in actual case of cleaning up

Column 2:

M : Heavy metals
H : Hydrocarbons
HH : Halogenated hydrocarbons
C : Cyanides

Column 3:

+ : Low energy need
0 : Moderat energy need
− : High energy need

Column 4:

Positive factors: − Large applicability, with respect to number of contaminants
− Low energy need
− Few bottlenecks to be solved
− Low amount of residual waste
− Low investment costs
− Universal applicability, with respect to number of sites
− Many allocation alternatives for the treated site
+ : Positive balance
0 : Positive and negative factors are about equal
− : Negative balance

Column 5:

+ : Good
0 : Moderate
− : Poor

Specific advantages of on-site cleaning techniques compared with macroencapsulation (or stabilization, immobilization) are:

-   The soil or site is re-usable without any special measures immediately after the treatment as well as in the long term.

-   Maintenance or monitoring is not required.

-   There is no serious risk of failure or ineffective remedial action.

-   Cleaning techniques will be more easily accepted from a psycho-social point of view.

-   Less expensive for relatively small contaminated sites.

Specific disadvantages of on-site cleaning techniques compared with macroencapsulation are:

-   More expensive for relatively large contaminated sites.

-   In general, certain waste streams have to be dealt with or disposed of after treatment.

-   Special safety measures are often needed during treatment.

-   The capacities of on-site treatment installations are generally limited.

Some further general remarks can be made regarding the potential processes for decontaminating soil:

-   Some of the processes mentioned in Sections 2 to 9 have already been applied for treating hazardous waste materials. This is especially the case with thermal and some chemical treatment processes. The experiences obtained with these processes can be valuable in developing processes for the treatment of contaminated soils.

-   In most cases of soil contamination very large amounts of soil have to be treated. This means that only short residence times of the soil in the on-site treatment installation are likely to be acceptable.

-   The specific properties of soils such as particle-size distribution, and the total percentage of organics, have a very strong influence on the type of treatment process that can be applied and the type of treatment that is likely to give the best results.

–   Except for a few, most of the treatment processes mentioned
     in Table 3.5 result in a residual hazardous waste. In most
     cases the amount of this waste fraction is only a few percent or
     less of the total amount of treated soil. If a treatment
     process produces a larger amount of waste the process becomes
     proportionally less attractive, since the waste will have to be
     disposed of in some way.

–   From an economic point of view each type of soil contamination
     will require its own specific treatment process, adapted to the
     specific properties of soil and contaminants. This means that
     for a particular clean-up situation an investigation into
     optimum treatment process and process conditions has to be
     carried out.

13   CONCLUSIONS AND RECOMMENDATIONS

 (i) A large number of processes are available for the on-site
     treatment of contaminated soil. These processes mainly
     originate from experiences already gathered in, for example
     hazardous-waste and waste water treatment, and in ore
     processing and soil handling.

(ii) 'On-site treatment' refers to treatment methods applied on-
     site after excavation of soil. Treatment on site may not
     always be practicable for operational, environmental or cost
     reasons. The term 'on-site treatment' is therefore usefully
     intepreted to encompass operations in which the contaminated
     soil is treated off-site eg at a central facility.

(iii) The following aspects should be thoroughly considered before
     an actual on-site treatment is applied:

     –   the impact of the treatment upon environment and human
         beings (including phsycho-social aspects);

     –   the reuse of the site in relation to some residual
         contamination and/or long-term effects;

     –   relevant legislation;

     –   waste streams resulting from the treatment and the way of
         disposing of such streams.

(iv) Although little work has yet been done on the evaluation of
     the different methods of treatment, it is often possible to
     predict whether particular processes promise good possi-
     bilities for application. The expected applicabilities of the
     different processes have been summarised in Table 3.5. Other

relevant qualitative information has also been included in this table. Much of the information is of necessity rather qualitative; there is little quantitative information available at the moment.

(v)   Only thermal treatment (evaporation by direct heating and incineration) and some methods involving treatment of soil with a liquid phase (extraction, classification and flotation) are in such an advanced stage of development that application will be possible in some cases of soil contamination at present.

(vi) The fact that only a little information is available concerning the other modes of soil treatment does not mean that they do not offer interesting potential applications. On the contrary, many principles of operation are interesting enough to merit more detailed investigation and further evaluation. The most interesting options should be developed to practical applications to make it possible to treat sites having different kinds of contamination profiles at reasonable costs.

(vii) Treatment plants can be most cost-effective when they can be easily adapted or converted for use on a large variety of contaminants and soils.

(viii) The development of on-site processes should be the subject of an intensive study in the near future in order to solve the numerous problems of severe soil contamination we face at the moment.

14   REFERENCES

1   R.H. Perry, C.H. Chilton, Chemical Engineers Handbook, McGraw-Hill Ltd, London, UK.
2   L.Ricci, the Staff of Chemical Engineering, Separation Techniques (2): Gas/Liquid/Solid Systems, McGraw-Hill Publications Co., New York, NY, USA.
3   D.J. De Renzo, Unit Operations for Treatment of Hazardous Industrial Wastes, Noyer Data Corporation, New Jersey, USA (1978).
4   Ontgiften, neutraliseren en ontwateren van afgewerkte concentraten (Detoxifying, neutralising and dewatering of spent concentrates). Report, edited by TNO and SVA, The Netherlands (1976).
5   Dr Ludwig Hartinger, Taschenbuch der Abwasserbehandlung (Handbook of sewage treatment), Ed Carl Hanzer Verlag Munchen, Wien, Austria(1976).

6   Ph Duchaufour, Pedology, Pedogenesis and Classification, George
      Allen and Unwin, London, UK, (1982) Translated by   R. Paton.
7   R. Hamnett, A study of the process involved in the electro-
      reclamation of contaminated soils, Dissertation, University of
      Manchester, UK (1980).
8   T.F. Stanczyk, B.C. Senefelder and J.H. Clarke, Solidication/
      stabilisation processes appropriate to hazardous chemicals and
      waste spills, in: Proc Conf Hazardous Materials Spills,
      Milwaukee, Wis, USA, 1982. J. Ludwigson (ed).
9   R.E. Landreth, Physical properties and leach testing of
      solidified/stabilised industrial wastes. US Environmental
      Protection Agency - order No PB 83-147983 MERL, Cincinnati,
      USA.
10   Inventarisatie Bodemsaneringstechnieken (Inventarisation of Soil
      Treatment Techniques), LGM (Delft Soil Mechanics Laboratory),
      Staatsunitgeverij, The Hague, Netherlands.
11   Handboek Bodemsaneringstechnieken (Handbook of Remedial Action
      Technologies for Contaminated Soils), Publ. Staatsuitgeverij,
      ISBN 90 12 04404 9, Den Haag, Netherlands (1983).
12   R.E. Edwards, N.A. Speed and D,E,Verwoert, Clean up of
      chemically contaminated sites, Chemical Engineering, 90 (4): pp
      73-81, (1983).
13   J.E. Brugger et al, the EPA-ORD Mobile Incineration System:
      Present Status, Oil and Hazardous Materials Spills Branch;
      Report No 167 MERL, EPA Edison, New Jersey, USA.
14   R. Scholz and J. Milanowski, Mobile system for extracting
      spilled hazardous materials from excavated soils, In: Proc Conf
      Hazardous Materials Spills, Milwaukee, Wis, USA, (1982). J.
      Ludwigson (ed).
15   G. Muller, S. Riethmayer, Chemische Entgiftung: das alternative
      Konzept zur problemlosen und entgultigen Entsorgung
      schwermetallbelasteter Baggerschlamme, Chemiker Zeitung, 106,
      (7-8): 209-292 (1982).
16   Anonymus, Schuimschiding voor bodemsanering (Froth flotation
      for cleaning soils), De ingenieur, 95, (4) 33 (1983).
17   Measures for Treating Soil Contamination Caused by
      hexavalent chromium in Tokyo, Bureau of Environmental
      Protection, Tokyo Metropolitan Government, November 1980.
18   E.E. Malouf and J.D. Prater, Role of bacteria in the alteration
      of sulfide minerals. Journal of Metals, 13 (5): 353-356,
      (1961).
19   Anonymous, Micro-biological leaching, a success story for
      Canadian research. Western Miner, 30-31,(March 1979).
20   E.E. Malouf and J.D. Prater, Technology of leaching waste dumps.
      Mining Congress Journal, 82-85, (November 1962).
21   P.G. Mikolaj and E.J. Curran, A hot water fluidization process
      for cleaning oil-contaminated beach sand In: Proc Conf Oil Spill
      Cleanup, USA (1972).
22   F. Colin, Evaluation de la fiabilite des procedes de fixation
      des boues utilises en France, Ministere de l'environment, Groupe
      de travail 'SOL', France, (December 1982).

23   C. Boulanger, La resorption in-situ d'anciens depots de dechets
      industriels, Ministere de l'environnement, Direction de la
      prevention des pollutions, France, (November 1982).
24   L. Gerschler, Mineralolhaltige Schlamme (Mineral Oil Containing
      Sludges), Copy from Series No 34 of Wiener Mitteilungen – Wasser
      – Abwasser – Gewasser (1980).
25   Handbook for Remedial Action at Waste Disposal Sites,
      Municipal Environmental Research Laboratory, Office of
      Environmental Engineering and Technology Office of Research and
      Development, USEPA, Cincinnati, Ohio 45368, USA, (1982).
26   J. Josephson, Hazardous Waste Landfills, Environmental Science
      and Technology 15 (3), 250–253, (1981).
27   D.R. Fussel (Edit), Revised inland oil spill clean–up manual,
      Concawe Report No 7/81, Den Haag, Netherlands, (1981).
28   B. Tramier (Edit), A field guide to coastal oil spill control
      and clean–up techniques, Concawe Report No 9/81, Roepers
      drukkerij, Den Haag, Netherlands, (1981).
29   J.A.C.M. van Oudenhoven (Edit), Disposal techniques for Spilt
      Oil, Concawe Report No 9/80, Den Haag, Netherlands, (1980).
30   National Conference on Management of Uncontrolled Hazardous
      Waste Sites, Washington, DC, USA, 1981. HMCRI, Silver Springs,
      Maryland (1981).
31   Conference on Management of Uncontrolled Hazardous Waste Sites,
      Washington DC, HMCRI, Silver Springs, Maryland, (1982).
32   O.W. Albrecht, Costs of Remedial Response Action at
      Uncontrolled Hazardous Waste Sites, MERL, USEPA (No 83–164 830),
      Cincinnati, USA (1983).
33   H.U. Wiedeman, Verfahren zur Verfestigung von Sonderabfallen
      under Stabilisierung von verunreinigten Boden, Berichte 1/82,
      Umweltbundesamt, Erich Schmidt Verlag, Berlin,(1982).

4

IN-SITU TREATMENT

D E Sanning

US Environmental Protection Agency

1   INTRODUCTION

This Chapter is concerned with the feasibility of rendering
harmless, or stabilising, contaminants in the ground and of
solidifying contaminated materials already placed in the ground.
Most of the processes available are analogous to those that might be
applied to hazardous wastes before they are placed in a landfill site
(commonly called stabilisation and solidification processes) or to
material excavated from the site for treatment on-site (see Chapter
3) or off-site followed by re-deposition.

2   SCOPE

The term in-situ treatment of contaminated ground means that the
treatment is applied without excavation of the material to be
treated (irrespective of whether the waste is excavated as a total
mass or removed in increments and irrespective of whether the
treated product is replaced or otherwise disposed of).

Conceptual possibilities include: grouting; liquid extraction
using water or some other solvent; microbial degradation; chemical
neutralisation, fixation or degradation; and electro-osmosis[1-3].
Most of these treatment systems have in common the need to inject
fluids into the ground in one way or another, although less complex
treatments such as deep ploughing to invert strata where contamination goes only to shallow depths is also a form of in-situ
treatment. Some systems have the added potential advantage of
improving some other property of the ground: for example, grouting
may improve engineering properties.

3   CLASSIFICATION OF TREATMENT METHODS

The potential methods of in-situ treatment can be classified in
terms of their objectives:

(i)   To render the contamination harmless. This embraces such
       processes as microbial or chemical treatment to destroy the
       contaminants and ground leaching to remove them physically.
(ii)   'Stabilisation' in which the hazardous contaminants are
       rendered insoluble or otherwise immobile or their hazardous
       characteristics (eg toxicity or combustibility) are
       neutralised. The implication is that the converse may occur
       and that changing site conditions may in time 'destabilise'
       the contamination.
(iii) 'Solidification' involving the conversion of the contaminated
       ground into a solid mass - one beneficial effect of which may
       be stabilisation of the contamination and a second an
       improvement in the engineering properties of the ground.

Some specific chemical constituents in wastes may be neutralised,
immobilised, or rendered less harmful by specific additives that
react chemically with the hazardous constituent.

The description of a chemical treatment process as one 'to
render harmless' or of 'stabilisation' depends essentially on the
extent to which the process might be reversible.

A strongly acidic (or alkaline) waste could be neutralised by
the addition of an alkaline (or acidic) material, or its level of
acidity (or alkalinity) could be controlled by an appropriate
buffering agent. A simple toxic species such as a cyanide might be
detoxified by reaction with a strong oxidising agent such as the
hypochlorite ion. Heavy metal ions may be immobilised by reaction
with an additive that forms a precipitate of very low solubility, or
may be rendered less harmful by reaction with a sequestering or
chelating agent, although the resulting chelate might pose separate
problems.

In general, each individual pollutant (or in some cases,
classes of pollutants such as heavy metal ions) in the waste could
require different chemical agents to render it non-hazardous.
Moreover, a particular chemical added to cope with one hazardous
constituent might cause antagonistic or counter-productive reactions
with other constituents. Thus, the addition of an oxidising agent
intended to destroy a specific organic compound might also change the
oxidation state of a metallic ion making the metal more toxic and/or
mobile.

Classifications other than by objective as defined above are
possible. For example:

   (i)   in terms of the agent applied as follows;
- chemical,
- microbial,
- solidification,
- thermal.

  (ii) in terms of method of application:
- surface
- ground leaching
- ground injection
- other (eg fusion/vitrification).

The discussion below involves all three classifications.

Two other aspects of in-situ treatment need to be dealt with. One occurs when the treatment is effected by injection of an agent into the buried mass or by ground leaching producing a polluted leachate while leaving the original waste in a non-hazardous condition. The subsequent treatment of the leachate (or perhaps gaseous by-product), if pumped or otherwise removed from the fill, is not considered here although it may form an essential part of the overall in-situ treatment. Such leachate treatment processes are not unique to in-situ treatment and existing appropriate technologies can be applied.

The second important aspect is that methods designed to isolate the contamination from any driving force that would cause the release or dispersal of hazardous components to the environment, ie 'macro-isolation' or 'macro-encapsulation'[3-5], could be seen as a form of in-situ stabilisation. This concept has not been used for the purposes of this study but it is important to recognise that the isolation of the site by the means of cut-off barriers, etc, may be an essential prerequisite of some in-situ treatments – for example to confine a treatment agent within the site. The isolation/barrier approach is thus discussed as necessary to provide a complete picture of the treatment processes discussed here but for a detailed discussion reference should be made to Chapters 6 to 9.

Isolation of the hazardous materials can also be approached on a less massive scale, in the form of 'surface encapsulation' and 'micro-encapsulation' techniques. These techniques will be considered as forms of solidification, and are discussed further in that context.

4   METHODS OF IN-SITU TREATMENT

The methods identified as having potential application for the in-situ treatment of contaminated land are listed in Table 5.1 together with their principal advantages and disadvantages and an and an indication of their current status, eg demonstrated in practice, experimental, conceptual.

The information in this table is entirely qualitative. Although individual vendors have information about the quantitative effects of specific interfering substances on the setting characteristics of grouts and the performance of specific solidification agents or techniques, information of this nature does not appear to be available in the open literature. A listing of possible contaminants that might adversely affect the use of thermal fusion/vitrification is given in Table 5.2.

This section outlines the conditions needed for effective treatment of contaminated ground by the methods identified in Table 5.1 and draws conclusions concerning the prospect for successful in-situ application. It proceeds from a discussion of general factors governing the feasibility of in-situ treatment methods involving application of a fluid to a consideration of processes based on:

(i)   ground injection,
(ii)   surface application,
(iii) ground leaching,
(iv)   ground injection including grouting,
(v)   processes using other techniques, such as the application of electrical energy.

4.1   Factors Governing the Feasibility of In-Situ Treatment

With the exception of the thermal treatment method by fusion described in 4.10 below, the available techniques for in-situ treatment all involve the contacting of a fluid with the contaminated ground. In the case of solidification this fluid then sets or gels. The fluid may be introduced by two main methods (i) by surface application and (ii) by injection under pressure. All the available processes are constrained by the chemical and physical heterogeneity of most contaminated ground and limitations on the depths at which treatment can be carried out (see 4.6). All must also be viewed in the light of difficulties in ensuring that the treatment has been properly carried out and probable high unit costs.

4.2   Methods of Applying Agents In-Situ

The number of in-situ application methods identified is very limited. They include:

(i)   Injection of agents in liquid, slurry, or possibly gaseous form into the mass of contaminated ground. Injection could be by means of porous tubes that penetrate the fill at strategic

Table 4.1   Advantages and Disadvantages of Selected Solidification/ Stabilisation Techniques

Advantages	Disadvantages

In-situ injection of neutralising chemicals

- When the chemistry and circumstances are such that a hazardous material is amenable to chemical control, chemical injection may be a cost-effective method to correct a problem due to leaching. - The method could potentially control a hazardous situation in which no other alternative is feasible.	- The fact that the source of the problem is buried deeply in the ground introduces many uncertainties such as the dimensions of the volume to be treated, the concentration gradients in the system, whether any causative material is retained in drums only to continue to propagate the problem,etc. - Some displacement of the pollutant, perhaps to environs outside of the landfill, will occur owing to the injection of the added volume of the chemical solution.

Grouting/ground injection*

- The technique is versatile, ie it can be used for a wide range of objectives. - It can be used for injecting any fluids. It can be used to reach specific targets - It is a well established, relatively straightforward technique with many experienced contractors. - Contaminated soil is treated in-situ. - It can be used to improve the effectiveness of other treatment and be used in their maintenance. - It can be used for applications at considerable depth. - It can be used in tight situations. - It is not likely to affect adjoining areas (compare with piling). - A wide range of ground conditions can be treated. - The technique is not likely to be limited by treatment volumes or site area. - There is a wide range of well established filler and chemical grouts readily available. - Applications can be staged (preventing over-design). - Equipment is not overly cumbersome or weighty.	- It is generally very expensive. - It cannot usually be applied at shallow depth (normally not less than 1.5 m). - Permeation cannot be guaranteed and repeated application may be required. - Soil and contamination regimes are likely to be heterogeneous and complex, presenting design problems, ie matching application strengths and viscosities to contamination and geotechnical conditions. - Applications (of grout) can be an impediment in subsequent groundwork (eg excavations for sewers and other services). - Durability of grouts is not proven in the potentially aggressive underground environments. - Robustness of application has to be proven. - Chemical grouts, solvents and other applications can present some operator hazards. - Some applications may create supplementary contamination.

Thermal fusion and vitrification methods

- The process is assumed to produce a high degree of containment of wastes. - The additives used can be relatively inexpensive (ayenite and lime). - Can be performed without close contact between workers and waste materials.	- Research method - Some constitutents (especially metals) can be vaporised and lost before they can bind with the molten silica if high-temperature processes are used. - The process is energy-intensive. The waste silicate charge must be heated (often up to $1350^{\circ}C$) for melting and fusion. - Specialised equipment and trained personnel are required for this type of operation. - No experience with organics, potential to create dioxin and other hazards.

*   Source: Reference 2

        locations to the depth required, with the injection liquid
        under pressure, as required.

(ii)   Application of the stabilising agent to the surface of a
        fill, with infiltration by gravity into the (shallow) fill.

(iii) Application of electrical energy by electrodes placed at
        strategic locations within the fill.

None of the referenced sources proposes methods of in-situ mechanical
mixing of a stabilising agent with buried solid wastes or
contaminated soil.

In order to carry out these types of treatment it is necessary
to know two things:

(i)   the degree and nature of contamination (this applies also to
     fusion/vitrification methods)

(ii) the physical characteristics of the site in terms of hydro-
     geology, geology and permeability of strata.

It is essential that there is intimate contact between the
treatment agent and the contaminants and the possibility of achieving
this will be largely governed by the physical characteristics of the
site.

Information on contamination is needed to determine whether
there will be any interference with the proper 'setting up' of a
solidification material by crystallisation, polymerisation, gelling,
adhesion, or other setting mechanisms. It is also needed to permit
formulation of the proper additives for neutralisation or detoxification, if this approach is used. The required level of detail
concerning contamination and physical characteristics differs for the
various processes.

In the disussion that follows, it is assumed that the required
information about the contaminated land, surrounding rock and soil,
and groundwater, is available or can be determined at the required
level of detail by established methods of sampling and analysis.
This information may include chemical composition and physical
characteristics of the waste or soil; quantities and spatial
distribution of the contaminants; types and characteristics of
surrounding and intermixed soils, rocks, and other geological
materials; and rates and directions of groundwater movement. Methods
for acquiring these types of information are described in a variety
of manuals, handbooks, and other sources[7,8].

The second requirement, that of providing a sufficient degree of
mixing or contact between the contaminants and the treatment
agent(s), presents a more fundamental and difficult problem in the
general case of multiple-constituent wastes buried and intermixed
with soils of different textures and hydraulic conductivities. A

considerable amount of information on the extent to which the degree
of contact between reactive agent(s) and material to be treated is
important has been established in trials of various solidification
and stabilisation processes for hazardous wastes. For example, in
some processes such as containment of the solid within a bituminous
material, the efficiency of the process (in terms of amount of waste
contained per unit of solid formed) depends on the thoroughness of
the mixing. In cases where the waste participates directly in the
chemical reactions that form the solid product (as in providing water
of hydration in a cement-setting process), or when two or more
additives must be blended (as with a polymerising agent and a monomer
in addition to the waste), the thoroughness of the mixing and
blending is critical to the formation of a solid with the desired
properties. Similarly, as discussed in Chapter 3, intimate mixing is
essential to most of the chemical on-site treatment processes.

The degree of mixing or contact that can be expected to occur
when a treatment agent is applied in-situ to a given landfilled waste
or contaminated soil is dependent on a large number of parameters
including:

   (i)   viscosity of the applied agent
  (ii)   permeability* of the waste material
 (iii)   permeability* of surrounding soils (if they are
          contaminated)
  (iv)   porosity of the waste material and soils
   (v)   degree of saturation of waste material and soils
  (vi)   spatial distribution of the waste material relative to
          soils, rocks, and other surrounding materials
 (vii)   setting time for solidification agents.

4.3 Surface Application

The stabilising agent can be applied to the land surface in
solid or liquid form and allowed to penetrate the waste materials or
contaminated ground by infiltration, provided the ground is
sufficiently permeable (and provided a chemical applied to the
surface in solid form is dissolved or otherwise transported by
rainfall or wetting). Surface application would be appropriate
primarily for very shallow, permeable, loosely-packed fills in which

---

* It is recognised the 'permeability' involves the penetration of
one substance (soil or waste) by another (usually a liquid). In
standard soil determinations, the liquid is considered to be water.
Permeability is sometimes reported in different units, depending on
the method of determination. Here, permeability is considered
synonymous with hydraulic conductivity and is reported in units of
distance/time.

simple chemical reactions can be used for rendering a specific waste
component less hazardous. It does not appear generally useful where
a buried waste is to be incorporated or enclosed in a rigid solid
mass.

4.4 Ground Leaching

This approach involves application of a liquid to a contaminated
soil mass, followed by collection of the liquid for disposal or
reuse. The liquid may be water to leach out soil contaminants, or
may contain chemical or biological agents that would react with
contaminants to form innocuous products. This technique is being
used in Sweden to reclaim a former herbicide factory site over a 5 to
6 year period.[12]. Here, water is applied by means of perforated
pipe laid in a ditch through the contaminated area. The water
permeates the affected soil, is collected by drainage pipes, then
treated by activated carbon.

Other methods for applying the water or treatment liquid
include (a) direct application to the surface, and (b) underground
pressure injection by use of grouting equipment. The latter is of
greatest interest in this study because it offers the potential for
faster, more localised application, and can accommodate gaseous
treatment fluids (eg air or oxygen for aeration) as well as liquids.
In this approach, as with many grouting applications, injection at
shallow depths may be a problem because of inadequate overburden
pressure. It may be possible to enhance penetration by the
application of electro-kinetic techniques (see 4.12) such as in the
work described by Esrig.[11]

4.5 Ground Injection

The injection of gelling and solidifying agents for engineering
purposes is known as 'grouting'. Grouting is frequently used to
reduce the permeability of a mass of material or to form cut-off
barriers to eliminate water flow beneath dams or adjacent to
engineering structures. The technical difficulties of grouting
contaminated sites, are indicative of the problems of ground
injection techniques in general as most employ similar equipment and
must cope with similar physical difficulties.

Conditions under which chemical agents can be successfully
applied by injection are limited by the extent to which the waste
and/or soil will be penetrated by the injected agent. There are
quantitative guidelines for determining the degree or extent of
penetration from a knowledge of properties of either the fill or the
surrounding (or intermixed) soils.

4.6 Grouting in Contaminated Soils

When grouting is applied to treat contaminated soils, it is often necessary to inject at shallow depths (less than 2 m), since the contaminated volume is generally in the soil mass immediately below the surface. One of the basic characteristics of grouting is the need for adequate overburden pressure to ensure that the grout does not escape to the surface through fissures before it permeates the target areas. To overcome this problem, two solutions are suggested:[2]

(i)   Applying a temporary surcharge, such as a suitable concrete block (placed on the surface and moved about, as necessary, by crane) or a mound of clean fill material.

(ii) The use of electro-osmosis to improve the receptivity of the soil to grouting, or as part of the grouting process. It has been shown that the presence of an electric field reduces the grouting pressures necessary and induces movement of the grout. The electric energy requirements are not well established but appear modest — in the range of 2 to 8 kWh per cubic metre of material.[11]

4.7 Grouting in Landfills

Problems arise from the heterogeneous nature of many hazardous waste disposal (and other contaminated) sites and the on-going chemical and biological activity that may be occurring within the fill. The potential for leachate production and gas generation must be taken into account.

The principal potential applications of grouting in landfills are considered to be:[2]

(i)   control of groundwater entry
(ii)   control of leachate egress
(iii) extinguishing fires in the landfill
(iv)   inducing anaerobic conditions (thus reducing potential for fires)
(v)   maintaining top temperatures (thereby improving bacterial activity — this can also be achieved by provision of cover)
(vi)   control of gas migration
(vii) improvement in ground stability.

The principle disadvantages of grouting are:

(i)   application may be difficult and uncertain depending on composition of the fill material
(ii) ultimate stabilisation of the fill may be delayed.

A related observation, made in the context of a different application of grouting, is that when the end use of the site is for construction, the presence of hard, cementitious grout in the soil may interfere with the installation of foundations and underground water pipes, electric lines, and other utility conduits; also, the presence of certain chemical grouting material may corrode such conduits. Advantages and disadvantages of grouting are summarised in Table 5.1: most of these apply equally to other forms of ground injection. Chemical grouts can suffer dissolution from water passing through the structure and soluble contaminants may be leached from it. The life of a grout depends on its solubility, its dimensional stability, its resistance to biodegradation, permeability of the gel, groundwater movement/hydrostatic head, and groundwater chemical composition.

4.8 Stabilisation by the Action of Ion Exchange Resins

Wastes containing heavy metals and certain organic compounds can produce leachate contaminated by toxic cations (eg heavy metals) and anions (eg cyanide). When such leachates contact selected ion exchange resins under appropriate conditions, the toxic ions in the leachate will attach to the resin while releasing an innocuous ion into the leachate. A related approach that might prove useful for contaminated land is to inject into the contaminated ground, a fluid containing ion exchange materials and a material that will polymerise or form a gel. Under appropriate conditions, the polymer or gel will form in-situ and will be embedded with the ion exchange resin. The solid component (eg the polymer or gel, or a matrix of these materials and soil) will tend to impede movement of leachate through the soil while immobilising toxic species in the leachate through the mechanism of ion exchange. Laboratory experimentation with this process indicated it to be technically feasible, but field demonstrations have not yet been reported[9].

4.9 Stabilisation by the Action of a Solidifying Agent

In a study[10] conducted for the US Environmental Protection Agency no situations were identified [6] where in-situ solidification of a landfilled waste by the action of crystallising, polymerising, or gelling agents appeared technically feasible. This approach to solidification requires intimate mixing or blending of the waste and the agent, or thorough dispersion of the waste throughout the solidifying agent. Thorough mixing or dispersion is difficult to ensure through in-situ application.

The two principal impediments to the use of this type of in-situ stabilisation are (i) non-homogeneity of the landfilled wastes, with respect both to physical properties and to chemical properties, and

(ii) the possible presence of waste components (not necessarily
pollutants) that would interfere with proper setting of the
solidification agent.

4.10 Thermal Fusion and Vitrification

Wastes and contaminated soil may be solidified by application of
sufficient heat to cause vitrification.[13]   In the case of in-situ
applications, the heat could be generated by passing an electric
current through the ground. At sufficiently high temperatures, any
soil or rock components will melt, most organic materials will
decompose, and many metallic components will either fuse or vaporise.
Gases and vapours may require recovery and/or treatment, depending on
the composition of the wastes. The compatibility of contaminants
with this form of treatment is summarised in Table 5.2. Upon
cooling, the fused mass will solidify into a glassy or crystalline
product that has about the same chemical stability as granite;[10]
however, the physical and chemical properties of the solid product
might reasonably be expected to depend on the composition of the rock
and soil.

The energy is applied through electrodes inserted in the site on
either side of the area to be melted. The electrodes are placed in
the ground or fill by drilling or other appropriate means, and a
strip of graphite in contact with the fill material is connected
across the electrodes to act as a 'starter' in melting the fill. A
cover is placed over the surface of that portion of the fill that
will be fused at a given placement of electrodes. The cover is
intended to capture gases released during the fusion. These are then
ducted to a treatment unit as necessary. The real world limitations
of this technique have not been ascertained at this time. However,
the description of this method in a vendor brochure[13] indicates
that it has been demonstrated in tests involving the fusion and
solidification of 45 kg contaminated soil. The method was later
demonstrated in tests involving over 8 tonnes of soil.[14] Although
the results of the latter tests were not available for review at the
time of writing, it is the tentative assessment of the author that
tests of this magnitude cannot be conservatively interpreted as
demonstrating that this method is suitable for large-scale
applications involving complex mixtures of wastes.

The estimated cost of fusion/vitrification falls in the range of
$125 to $300 per cubic metre (Eastern USA, humid site, is near to
$300/m^3) (1982 US dollars), of which about 30-46% is for energy,
10-13% for equipment, 36-45% for labour   and 5-10% for materials
(electrodes). As with many developing technologies, the cost is
expected to decrease greatly as experience is accumulated and
improved techniques and equipment are developed. Comparison of these
estimated costs of fusion/vitrification with those of several

Table 4.2 Compatibility of Selected Contaminants With Thermal Fusion/Vitrification

Waste component	Thermal fusion and vitrification
Organics	
1 Organic solvents and oils	Wastes decompose at high temperatures, may form undesirable pyrolysis products.
2 Solid organics (eg plastics, resins, tars)	Wastes decompose at high temperatures.
Inorganics	
1 Acid waste	Can be neutralised and incorporated.
2 Oxidisers	High temperatures may cause undesirable reactions.
3 Sulphates	Compatible in many cases.
4 Halides	Compatible in many cases.
5 Heavy metals	Compatible in many cases.
6 Radioactive materials	Compatible.

Source: Reference 3

alternative methods for controlling pollution from disposal sites (as presented in Reference 15 and updated to 1981 dollars) suggests that most of the alternatives, with the possible exception of complete excavation-treatment-reburial, are substantially less costly. The cost differential depends strongly on types of waste and other site-specific factors (including labour costs).

If, for either technical or economic reasons, the fusion/ vitrification method should not prove feasible for use with large-scale contaminated sites containing complex mixtures of chemicals, it may nevertheless prove to be appropriate and cost-effective for treatment of lesser quantities of highly contaminated land, especially if the removal of the wastes would present severe problems of environmental contamination or threats to the health of personnel involved in excavation and treatment.

4.11 Microbial Treatment

Bioreclamation techniques can be effectively used to decontaminate soils and/or a polluted aquifer. The applicability of the process has been demonstrated in both the United States and Europe. The indigenous microorganisms are usually stimulated via nutrient addition and pH and redox control. Only microbial degradable organic contaminants can effectively be treated by these processes. In some instances, various cation contaminants may be oxidized to a less environmentally hazardous state.

4.12 Application of Electrokinetic Techniques

The effects associated with the flow of electricity through porous media are termed electrokinetics.[15] The theoretical potential of electrokinetics in the treatment of contaminated land lies in two areas, (a) the extraction of toxic elements from the soil and (b) an aid in the permeation of grouting or leaching materials in soils of low permeability.

The first application of an electrokinetic technique to clean up contaminated soil appears to have been in 1936 when it was used in India to treat alkaline soils. It has more recently been employed in the USSR to remove salts from agricultural soils, with a beneficial side effect of improved soil structure. The process has also been investigated by the US Bureau of Reclamation. The use of electricity to remove water and to achieve the consolidation of soils is a method that has been used successfully for many years in Civil Engineering.

In a typical application, a series of anodes and cathodes would would be installed, an amount of water applied, and the salt-laden water flowing under the induced current collected at the cathode.

Application of electrokinetic techniques to soils containing a range
of contaminants is conceptually attractive, but there is no evidence
of any field applications. The process merits further investigation.

4.13 Vegetational Uptake

Plant uptake of an element is a function of four processes and
their interrelationships. These processes include availability of
the element in the soil, movement of the element to the root,
absorption of the element by the root (roots absorb only ionic forms
of elements) and translocation of the element in the plant.

There is some evidence from research studies that certain
contaminants in soil will be taken up and translocated into growing
plants. Repeated growth, harvest, and regulated disposal of the
plants may have the potential for removing contaminants from the
soil. Removal is highly variable and depends on the soil type, the
specific contaminant and concentration level, and the crop used. In
various parts of the world plants can be found that have developed
the ability to accumulate specific elements to an extent that would
generally be lethal to most plants. The ash of the plant may, in
rare instances, contain percentage amounts of the accumulated
element, but in most cases the degree of concentration is much less.

The first steps in the development of a vegetative uptake system
are as follows:

(i)   Determine whether vegetative extraction from the contaminated
      soil has a high probability of being the most technically and
      cost effective approach at a specific site, realising that
      this approach will require a substantial time period and
      intensive agronomic management over that time.
(ii) Determine whether suitable plant species (or varieties
      withinspecies) are available to accomplish the desired
      contaminant extraction.
(iii) Determine whether the site posseses, or can be readily
      modified to possess, soil conditions which will support
      optimal or better growth of the selected plant materials.
(iv) Conduct greenhouse-scale confirmatory uptake tests.
(v)   Confirm that the plant materials which have extracted soil
      contaminants can be adequately disposed of in an environ-
      mentally safe manner, and that the plant mass and harvesting
      mechanics are realistically manageable.

Although there are plant species that are capable of measurable
extraction of numerous inorganic soil contaminants, such as heavy
metals and elements metabolised similarly to nutritional elements,
there is little evidence as yet regarding the potential of this
approach in soils containing contaminants such as hydrocarbon

solvents and refractory organic compounds. A few investigators have proposed studies parallel to those which have confirmed the uptake and metabolism of soil-applied herbicides such as those commonly used in agricultural practice. However, since many organic pollutant molecules are substantially different in structure and reactivity from agricultural herbicides, further research is essential to confirm the utility of this concept.

5   FEASIBILITIES

None of the treatment methods considered here will be generally applicable to all situations. Most will, however, have some specific applications. Particularly promising are chemical leaching and treatment to achieve neutralisation, precipitation, etc, and grouting to achieve physical stabilisation. There are, however, serious constraints as to when even these may be applied. The penetration of grouts and chemical agents may be improved by the application of electrokinetic techniques.

5.1   In-situ Application of Solidification/Stabilisation

Current methods for hazardous waste solidification or stabilisation can not be applied to the in-situ treatment of contaminated ground and, as discussed in Chapter 3, there are limitations to their application to materials excavated from contaminated sites.

5.2   Injection of Chemical Agents

Treatment by injection of chemical agents into a contaminated site appears to be a technically and economically feasible approach to pollution control only in specific, limited cases where the following conditions exist:

(i)   The land contains a contaminant homogeneously distributed and that can be neutralised, chemically decomposed, or otherwise rendered harmless by relatively simple reactions with a single chemical agent.

(ii) The pollutant components are either in solution or in the form of a solid that is highly permeable* to the stabilising agent (presumed to be in liquid form).

---

* Here, 'highly permeable' is used to mean a permeation rate similar to that of clean sand to water - about 10^1 to 10^3 cm/second. The term is used only as a rough indication of the soil or waste permeability to the stabilisation agent.

(iii) Soils intermixed with the waste material (and contaminated
       soils which surround the waste and which must also be
       neutralised) are highly permeable to the stabilising agent.

The cost of stabilising a hypothetical 4 hectare site containing about 180,000 m^3 of cyanide-polluted waste by injection of
sodium hypochlorite solution has been estimated to be about 15 to 30%
of the cost of excavation alone,[16] ie excluding the cost of
neutralisation, transporting, or reburial subsequent to excavation.

In cases where chemical injection is considered as a feasible
method of stabilisation, the following restrictive factors should be
considered. First, the procedure is pollutant-specific. Second,
long-term effectiveness of the procedure is predicted on the
conditions that the pollutant neutralised is not regenerated by
continuing chemical or biological processes. Third, if the specific
pollutant is in an undissolved solid form, the concentration of the
stabilising agent and the contact time must be adjusted to allow the
reaction to proceed to completion. Fourth, the neutralising agent, if
not completely reacted, may itself be considered a pollutant.

5.3 Fusion/Vitrification

In-situ vitrification of landfilled hazardous inorganic wastes
appears technically feasible in concept, but expensive relative to
the other methods (with the possible exception of excavation-
treatment-reburial, which is not considered an in-situ method). Its
technical feasibility for use with wastes containing materials that
form toxic gases or vapours at high temperatures (for example,
arsenic, mercury or organic materials that produce dioxins) depends
on the effectiveness of the vapour-capture cover and vapour/gas
treatment equipment.

6   PRACTICAL EXAMPLES

There are few published examples of in-situ reclamation. Those
that do exist tend to involve the treatment of soil to limited depth
by such techniques as admixture of lime to control plant availability
of metals, deep ploughing to strata or soil leaching/solution
mining. This is to be expected since as discussed above the
technology has ingeneral yet to be tried on a pilot scale. Of six
published examples looked at in this study, one involves admixture
of ion-exchange resins,[17] two used oxidation[18] and reduction
techniques[19] and two are really examples of groundwater treatment
that illustrate chemical approaches (precipitation[20] and injection
of silica gel[21]) that may be feasible for treatment of bulk
contaminants.

6.1 Admixture of Ion-Exchange Resins

Heller[17] has described the use of selective cation exchange
resins to reduce the availability of certain toxic metals to plants.
Resins desorbing calcium, magnesium and potassium were employed in
tests on contaminated soils subject to continuing emissions from
metallurgical works. The tests were performed on soils containing up
to 1900 mg/kg Cu, 1500 mg/kg Pb and 530 mg/kg Zn.

6.2 Oxidation

A Federal Republic of Germany report,[18] describes the
injection of oxygen-enriched water to improve groundwater quality and
protect against pollution. The suggested mechanism is that concentration of certain ionic species in groundwater varies with oxidation
and reduction reactions. Biological decomposition of organic matter
in soil and groundwater consumes oxygen and decreases redox potential.
The redox potential can be increased either by adding oxidising
agents (usually atmospheric oxygen) to the contaminated soil, or by
injecting oxygenated water into the groundwater. In addition to this
basically inorganic mechanism, the increased level of oxygen in soil
and groundwater can also promote biological reactions beneficial to
water quality.

6.3 Reduction

In Japan[19] the Bureau of Environmental Protection of the Tokyo
Metropolitan Government identified a number of sites that had become
contaminated as the result of the disposal of chromite roast residues
from the manufacturing of sodium bichromate. The remedial actions
consisted of excavation of the primary contaminated areas and
treatment of the secondary contaminated areas with the reducing
agent, ferrous sulphate. The soluble, hexavalent chromium was
thereby reduced to the insoluble trivalent chromium. Further
protective measures included surrounding the site with steel sheet
piles to the depth of an impermeable layer. In some cases the sheet
piles were used in tandem with a bentonite slurry wall and/or a wall
of slow-effective reducing agent with lignite and Japanese acid clay.
The surface was then graded and covered.

6.4 Precipitation

An oxidising agent, potassium permanganate, was used to reduce
arsenic concentration in groundwater in the vicinity of a zinc ore
smelter near Cologne[20]. Arsenic concentrations of as high as
56 mg/l were detected in groundwater at 20 m depth. Monitoring data
revealed that the arsenic was in trivalent form in the regions of

higher concentration (less than 0.1 mg/l). These results, together
with observations of arsenic compounds precipitated on soil samples,
suggested that most of the dissolved arsenic was present in the
trivalent state and that transformation into pentavalent species in
the presence of calcium and ferrous ions would cause an appreciable
fraction of the arsenic to precipitate.

From December 1976 to May 1977, a solution of potassium
permanganate was injected into 17 wells and piezometers. A total of
2900 kg $KMnO_4$ was injected. The average arsenic concentration in
groundwater samples decreased from 13.6 mg/l in 1975 to 0.06 in 1977.
However, this average value had increased to 0.4 mg/l by 1979.

6.5 Injection of Silica Gel

The impacts of silica gel injection on groundwater have been
reported[21] in quantitative form[3] and are summarised briefly as
follows:

–   Alkalinity increases as a result of the alkaline components
     of the water glass (an aqueous solution of sodium silicates –
     Na_2SiO_3 and Na_2SiO_4) or alkaline and alkaline–earth
     precipitants.
–   Organic content increases when organic precipitants are used,
     causing O_2 consumption and reduction of O_2 – containing
     compounds, resulting in the temporary occurrence of a strongly
     reducing environment.
–   Heavy metal content increases because of impurities of the water
     glass (an aqueous solution of sodium silicates – Na_2SiO_3
     and Na_2SiO_4) or dissolution from the sediment by the action
     of CO_2.
–   Heavy metals precipitate as sulphides in a reducing environment
     only.

7   LONG–TERM EFFECTIVENESS

The long–term effectiveness of in–situ treatment methods has
been discussed in detail in Chapter 2. Some, such as the admixture
of lime to metal–contaminated soils, are obviously only transient in
their effect. Others such as the example of soil leaching[12]
described in 4.4 above are designed to be fully effective at the time
of application. In general, the application of the various
techniques to contaminated lands is so new that little data on long–
term effectiveness yet exist.

8   CONCLUSIONS AND RECOMMENDATIONS

In-situ treatment can be an attractive alternative to removal of wastes and contaminated soil. In those cases where it can be determined to be a feasible approach, it will usually be implemented as a part of an overall remedial action scenario involving the use of other methodologies (eg containment) in parallel. The application of in-situ techniques to uncontrolled hazardous waste sites is subject to many restrictive factors and the final decision on their applicability should only be made after careful deliberation. None of the in-situ treatment methods appear to be generally applicable to large areas of contaminated land containing mixed industrial wastes, but several of the methods appear promising for some specific type applications. The individual methods discussed here, such as fixation, chemical injection, and stabilisation, etc, are not new techniques or processes but their application to in-situ treatment at contaminated sites constitutes a major change in the thrust of application. In fact, the mode or technique of effectively utilising existing chemical principles and adapting, modifying or developing new equipment, presents the most formidable obstacle to the present day application of in-situ treatment techniques to the treatment of contaminated land. Methods to increase or decrease the permeability of waste materials and soils and improved diffusion techniques for intimately mixing chemical additives with waste materials and contaminated soils are needed.

Cost and economic restrictions to innovative, techniques can be lessened by reducing energy requirements through the development of new and better equipment and improved procedures for the application of these techniques.

Future technology advancements will depend in a large part on the ability to recognise, the limitations of existing technology to deal with the problems of contaminated sites. The commonly held belief that resources spent on research means less for cleanup actions is totally without merit. The boot strap approach to dealing with the problem of contaminated lands is contrary to the basic, fundamental reasons for the success of high technology endeavours in the world today.

9   ACKNOWLEDGEMENT

The assistance of Mr Ihor Melnyk in the preparation of this Chapter is most appreciated.

10   REFERENCES

1   J. B. Truett, R. L. Holberger and D. E. Sanning. 'In-situ
      treatment of uncontrolled hazardous wastes sites'. Proceedings
      of the Third Conference on the Management of Uncontrolled
      Hazardous Waste Sites, Washington, DC, 451-457; HMCRI, Silver
      Spring, Maryland (1982).
2   D. L. Barry. 'Treatment options for contaminated land'.
      Prepared by Atkins Research and Development, Surrey, UK, for the
      UK Department of the Environment's Central Directorate on
      Environmental Pollution, (1982).
3   US Army Waterways Experiment Station. 'Survey of Solidification/
      Stabilisation Technology for Hazardous Industrial Wastes',
      Report No 600/2-79-065, NTIS No PB 299206. Prepared for USEPA,
      Cincinnati, Ohio, July 1979.
4   Halliburtion Pressure Grouting Service. A brochure from
      Halliburton Services Division of the Halliburton Co, Duncan,
      Oklahoma,(April 1971).
5   R. J. Pojasek (Editor). 'Toxic and Hazardous Waste Disposal'.
      Volumes 1, 2, 3, 4. An Arbor Science Publishers, Inc, Ann
      Arbor, Michigan, 1979 and 1980.
6   P. Malone and L. Jones. 'Guide to the disposal of chemically
      stabilised and solidified wastes', Report No SW-872, PB 81-181-
      505. Prepared by the US Army Corps of Engineers, Waterways
      Experiment Station, for the USEPA, Cincinnati, Ohio, 1980.
7   Dennis, Fenn, et al. 'Procedures Manual for Groundwater
      Monitoring at Solid Waste Disposal Facilities', Report No SW-
      611, Prepared by Wehran Engineering Corporation and Geraghty and
      Miller, Inc. for the USEPA, Washington, DC, 1980.
8   US Environmental Protection Agency, 'Methods of Chemical
      Analysis of Water and Wastes'. Environmental Monitoring and
      Support Laboratory, USEPA, Cincinnati, Ohio, 1979.
9   J. F. Porter. 'Investigation of in-situ gelation control
      emissions from abandoned waste sites'. NTIS No PB 82-103508,
      Prepared by Energy and Environmental Engineering, Inc for the
      National Science Foundation, Washington, DC, (1981).
10   J. B. Truett and R. L. Holberger. 'Feasibility of in situ
      solidification/stabilisation of landfilled hazardous wastes'.
      Prepared by the Mitre Corporation for the USEPA's MERL,
      Cincinnati, Ohio, MITRE No MTR-82W182. McClean, Virginia,
      (1982).
11   Melvin I. Esrig, 'Applications of Electrokinetics in Grouting'.
      Journal of the Soil Mechanics and Foundation Division,
      Proceedings of the American Society of Civil Engineers,
      September 1968.
12   Lars-Gunnar, Lindfors. 'Reclamation of site of herbicide
      factory'. Proceedings of the Conference on Reclamation of
      Contaminated Land, Eastbourne, 1979, Society of Chemical
      Industry, London, 1980.

13   K. H. Oma, et al, 'In-situ Vitrification of Transuranic Wastes:
      Systems Evaluation and Applications Assessment'. Battelle-
      Pacific Northwest Laboratory, Richland Washington, PNL-4800
      (1983).

14   W. F. Bonner, 'Personal Communication'. Battelle-Pacific
      Northwest Laboratories, Richland, Washington) January 30, 1982
      and March 22, (1982).

15   R. Hamnett, 'A Study of the Processes Involved in the Electro-
      Reclamation of Contaminated Soils'. A dissertation submitted
      for the Degree of Master of Science in Pollution and
      Environmental Control to the University of Manchester, (1980).

16   Andrew J. Tolman, et al. 'Guidance manual for minimising
      pollution from waste disposal sites'. Report No EPA 600/2-78-
      142. Prepared by A. W. Martin Associates, Inc. for the USEPA,
      Cincinnati, Ohio, (1978).

17   H. Heller, 'The Use of Selective Cation Exchange Resins for
      the Fixation of Phytotoxic Heavy Metals in Cultivated Soils'.
      Proceedings, Conference Reclamation of Contaminated Land,
      Eastbourne, 1979, Society of the Chemical industry, London
      (1980).

18   U. Rott. 'Protection and improvement of groundwater quality by
      oxidation processes in the aquifer'. Proceedings of an
      International Sympsoium on Quality of Groundwater,
      Noordwijkerhout, The Netherlands, March 23-27, 1981. (Studies
      in Environmental Science, P Glasberger, Editor, Volume 17),
      Elsevier Scientific Publishing Co, Amsterdam, The Netherlands.

19   Bureau of Environmental Protection, Tokyo Metropolitan
      Government, 'Measures for Treating Soil Contamination Caused by
      Hexavalent Chromium in Tokyo'. October 1980.

20   G. Matthess. 'In situ treatment of arsenic contaminated
      groundwater'. Proceedings of an International Symposium on
      Quality of Groundwater, Noordiwijkerhout, The Netherlands, 1981,
      (Studies in Environmental Science, P Glasbergen, Editor, Volume
      17), Elsevier Scientific Publishing Co, Amsterdam, The
      Netherlands.

21   K. Aurand, et al. 'Groundwater impact of silicate injections'.
      Proceedings of an International Symposium on Quality of
      Groundwater, Noordwijkerhout, The Netherlands, 1981, (Studies in
      Environmental Science, P Glasberger, Editor, Volume 17),
      Elsevier Scientific Publishing Co, Amsterdam, The Netherlands.

COVERING SYSTEMS

G D R Parry and R M Bell

Environmental Advisory Unit
Liverpool University, UK

1   INTRODUCTION

This Chapter is concerned with the provision of covering systems
for the treatment of contaminated land. These may be used on their
own or in conjunction with vertical and horizontal in-ground barriers
or cut-offs (see Chapter 7) to achieve total isolation or 'macroencapsulation' of a site. This isolation process, in its extreme
form, may involve attempts to enclose completely the site and its
contents within an impermeable barrier which will prevent both
ingress and egress of water and pollutants from all directions. This
process is extremely costly, and therefore it will often be
considered to be sufficient simply to superimpose a covering layer on
the site which may be simple or complex, to attempt to isolate the
underlying material. The term 'covering layer' is preferred to
'capping'; the definition of the latter has tended to be restricted
to provision of a totally impermeable layer. This may not always be
desirable, even if possible, and the decision must take into account
local requirements for minimising pollutant migration.

For the purposes of the Pilot Study the term 'covering system'
is used to include covering layers which may be simple, sandwiches of
materials, or complex. Also included are barrier layers, chemically
and biologically active layers and impermeable cappings.

2   SCOPE

There has been only limited systematic investigation of the
broad spectrum of problems associated with covering and hence the

113

redevelopment of contaminated land sites. However, some individual
examples, eg treatment of metal mine sites contaminated with toxic
metals, have received considerable attention. In addition, pollution
of underlying strata and groundwater has been well studied but still
remains uncertain in interpretation. Until recently, only limited
attention has been paid to the design and construction of cover
material for landfill, hazardous waste sites and contaminated land
sites to provide physical and chemical isolation of the material
beneath. Some information is available on the properties of liners
for landfill and this is in part applicable to covering materials.

While many of the reclamation schemes carried out so far in the
United Kingdom and other countries have incorporated provision of
cover as a treatment, such schemes have been carried out largely on
an ad-hoc basis, viewing each site individually. There is a need for
a more rational basis for cover design along the lines of the
guidance produced in the United States, for the design and construction of cover for new landfill sites.[1]

As a prelude to the preparation of this report a bibliography[2] was produced that is available on request from the authors.

The aims of covering systems and the conflicts that can arise
between primary aims of containment or isolation and proposed final
land use are discussed below. Where information is available, the
potential causes of barrier failure have been investigated and the
performance of particular types of covering system reviewed. Areas
of uncertainty, and research needs, are identified.

3   FUNCTIONS OF COVERING SYSTEMS

There are two main methods for the treatment of contaminated
land currently in wide use (as discussed in Chapters 3 and 4,
techniques for in-situ treatment and for on-site treatment of
contaminated soil are not as yet well developed). The first is to
remove the contaminated material and rebuild the site with inert
fill. This of course infers that the contaminants will be disposed
of elsewhere, and by a 'safer' method. The second method is to
contain the contamination where it occurs. In its ultimate form this
requires the complete macroencapsulation of the contamination by the
provision of surface cover and in-ground vertical and horizontal
barriers. However, in many cases the provision of surface cover on
its own should provide adequate protection to human health and the
environment. To date this has been the approach largely adoped by
the United Kingdom.

The emphasis of reclamation must be on 'long-term' solutions,
but there are occasions when emergency containment and isolation of a
site may be required.

The choice of cover will be influenced by a number of factors relating to subsequent land use and the nature of the contaminants. These demands may be conflicting. For example:

(i)   control of moisture movement, ingress or egress, by low-permeability cover may inhibit successful gas dispersion and require additional specification for gas disposal;
(ii)   tree planting on a site where a clay seal has been used may result in loss of integrity by penetration of tree roots.

Primary functions of covering systems can be grouped under three broad categories.

(i)   To prevent exposure of the at-risk targets to potentially harmful contaminants.
(ii)   To sustain vegetation growth.
(iii) To fulfil an engineering role.

Each of these categories can be sub-divided into a variety of secondary functions. The following list is not in order of importance:

-   Control of gas movement.
-   Control of leachate production.
-   Control of soil fluid movement.
-   Limitation of surface water ingress.
-   Prevention of capillary movement of contaminants in soil solutions.
-   Minimisation of fire hazard.
-   Prevention of dust blow.
-   Erosion control.
-   Support of vegetation.
-   Inhibition of root penetration through the cover system and prevention of biological translocation of potentially harmful chemicals.
-   Improvement of structural properties of site, eg facilitate vehicle movement.
-   Improvement of aesthetic appearance.
-   Reduction or elimination of toxicity at the site surface.

A number of other requirements can be added to the above list which can be categoriesed as performance requirements. Examples of these are:

(a) Resistance to climatic change, ie cold or arid climate deterioration.
(b) Maintenance of cover integrity, ie robustness.
(c) Preservation of slope stability.

The above requirements are obviously inter-related and the design of covering systems requires a ranking of priorities to determine those of greater significance for a particular site. In addition, in order to satisfy the primary requirements, schemes should be designed to take into account their effect on other cover functions. For example, provision of an impermeable barrier to prevent water ingress into a site where gas production is likely, will also demand additional gas drainage and venting systems.

4   COVERING MATERIALS

The choice of cover material and its method of application is governed by site-specific characteristics and requirements, availability and cost. There are four main categories of materials that may be considered:

  (i)   Natural materials.
 (ii)   Modified soils.
(iii)   Synthetic cover.
 (iv)   Waste materials.

4.1   Natural Materials

The performance of soils as covering material is principally dependent on soil structure, in particular its particle size distribution and mineralogical characteristics. Well graded fine-grained soils tend to have acceptable low permeability values and are best suited for use where prevention of water ingress is of primary concern. However such soils are susceptible to wind blown erosion and cracking in drought conditions. Table 4.1 describes soil types in terms of their performance but it should be noted that the degree of compaction is also a significant factor affecting performance.

Where vegetation cover is required as a final or intermediate treatment, top soil is usually imported as final cover. Top soil quality is dependent on a wide range of physical and chemical factors which influence its ability to serve as a growth medium. Soil structure is seriously influenced by handling and compaction should be avoided. Top soil should be analysed to ensure that unacceptable concentrations of substances toxic to plants are not present.

4.2   Modified Soils

In the absence of well graded fine-grained soils, blending two soil types can broaden the grain size distribution and reduce percolation. This mechanical mixing also improves the reliability of mathematical predictions of performance[4].

Table 5.1   Ranking of Soil Types According to General
            Performance of Cover Functions

	Traffic-ability	Impeding water perco-lation	Impeding gas migration	Erosion control	Support vege-tation
Gravel	1-5	5-12	4-10	1-5	5-10
Sand	1-7	6-11	5-8	2-7	1-9
Silt	5-9	3-4	3	9-13	3-4
Clay	6-10	1-2	1-2	8-12	7-8

Where   1 = Very good at achieving function
       13 = Very poor at achieving function

Soil stability and mechanical strength can be modified by the addition of small quantities of Portland cement. An addition of as little as 1% cement or lime to granular soil has been shown to have stabilising effects. Incorporation of about 8% cement has been used to strengthen soils for road surfacing and embank-ments[2].

The pozzolanic properties (the ability to react with lime to form cementing compounds) of fly ash from coal-burning power stations have also been used to strengthen and stabilise soils. Lime is usually added to assist the cementing process[2].

Addition of 2-8% lime to clayey soils can be used for its effects as a flocculant or base exchanger. In fine cohesive soils it can have a strengthening effect over time because of its chemical reaction with clay minerals.[2]

Deflocculation of fine-grained soils containing clay minerals to reduce permeability, increase density and improve compaction, can be achieved by the addition of soluble salts such as sodium chloride and tetra-sodium pyrophosphate. Such dispersants however, tend to increase the shrink and swell behaviour of silts and clays and may make them more susceptible to erosion and cracking.

The primary constituent of bentonite is the clay mineral mont-morillonite. It swells between ten and fifteen times its original volume when hydrated. When it is mixed into soil, the swelling that occurs fills the pores that would otherwise permit seepage, enabling very low permeability to be achieved,[4,5]. There are many mechanical requirements for bentonite to work properly, these include proper compaction rates.[6]

Low permeability to water does not guarantee low permeability to other liquids. The properties of bentonite, and of other clays, can be severely modified by the presence of a wide range of potential contaminants. Research is in progress to develop chemically resistant bentonites.[7,8]

A wide variety of other soil additives are available and Lutton et al[2] provides a comprehensive list of over one hundred varying in their soil modification properties and effectiveness.

Freeze-thaw characteristics of soils are important considerations in certain climates. A signficant reduction in compaction can occur when soil pore water is frozen. Addition of calcium chloride changes a number of soil properties and may help soil placement and reduce adverse effects of freeze-thaw cycling.

4.3   Synthetic Cover Materials

A variety of commercially available materials are on the market for use as alternatives to soil-based cover.[9-11] They range from flexible membranes to rigid concrete. The use of these two forms is generally restricted to sites where there is a high risk of pollution because of cost, but the group of materials known as 'geotextiles' has useful applications on very many sites.

Flexible membranes which must be resistant to tearing should be placed over a smooth buffer of soil or sand in a relaxed state. Rigid structures must resist cracking. Both types of material may be liable to microbial and/or chemical attack.[12,13] Additional damage may be caused to membranes by burrowing animals, plant roots and exposure to sunlight.

It is important that cover over membranes is placed so that erosion does not subsequently expose the membrane.[14] In addition uneven settlement can cause severe stress on membrane cover. Experience in the use of membranes for surface sealing and lining sites is limited to about 20 years.[15]

Geotextiles embrace a wide range of mainly sheet materials including felts made from polypropylene and other fibres and plastics meshes ranging in mesh size from very small to several centimetres.[13] The fibrous mats may be used to improve the bearing capacity of soft ground (eg laid on mud and covered with stone), as filter layers and to inhibit root penetration. The meshes, may be used to inhibit root penetration, to stabilise the surface of slopes, and to reinforce embankments.

4.4 Waste Materials

There are a large number of waste products that can fulfil both the engineering and isolation functions of cover. Many, however, contain trace elements or other components that may be toxic or harmful to the environment and this needs to be investigated before use. The attitude of authorities to the disposal of these wastes differs from country to country and their use will be governed by the prevailing regulations. Some of the materials mentioned below were formed at high temperature and are either crystalline or glassy: leachability of toxicants is consequently generally low.

4.4.1 Fly Ash. Fly ash (pulverised fuel ash) is produced from power stations in large quantities in all member countries.[16] It has a silt-like nature, good engineering properties, pozzolanic properties and is able to sustain vegetation when suitably amended or treated. However, fly ash quality and composition will vary with the coal burnt. It is particularly important to distinguish between those derived from bituminous and from soft or brown coals. The latter often contain large amounts of 'free' or uncombined lime. Fly ashes frequently contain elevated levels of a number of potentially toxic elements, the significance of which has been widely studied, with particular reference to the potential for pollution of water and of uptake into vegetation growing on it.[17-20]

Bottom furnace ash and molten ashes are additional by-products of electricity generation.[16,17] They are coarser grained than fly ash, and their polluting potential should be similarly investigated before use.

4.4.2 Slags from Iron and Steel-making. The major wastes produced by the iron and steel making industry are discussed in Chapter 12. Blastfurnace slags arising from recent production are generally physically and chemically stable crystalline or glassy products of low pollution potential.[16,17,21,22] Old slags, such as may be found on steel works sites, may not be so stable.[23]

Slags derived from steel making vary greatly in composition depending upon their parent process. They tend to be both physically and chemically unstable and should only be used as fill with caution. They should not be used as hardcore beneath buildings.

4.4.3 Non-ferrous Slags. Slags from non-ferrous industries generally contain relatively high levels of toxic elements making them more a source of contamination than useful materials for site reclamation.

4.4.4 Domestic Refuse Incinerator Ash. Similar restrictions to the above also apply to domestic refuse incinerator ash which can be toxic to vegetation as a result of its heavy metal content.

4.4.5 <u>Overburden Materials</u>. Overburden from mines, quarries, coal and aggregate extraction areas can be used as soil cover substitutes, as can non-toxic sludges and dredged silts.[24]

4.4.6 <u>Dredged Silts</u>. Many dredged silts, eg those from major waterways and docks, are heavily contaminated with toxic elements and other substances, eg oil. This, and their very fine grain size greatly restricts their usefulness in site reclamation.

4.4.7 <u>Sewage Sludge</u>. Digested and composted sewage sludge can be a valuable resource both as a cover material and a surface ameliorant provided that its content of toxic elements and other toxic substances is not too high. It has been used in a variety of locations in the United Kingdom as a source of plant nutrients and organic material on mineral wastes. Imprudent use of the material in excessively thick layers can lead to problems of inhibition of plant growth as a result of anaerobic activity created by biological action in the sludge.

4.4.8 <u>Construction Rubble</u>. Builders' rubble is often in plentiful supply in or near urban areas. Its general alkaline properties and physical structure make it a useful product for use as a break layer provided that it is free of potentially aggressive contaminants (eg sulphate in the form of old gypsum plaster).

5   PERFORMANCE REQUIREMENTS

5.1 <u>General Principles</u>

The effectiveness of a covering system based upon soil or soil-related materials in performing the principle functions described in Section 3 above will largely depend on five interacting factors:

(i)   its effectiveness in controlling entry of water into the underlying contaminated ground;

(ii)   its effectiveness in preventing upward and lateral movement of contaminants through the ground;

(iii) the ability of the covering material to bind pollutants through chemical and physical absorption onto constituent soil particles;

(iv)   interactions between covering materials, the underlying contaminated materials and biological systems;

(v)   the engineering behaviour of the system and its component parts.

The multi-functional, sometimes conflicting, requirements of cover suggest that a layered system for covering contaminated sites has a number of advantages over trying to identify and place a single layer.[2] Such a scheme might include:

  (i)   top soil or sub-soil made amenable to support vegetation;

 (ii)   barrier layer or membrane to prevent passage of water, gas or volatile organic compounds;

(iii)   buffer layer to protect barrier by providing a smooth working base;

 (iv)   water drainage channel or layer;

  (v)   filter layer to control interaction between materials of different grain sizes, eg penetration of fines into a coarse porous material;

 (vi)   gas drainage layer or vents.

To this list may be added:

(a)   break layers to prevent upward movement of soil water by capillarity, which can also serve as gas dispersion layers;

(b)   chemical barriers;

(c)   barriers to biological processes.

In addition, the system as a whole will have to be:

- capable of effective installation given the constraints imposed by normal civil engineering tolerances, working methods and achievable standards of workmanship;
- remain effective for the design life (see discussion in Chapter 2);
- be resistant to intervention by man, deliberately or accidentally, and to foreseeable natural events such as flooding, extreme rates of precipitation or earth tremors (it is normal practice to allow for such factors in the design of many engineering structures);
- preferably not depend on continuing maintenance to ensure effectiveness, but it will require monitoring.

It should be noted that whilst cover systems may be effective in reducing migration of contaminants above the water table they will have little, if any, effect on those below the water table.

5.2 Control of Water Ingress

Prevention of water ingress is considered by a number of authorities[11,25,26] to be the primary function and most effective method of reducing leachate output from a hazardous waste site. The method of preventing percolation or infiltration into a site involves the use of natural or artificial impermeable layers together with alterations in site topography and drainage (see for example Dowiak et al[27]).

Evenly graded fine-grained soils with a low permeability value are one of the most useful means of providing cover to impede the

passage of water into the layers beneath. Desirable compaction of
the capping layer during its construction to engineering requirements
is a most important process to further reduce pore space.
Additionally, controlled compaction of contaminated wastes can
significantly reduce leachate production.[28] As stated above,
blending and increasing cover thickness are also involved in water
percolation control.

The use of membranes and artificial liners is restricted both by
economics and by handling difficulties.[17] The newer artificial
materials need long-term field trials before they become accepted as
permanent solutions.[15,29]

Compatability studies should also be carried out between
wastes, leachates and liners.[30] Such studies lasting up to nine
years on a variety of liners have reported only modest structural
effects although some membrane failure had already occurred.[31]

In order to improve the performance of water barrier layers,
discontinuities such as deep cracks or surface ponding effects
resulting from differential settlement must be avoided. This and
other factors are assisted by grading and increasing the slope of the
site surface. As rainfall run-off increases with slope, infiltration
of the soil surface decreases.[31]

Adequate drainage to intercept and direct all water from outside
the area is required to assist the effectiveness of the
impermeable barrier. Sub-surface drainage in the covering layer
itself can also be used to reduce water ingress.

5.3 Gas Control

The production, migration and dispersion of gases from contaminated sites are discussed at length in Chapter 10 with a particular
emphasis on toxic gas emissions.

The production of methane, carbon dioxide and other gases from
the degradation of organic matter is inherent in the disposal of
refuse to land and has received wide attention with regard to
predicting production rates and migration and developing means for
its control or beneficial use as an energy source. Management of the
problem can take two forms based on the ease with which the produced
gases will migrate sub-terraneously through continuous pore space in
soils.[32]

In areas where permeable cover is provided, landfill gas can
migrate to the surface layers of cover, resulting in anaerobic
conditions and causing significant damage to vegetation.[33,34]
Active control of gas from landfill and contaminated land generally

employs the use of layered covering systems[32,35] which might
include:

  (i)   Placement of impervious liner material.
 (ii)   Selective placement of granular material.
(iii)   Evacuation and venting of the gas.

     A suitable gas barrier in a layered system will usually
consist of damp clay optimly compacted, or an artificial membrane.
Some limited research has investigated a gelatinous soil barrier in
limiting radon gas emissions.[36] The clay or other impermeable
layer will also act to prevent water movement into the site. Beneath
this there is the requirement for a gravel layer which can be used to
channel the gas to vents. This is required to prevent positive
pressure directing the gas into adjacent land or buildings. The
surface layer can be of top soil to encourage vegetative cover (the
relationship between plants and gas is discussed further in 5.6
below).

     Impermeable layers in covering systems can also be used to
contain gas in order that it may be pumped from wells or channels.

5.4   Buffer Layers and Filter Layers

     Buffer layers, usually a sandy material or synthetic fabric
(geotextile), can be placed above and below impermeable layers to
protect clays from drying and cracking or synthetic layers from
punctures or tears.

     Filter layers are used to prevent the fine particulate material
of the impermeable layer or buffer layer from entering coarser
grained layers such as gas or drainage channels below.

5.5   Break Layers

     The principal guidance available in the literature concentrates
on prevention of water ingress into a site, or of control of gas out
of a site. Relatively few examples are recorded of recognition of
the equally important possibility of upward migration of
contamination into the upper parts of covering systems by a variety
of chemical, physical and biological mechanisms.[37,38] The factors
governing such upward migration are in need of further
investigation.

     It is believed that soil suction (capillary migration of fluids)
can play a major role in movement of contaminants upwards through
covering layers[39] under some climatic conditions. In general, clay
has always been assumed to have a very low permeability, hence its

common use as a sealing layer. This is only true, however, when the
clay is water saturated. When placed above the water table, or in a
particularly harsh environment, it is quite possible for a clay cover
to dry out and crack.[40-42] When cracked it is no longer effective
at stopping water flow and can then allow significantly greater
quantities of fluid to pass than sand.[43] Clay barriers can also
lose integrity by the action of burrowing animals and decaying plant
roots.[44]

Inclusion of a clay layer in a covering system can make use of
its cation exchange capacity where it is recognised that heavy metals
may be bound from soil water onto the clay.[45] It is important to
note that both pH and type of leachate solution affect adsorption
processes.[46] In general terms adsorption of cationic heavy metals
increases while adsorption of the anionic heavy metals decreases as
pH increases.

Clay layers can also be affected by aggressive chemicals
and solvents, allowing the passage of materials they were designed to
contain.[47] It is therefore suggested that the permeability of
prospective clay liners is tested using the leachate to which they
will be exposed. It has been proposed that the time needed for such
tests can be decreased by the use of elevated hydraulic gradients. A
gradient of between 50 and 100 seems optimal in that the tests can be
completed in 2-3 months.[48] Where clay covers are designed to
minimise precipitation infiltration, it is normally not necessary to
utilise anything other than rain water as a permeant. Distilled
water is not recommended.[49]

In some cases a reaction between the leachate and the barrier
can actually reduce the permeability of the barrier.[36]

To counter these problems the concept of a break layer has been
developed. A granular material (eg sand, grit or building rubble) is
used beneath the seal to reduce capillary movements of water-soluble
pollutants up the soil column. Such layers would require the
protection of a filter layer to prevent downward movement of fine
material into the pore spaces and increase the potential for
capillarity. An example of the usefulness of this approach is
discussed by Cline et al.[50]

Biological activity, that is earthworm or rodent burrowing, or
plant root translocation, can play a significant role in the movement
of contaminants from below to above a covering system. They can also
destroy the integrity of many natural and synthetic sealing
layers.[40,41] Similar to the need to protect sealing layers from
below, there is the need to protect them from biological activity
above.

One of the best examples of this type of barrier design is provided by Cline et al.[50] A layer of loose rock placed between buried waste and top soil successfully prevented burrowing animals, plant roots and insects from translocating waste contaminants into the surface layers. Air space between the rocks was maintained with stone sizes of between 3.8 and 7.6 mm diameter and a filter layer was incorporated.

Limited information is available on the use of layers as chemical barriers. Limestone layers have been used in a number of cases to control acid generation and upward movement of heavy metals.[24,38] The use of this material is particularly attractive because of its (a) wide geographic distribution, (b) particle size and (c) relatively low cost. Limestone affects the mobility of hazardous metals by (a) adsorbing metal ions directly, (b) forming less soluble calcium and carbonate compounds and/or (c) raising pH levels.[53] Layering appears to be a more effective and efficient way of using limestone than intermixing with soil since soil acids react readily with limestone resulting in its decomposition.[4] Revegetation of land polluted by chromate smelter waste required a more complex chemobarrier.[51,52] Chromate smelter waste is phytotoxic as a result of water-soluble chromium and high pH. The solubility of the chromate anion facilitates movement into the rooting zone of the covering material and a layer of granular material below the soil was needed to form a capillary break. In addition ferrous sulphate was added to the surface after covering to chemically reduce traces of chromate waste being spread by the wheels of earthmoving equipment.

5.6 Vegetation Support Layers

Development of vegetation on the surface of a covering system is recognised as being extremely important for:

 (i)   increasing evapotranspiration
 (ii) reducing wind and water erosion
(iii) strengthening and increasing soil stability
 (iv) improving aesthetic appearance
 (v)   restricting access to the site.

Consideration of soil fertility, pH, climate and species selection is required if rapid establishment of perennial vegetation is to be achieved.[54] Invariably because of the prohibitive costs of importing good quality top soil, low productive soils, sub-soils or even quarry or mine waste may be the only surface layer available. Even if top soil is specified it is not always provided[55] and invariably it needs to be assessed before use.[56] The problems are compounded where the cover contains an impermeable layer. This may cause excessive water retention in the vegetation zone, or excessive water loss during dry periods.

Soils imported to support vegetation growth must contain sufficient nutrients to allow such growth and not contain unacceptable concentrations of toxic substances. If sufficient nutrients do not occur naturally, a soil conditioner such as organic compost, properly balanced chemical fertiliser or even digested sewage sludge should be added to attain a satisfactory nutrient level.[56]

The depth of any added top soil required is dependent upon its physical and chemical structure as well as the type of vegetation it is to support.[39]

Vegetation that develops a dense but shallow root system is more effective in surface stabilisation and reducing water penetration than vegetation with deeper less dense roots. They are also less likely to puncture or penetrate any barrier layer.

On sites where gas is being produced further precautions, including additional protection techniques and the careful selection of plant species may be required to ensure no vegetation loss of deep rooting species. Plastic sheeting underlain by gravel and vented by vertical PVC pipes, a 1 m mound underlain with 30 cm of clay, and a 1 m mound with no clay barrier, have all proved effective in preventing penetration of gas into the root systems of test species.[57] More research is required to investigate the performance of some species in the absence of permanent landfill cover.

5.7 Engineering Requirements

The 'engineering properties' of a covering system and its component parts are important in two respects:

  (i)   the handling, placement and trafficking whilst work is in
         progress;
 (ii)   the behaviour of the system once complete – for example to
         meet a need to support roads, buildings, etc.

Conflict between these requirements and certain of the others discussed above can again arise. The high silt or clay content conducive to good resistance to movement of metals or of water is likely also to produce poor working conditions to such an extent that work may only be possible under extremely good weather conditions.

A further possible conflict concerns site stability. The reduction of settlement and the provision of a good platform on which to build generally requires a high degree of compaction. In contrast it is generally necessary to avoid compaction for healthy vegetative growth.

Another practical requirement is that the requirements of the system, for example in terms of grading of materials, thickness required and the variations in these that can be 'tolerated', are consistent with what can be realistically achieved.

The engineering processes for the movement, placement and consolidation of materials are well established and will not be discussed further here. However it is very important that those charged with carrying out these works are made properly aware of the functions of the system and the features that are essential to its effective performance.

6   LONG-TERM EFFECTIVENESS OF COVERING SYSTEMS

The covering of offensive or potentially offensive materials is an instinctive reaction that has been practised throughout history. There are without doubt many examples of sites treated in this way in recent decades, the most common of which will be controlled landfills, dumps and tips used for the disposal of domestic refuse. Despite their number (some of them are described below in Section 7) they will rarely have been approached in the rational way that has been outlined above. However, this past experience provides some guidance to the design of covering systems in the future and limited research studies are in progress involving investigation of already reclaimed sites.[58]

The particular characteristics of covering systems in relation to long-term effectiveness are discussed in Chapter 2. The important point to note is that many of the processes that protection must be provided against are very slow and the most acute situation does not necessarily occur at the time at which the remedial actions are carried out. For example, tree roots will grow over a period of time whereas the synthetic membrane that may be threatened may become progressively weaker due to chemical or microbial action. If trace elements are to migrate upwards through either upward moisture movement or plant uptake and decay the time-scale for a significant effect may well be decades or longer.

Thus the lack of historic evidence, the changing properties of the materials of construction and the long time scales involved, make prediction of performance difficult. Something can of course usually be said on the basis of the experience and research in the literature cited here but continuing long-term research is required.

Attempts have been made to produce evidence on the probable performance of covering systems using enhanced environmental conditions judged likely to increase the rate at which the 'failure' processes are likely to occur and the establishment of trials under controlled and carefully monitored conditions. One such study in the UK is making use of columns of contaminated and covering materials kept under conditions conducive to the upward migration (wicking) of moisture.[43,58-62] A separate parallel study has involved field trials based on growing food crops and other vegetation on systems involving various thicknesses and types of covering materials.[63-65] These studies are of only limited duration (2-5 years) and cover only a small set of variations. There is a clear need for further long-term research including the instrumentation and monitoring of sites that are to be reclaimed.

7   SELECTED PRACTICAL EXAMPLES OF COVERING SYSTEMS

The information provided below has been extracted from the available literature and the authors' personal experience. Selection was made partly on the basis of the availability of documented data and partly to cover the range of techniques described elsewhere in earlier sections. The examples include practical application of simple break layers, chemical barriers and complex covering systems. Also included is a description of the use of vegetation as an integral part of cover design.

Information on the success of the treatments is very limited. The absence of detailed monitoring programmes compounds the problem and it is generally difficult to comment on the long-term success of the treatments described or to know what effort, if any, went into their initial design. They should not be viewed as necessarily good examples of the 'art' of covering system design and implementation – they are simply examples of current and recent practice.

It is clear from studies[58] in the UK of already reclaimed sites that, whatever the intentions and merits of the original design, they have, in a number of cases, been compromised by poor execution. This seems to have arisen in part from a simple lack of care but also possibly from a failure of the persons carrying out the works to be sufficiently aware of the objectives of the treatment or of the critical nature of certain aspects of it.

7.1 Site A - Mine Tailings, Canada

The presence of high salt concentrations and acidity in the mine tailings at this site facilitated acid migration into soil cover. A two layer cover system was developed incorporating a gravel 'break' layer positioned directly on the mine tailings and a top layer of soil. Studies have shown that gravel was an effective barrier to acid migration and inhibited the movement of iron, aluminium, zinc

and copper into surface soil. Cover not protected by a gravel layer
became contaminated by the above elements.[37]

7.2 Site B - Revegetation of Mine Wastes, Australia

Mine wastes (slimes) rich in zinc and copper have been vegetated
at this site using a variety of covering systems employing both
physical and chemical barriers. Greatest success was achieved when a
layered system of cover was used incorporating gravel and soil.
Chemical techniques were also incorporated into the system where
crushed limestone at 63 t/ha was mixed into the waste surface to
reduce the potential for capillary movement of toxic ions from the
slimes into the covering system. The final system adopted consisted
of three layers. The slimes were sealed with clay, superimposed by a
layer of gravel, and topped with top soil.[38]

7.3 Site C - Domestic Refuse Landfill Site, London, UK

A former domestic refuse site intended for redevelopment was
found to be contaminated with heavy metals. A layering system of
cover was selected and modified to perform different functions
dependent upon final land use. The site was covered with a 150 mm
base coarse layer of hardcore to deter upward movement of salts and
to provide a drainage layer. A 25 mm layer of fine porous material
was superimposed on the hardcore to prevent ingress of top soil into
the pore space. A final layer of 150 mm of top soil was provided in
those areas scheduled for open space. Where tree planting was
required 550 mm of top soil was provided. In those areas scheduled
for use as gardens a minimum of 550 mm of clean top soil was
specified as the final layer[66] but at the time of writing none of
the residential areas have yet been developed.

7.4 Site D - Metalliferous Mine, North Wales, UK

Former metal mining of the North Wales ore body during the last
century has resulted in a number of abandoned tailings spoil heaps
rich in lead and zinc. At this site 260 000 tonnes of tailings spoil
had been deposited on the banks of a tributary of the river Conwy.

Continuous stream and gulley erosion of the unstable surface
resulted in serious damage to some 6 ha of agricultural land in the
flood plain 2 km distant from the mine due to contamination with
lead, zinc and cadmium. The contamination was sufficient to inhibit
cereal production and grazing of this land. In addition shellfish
production was also adversely affected by heavy metals from this
source 20 km distant at the estuary mouth.[58,67-69]

The total area of the site covered 2.2 ha and its visual disamenity was insignificant in the context of its surroundings. However, some rapid and permanent surface stabilisation was required to prevent recurrence of the transported pollutant problem.

Steep tip slopes prohibited the use of conventional soil covering because of the potential for slippage and renewed erosion. Direct development of a metal tolerant grass sward on the tailings surface was also rejected because of the serious consequences of even small areas of sward failure. A combination of the two systems was eventually selected. A covering layer of readily available quarry overburden of 100 mm and 5 tonne/ha of crushed limestone was spread over the site and seeded with a glass clover sward containing 60% Festuca rubra Merlin. This is a grass developed for its tolerance of high soil concentrations of lead and zinc.

Unlike many covering layers that are designed to inhibit root penetration, the quarry overburden allowed the tolerant grasses to root into the underlying contaminated material. Thus the surface amendment was bound to the mine tailings beneath. This, together with the growth of non tolerant species on the overburden alone, has resulted in a stable surface and control of the erosion problem which was judged effective 3 years after its establishment.[69] Subsequent investigations of the site have been reported by Jones et al.[58]

7.5   Site E - Gas Works Waste, London, UK

Former sites of town gas production from coal and the by-products industry present a major source of severely contaminated land. These sites are usually associated with dense urban areas in the UK where land is at a premium and redevelopment pressures are high. One such site in London operated for over 100 years. By-products and wastes were disposed of on site during the working life of the gas works resulting in a waste tip covering an area of 5 ha with an average height of 16 m. The estimated 430 000 cubic metres of waste was made up mainly of boiler ash, clinker, iron oxides and lime residues. Associated with these deposits were potentially dangerous concentrations of cyanides, phenols, sulphides and other compounds. Large quantities of solid and liquid tarry wastes had also been disposed of on this tip.

Much of the surrounding site was suitable for conventional redevelopment to industry and housing, but redevelopment of the tip itself was scheduled for development as open space. Removal of the waste materials was considered uneconomic because of handling and subsequent disposal problems.[35]

Unlike the former example, open space development on this site presents a series of after use problems. Firstly, there is the major

difficulty of establishing hard wearing vegetation cover on material
that is extremely inhospitable to plant growth. Secondly, there is a
need to isolate the toxic materials from site users and services.
Provision of a cover layer in this case has required specification of
a multi-purpose system using available natural material.

The decision to open the site to public access requires
sufficient thickness of protective layer to prevent toxic materials
reaching the surface either directly or via the biological activity
of the plants that are to cover the final surfaces. This is to be
achieved by a seal of London Clay of 1.2 m covered with 0.3 m of top
soil.

Other requirements of the covering system necessitate exclusion
of percolating water into the tip mass. A 225 mm thick gravel
drainage blanket has been proposed immediately beneath the clay
together with a surface drainage system. A further problem arises
because of the complex nature of the wastes. It is likely that
earthmoving and tip moulding will result in mixing of the wastes,
resulting in a potential for gas generation. The permeable drainage
blanket has been designed to also act as a gas drainage channel to a
strategically located venting system.

7.6   Site F - Heavy Metals Wastes, Cheshire, UK

An alternative approach to physical separation of contaminated
material from the final restored surface is provided by the use of
chemical barriers. At this site over 50 years of copper refining and
associated waste disposal resulted in some 10 ha of tip seriously
contaminated with copper, zinc and cadmium, and significant
quantities of barium. On closure of the works, the waste tip area
was acquired by the local authority for redevelopment as a golf
course.

Concentrations of copper in excess of 14 000 mg/kg, together
with a variety of other readily available heavy metals resulted in a
surface which was extremely inhospitable to plant growth. The
presence of barium further necessitated the provision of an adequate
isolating or covering system.

Top soil sources and other covering materials were not available
but the waste product of a local soda ash industry, consisting
largely of calcium hydroxide and calcium sulphide was.

The alkaline waste (pH in the range 9 to 12) has been used to
provide a 0.5 m cover over the metal rich waste. The high alkalinity
and salinity of the cover material in itself will inhibit plant
growth, but will also reduce the solubility of the metals in the
original waste and reduce their potential for movement. The

revegetation requirements for the golf course have been overcome by provision of a further layer of 150 mm of top soil.[58]

7.7 Site G - Landfill, Connecticut, USA

This landfill was closed in 1978 because it was polluting the groundwater and posed a threat to a public water supply reservoir located near it. After close evaluation of the analytical monitoring data and the corrective options that were available, an engineered placement of a PVC membrane was selected. Refuse was regraded so that the slope was at least 1% and the side slopes 9%. A minimum of 150 mm of compacted borrow material consisting of sand and gravel was placed over the refuse and a drainage ditch was constructed around the site to divert surface water runoff. A 100 mm blanket of sand washings was spread above the borrow material in order to protect the top seal.

Several materials were field evaluated and PVC was selected as being the most suitable and cost effective. The seams of the PVC were field cemented and a gas venting system was installed. The top seal covered the entire landfill surface and the drainage ditch. It was covered with at least 450 mm of borrow material into which decomposed leaves and sewage sludge were mixed. The site was vegetated with grasses. A soil moisture/groundwater monitoring network was also installed to determine the effectiveness of the top seal.[70].

It is anticipated that up to 98% of the natural infiltration of precipitation into the landfill will be intercepted and routed away from the landfill. Leachate generation will be significantly decreased and its impact on the groundwater will be minimised.[71]

Monitoring of the landfill continued for several years before closure, establishing complete baseline data. It will continue for at least two years following the installation of the remedial action to determine the effectiveness. In 20 months since the implementation of the remedial action, the groundwater quality has improved and the leachate plume has retreated to the landfill.[72]

The need for subsequent inspection of such sites has been shown by the fact that erosion had removed some of the soil from the sides of the landfill exposing the PVC membrane. These erosion gulleys have since been repaired.[14]

7.8 Site H - Site Contaminated with Hexachlorocyclohexane, Germany

During the years 1949-1945 residues from hexachloro-cyclohexane

(HCH) production were dumped on a private company site at Gendorf, Bavaria. In consideration of possible leaching and diffusion and the consequent risk of hazard, the residues were encapsulated in a water-tight manner on the site. In addition a surface seal was used to create an impervious surface and provide for run off of precipitation.

Following encapsulation of the site with a steel pile barrier and PVC film additional surface covering layers were deposited. Immediately above the PVC film 100 mm of sand were placed. This was superimposed by 200 mm of gravel and a further layer of 80 mm of asphalt. A final surface of 30 mm of alsphatic concrete provided the final seal. The site was provided with a 3% surface fall in order to shed precipitated water. No results are available for its effectiveness.

8   CONCLUSIONS

(i)   A number of factors influence the choice of covering systems for contaminated land. Primarily the function and performance requirements are governed by occurrence of hazard, the local geology and hydrogeology, and the projected use of the site.

(ii) The local availability of materials often imposes a particular cover system on a site. This is indicated by the practical examples described in the text.

(iii) Where cover systems are described, their structure is often theoretically based on the chemical and physical properties of the component parts rather than practical tests or experience.

(iv) There is a need for an assessment programme to be developed involving site monitoring to evaluate cover performance.

9   REFERENCES

1   R. J. Lutton, G. L. Reagan and L. W Jones, Design and construction of cover for solid wastes landfill. US Environmental Protection Agency EPA 600 79-165.

2   R. M. Bell and G. D. R Parry, Environmental Advisory Unit, Liverpool University. Bibliography on Covering and Barrier Systems for Contaminated Sites.

3   W. H. Fuller, 1980, Soil modifications to minimise movement of pollutants from solid waste operations. in: 'CRC Critical Reviews in Environmental Control', March 1980, pp 213-270.

4   A. Kinard and F. P. Tuenge, 1979, Environmental safe landfill handles hazardous waste. Solid Wastes, 69, (2): pp 562-566 (1979).

5   A. Kinard and F. P. Tuenge, 1979, Landfill cell designed for hazardous waste disposal. Public Works, 110, (5): pp 92-94 (1979).

6   K. Garner, 1980, Sealing landfill sites 2. Bentonite.
     Pollution, Energy and Safety Monitor, (No 55): p 5,
     (1980).

7   H. Hass, Grouting and cut-off walls for the total
     encapsulation of contaminanted sites as long-term
     remedial measures - problems and research activities
     (private communication).

8   Prof Muller-Kirchenbauer, Hannover University (private
     communication) (presentation to CCMS Study Group in
     association with Hass ref 8).

9   I. F. Knipschild, Sealing landfill sites. I. The membrane
     lining. Pollution, Energy and Safety Monitor, (No 55):
     pp 3 and 16 (1980).

10   D. E. Sanning, Surface sealing to minimise leachate
     generation at uncontrolled hazardous waste sites. in:
     Proc Conf Management of Uncontrolled Hazardous Waste
     Sites. Washington DC, October 1981, HMCRI Silver Spring,
     Maryland, pp 201-205 (1981).

11a H. E. Haxo, Evaluation of selected liners when exposed to
     hazardous waste. in: Residual Management by Land
     Disposal, Proc of the Hazardous Waste Research Symp,
     W. H. Fuller, ed. US Environmental Protection Agency
     EPA 600/9-76-015, pp 102-111 (1976).

11b H. E. Haxo, Assessing synthetic and advanced materials for
     lining landfills. in: Gas and Leachate from Landfills:
     Formation Collection and Treatment, E.J. Genetelli., and
     J. Cirello, eds US Environmental Protection Agency
     EPA 600/9-76-004, pp 130-159 (1976).

12   S. Dakessian, M. Fong, and R. White, Lining of waste
     impoundments and disposal facilities. US NTIS PB81 -
     166365, p 411 (1980).

13   P. R. Rankilor, Membranes in Ground Engineering, John Wiley
     and Sons, Chichester/New York (1981).

14   R. J. Lutton, V. H. Torrey, and J. Fowler, Case study
     of repairing eroded landfill cover. 8th Ann Res Symp
     on Land Disposal of Hazardous Wastes. Ft Mitchell,
     Kentucky. March 1982, D. W. Shultz, ed, US Environmental
     Protection Agency. pp 486-494. (1982).

15   H. E. Haxo, Interaction of selected lining materials with
     various hazardous wastes. II. Disposal of Hazardous
     Waste. Proc 6th Ann Res Symp, Illinois, US Environmental
     Protection Agency EPA 600/9-80-010, pp 160-180 (1980).

16   W. Gutt and P. J. Nixon, Use of waste materials in the
     construction industry. Materials and Structures, 12
     (70): pp 255-306 (1979).

17   R. D. Stephens, Report of expert seminar on problems of
     hazardous waste sites. Waste Management Policy Group,
     Environmental Committee, Organisation of Economic Co-
     operation and Development, Paris 1980.

18   E. G. Barber, G J. Jones, P. G. K. Knight and M. H. Miles, PFA utilisation, Central Electricity Generating Board, London (1972).

19   C. J. Santhanam, R. R. Lunt, S. L. Johnson, C. B. Cooper, P. S. Thayer and J. W. Jones, Health and Environmental Impacts of Increased Generation of Coal Ash and FGD Sludges. Environmental Health Perspectives, 33: 131-157 (1979).

20   A. L. Page, E. A. Elseewi and I. R. Straughan, Physical and chemical properties of fly ash from coal-fired power plants with reference to environmental impacts. Residue Rev, 79: 83-120 (1979).

21   J. J. Emery, Assessment of ferrous slags for fill applications, in: Proc Conf Reclamation of Contaminated Land, Eastbourne 1979, Society of the Chemical Industry, London, pp F1-1-10 (1980).

22   J. J. Emery and D. Matchett, Use of stabilized industrial sludge for land reclamation, Ibidem, pp F2-1-8.

23   G. H. Thomas, Properties of iron and steel slags, Proc Conf Reclamation of former iron and steelworks, Windemere Durham County Council, Durham pp C1-8 (1983).

24   M. G. Carter, Restoration of a toxic waste quarry. J of Environ Management, 9: pp 12-129 (1978).

25   W. J. Mikucki, E. D. Smith, and R. Fileccia, Characteristics, control and treatment of leachate from military installations. NTIS AD-A097-935/1, p 103 (1981).

26a   R. J. Lutton, Predicting percolation through waste cover by water balance. Disposal of Hazardous Waste. Proc 6th Ann Res Symp, Illinois. US Environmental Protection Agency EPA 600/9-80-010, pp 118-122 (1980).

26b   R. J. Lutton, Evaluation cover systems for solid and hazardous waste. NTIS PB81-166340, pp 68 (1980).

27   M. J. Dowiak, R. A. Lucas, A. Nazar and D. Threlfall, Selection installation and post closure monitoring of a low permeability cover over a hazardous waste disposal facility. Proc Nat Conf Management of Uncontrolled Hazardous Waste Sites, Washington DC, HMCRI Silver Spring, Maryland, pp 187-190 (1982).

28   A. Schniering, Achievements and prospects for controlled compaction of fills of stored material. Lecture to the Committee for Communal Technology, Community Association of the Ruhr District Ltd, Oberhausen, Havensteinstrasse 50 (12-6-1981).

29   R. Harrington, and A. Sayers, Lining and capping of landfill sites. Munic Eng Environ Technol, 158, (16):pp 376-379 (1981).

30   W. H. Fuller, C. McCarthy, B. A. Alesii, and E. Niebla, Liners for disposal sites to retard migration of pollutants. In Residual Management by Land Disposal, Proc

of the Hazardous Waste Research Symp, W.H. Fuller., ed.,
US Environmental Protection Agency EPA 600/9-76-015,
pp 112-126 (1976).

31   H. E. Haxo, Effects on linear materials of long-term
exposure in waste environments. Proc 8th Ann Res Symp on
Land Disposal of Hazardous Wastes. March 1982, US
Environmental Protection Agency EPA 600/9-82-002,
pp 191-211 (1982).

32   J. Pacey, Controlling landfill gas, Waste Age, 13 (3):
pp 32-36 (1981).

33   E. S. Pankhurst, The effects of natural gas on trees and
other vegetation. Techniques No 35. Landscape Design,
(Feb): pp 32-33 (1980).

34   E. F. Gilman, F. B. Flower, I. A. Leone, and J. J. Arthur,
Vegetation growth in landfill environs. In Muncipal
Solid Waste: Land Disposal. Proc 5th Ann Res Symp
Florida, 1979. M. P. Wanielista, and J. S. Taylor, Eds,
pp 192-208 (1979).

35   D. N. Netherton and B. I. Tollin, Reclamation of the Beckton
Alps and adjacent areas. Public Health Engineer 10 (4):
pp 202-221 (1981).

36   B. E. Opitz, W. J. Martin and D. R. Sherwood, Gelatinous
soil barrier for reducing contaminant emissions at waste
disposal sites. Proc Nat Conf Management of Uncontrolled
Hazardous Waste Sites. Washington DC, Nov 1982, HMCRI
Silver Spring, Maryland, pp 198-202 (1982).

37   S. Ames, Reclamation of land disturbed in mining. in:
Proc 3rd Annual British Columbia Mine Reclamation
Symposium, Vernon, BC, pp 311-324 (1979).

38   B. Craze, Investigation into the revegetation problems
of Captains Flat Mining Area. J. Soil Conserv, N Wales,
33: pp 190-199 (1977).

39   R. C. Barth and B. K. Martin, Reclamation of Phytotoxic
tailing. Minerals and the Environment, 3, (2):pp 55-65
(1981).

40   J. Pacey and G. Karpinski, Retrofitting existing landfills
to meet RCRA standards for leachate control. Solid Waste
Management, 22 (2): pp 46-49 (1979).

41   J. Pacey and G. Karpinski, Selecting a landfill liner.
Waste Age. 11 (7): pp 26-28 (1980).

42   S. C. Reinfeldt, Wisconsin landfill closure requirements.
Solid Waste Management, 24, (12): pp 38-39 (1981).

43   T. Cairney, In-situ reclamation of contaminated land: The
problem of a safe design, Public Health Engineer
10 (4): pp 215-218 (1982).

44   E. Rumbery, Investigation of the behaviour of sealing
membranes towards rodents. Research Report 102-03-401.
Water Economy Federal Environment Office (1982).

45a   R. A. Griffin, R. R. Frost, and N. F. Shrimp, Effect
      of pH on removal of heavy metals from leachate by clay
      minerals. In Residual management of land disposal. Proc
      of the Hazardous Wastes Research Symp., W.H. Fuller.,
      ed., US Environmental Protection Agency, EPA 600/9-76-
      015, pp 259-268 (1976).
45b   R. A. Griffin and N. F. Shrimp, Leachate migration
      through selected clays. In Gas and Leachate from
      Landfills: Formation Collection and Treatment,
      E.J., Genetelli and J. Cirello., Eds., US Environmental
      Protection Agency, EPA 600/9-76-004, pp 92-96 (1976).
46   B. A. Alesii and W. H. Fuller, The mobility of three
      cyanide forms in soils. In Residual management of Land
      Disposal, Proceedings of the Hazardous Waste Research
      Symposium, W. H. Fuller, Ed, US Environmental Protection
      Agency EPA-600/9-76-015, pp 213-223 (1976).
47   A. Morrison, Can clay liners prevent migration of toxic
      leachate? Civil Engineering ASCE, (July): pp 60-63
      (1981).
48   D. C. Anderson, K. W. Brown and J. Green, Organic leachate
      effects on the permeability of clay liners. Proc Conf
      Management of Uncontrolled Hazardous Waste Sites.
      Washington DC, HMCRI, Silver Spring, Maryland, pp 223-
      229 (1981).
49   J. C. Evans, and H. Y. Fang, Geotechnical aspects of
      the design and construction of waste containment systems.
      Proc Nat Conf Management of Uncontrolled Hazardous Wastes
      Sites. Washington DC, HMCRI, Silver Springs, Maryland,
      pp 175-182 (1982).
50   J. F. Cline, K. A. Gano and L. E. Rogers, Loose rock
      as biobarriers in shallow land burial. Health Physics,
      39: pp 497-504 (1980).
51   R. P. Gemmell, Revegetation of derelict land polluted
      by a chromate smelter, Part 1. Chemical factors causing
      substrate toxicity in chromate smelter waste. Environ
      Pollut, 5: pp 181-197 (1973).
52   R. P. Gemmell, Revegetation of derelict land polluted
      by a chromate smelter, Part 2. Technique of revegetation
      of chromate smelter waste. Environ Pollut, 6: pp 31-37
      (1974).
53   W. H. Fuller and J. Artiola, Use of limestone to limit
      contaminant movement from landfills. 4th Ann Res Symp
      of Land Disposal of Hazardous Wastes, San Antonio, Texas,
      March 1978, D. W. Shultz, ed, US Environmental Protection
      Agency pp 282-298 (1978).
54   W. A. Duvel, Solid waste disposal: landfilling. Chemical
      Engineering, 86 (14): pp 77-86 (1979).
55   H. E. Bloomfield, J. F. Handley and A. D. Bradshaw, Topsoil
      quality. Landscape Design, (August) pp 32-34 (1981).

56   A. L. Tolman, A. P. Ballestero, W. W. Beck and G. H. Emrich,
      Guidance manual for minimising pollution from waste
      disposal sites. US Environmental Protection Agency EPA
      600/2-78.142. PB 286905 (1978).

57   I. A. Leone, F. B. Flower, E. F. Gilman and J. J. Arthur,
      Adapting woody species and planting techniques to
      landfill conditions. US Environmental Protection Agency
      EPA 600/2-79-128. PB 80-122617 (1979).

58   A. K. Jones, R. M. Bell, L. J. Barker, and A. D. Bradshaw,
      Coverings for metal contaminated land, in: Proc Conf
      Management of Uncontrolled Hazardous Waste Sites, HMCRI,
      Silver Spring, Maryland pp 183-186 (1982).

59   A. K. Jones, M. S. Johnson, R. M. Bell, and A. D. Bradshaw,
      1980, Biological aspects of the treatment of heavy metal
      contaminated land for housing development schemes. in:
      Proc Conf on Reclamation of Contaminated Land, Society
      of the Chemical Industry. London, Paper C3 (1980).

60   A. K. Jones, Journal Royal Society of Health, 102: pp 73-
      79 (1982).

61   R. M. Bell and G. D. R. Parry, The upward migration of con-
      taminants through covering systems, in: Proc Conf Management
      Uncontrolled Hazardous Waste Sites, Washington DC, 1984,
      HMCRI Silver Spring, Maryland (1984).

62   T.C. Cairney, A rational approach to the design of cover for
      contaminated sites, in: Proc Conf Contamination in the
      Environment, London 1984, pp 294-299, CEP Consultants Ltd,
      Edinburgh (1984).

63   N. W. Lepp and M. R. Harris, A strategy for evaluation
      of soil covering techniques to reduce metal uptake by
      soft fruits and vegetables. in: Proc Conf on Reclamation
      of Contaminated Land, 1979, Society of the Chemical
      Industry, London, pp C71-11 (1980).

64   M. R. Harris, S. J. Harrison, N. J. Wilson and N. W. Lepp,
      Varietal differences in trace metal partitioning by six
      potato cultivators grown on contaminated soil. in:
      Proc Conf Heavy Metals in the Environment, Amsterdam
      1981, CEP Consultants Ltd, Edinburgh (1981), pp 399-402.

65   M. R. Harris, S. J. Harrison and N. W. Lepp, Seasonal variations
      in the metal content of amenity grass and its use as an
      indicator of reclamation treatment performance, Science of
      the Total Environment, Vol 34, 267-278 (1984).

66   M. J. McCarthy, Reclamation of a refuse tip for open
      space and housing development. in: Proc Conf on
      Reclamation of Contaminated Land, 1979. Society of the
      Chemical Industry, London 1980 Paper B8.

67   J. N. M. Firth, M. S. Johnson and I. G. Richards, The
      reclamation of lead mine tailings at Parc Mine, N Wales.
      in: Trace substances in Government Health, Vol XV,
      University of Missouri, ed, D. D. Hemphill, pp 333-339
      (1981).

68   M. S. Johnson, Options for treatment of metal contaminated
      mine workings. Proc Conf Reclamation of Contaminated
      Land 1979, Society of the Chemical Industry, London
      Paper C9.
69   M. S. Johnson, T. McNeilly and P. B. Putwain, Revegetation
      of metalliferous mine spoil contaminated by lead and
      zinc. Environ Pollut 12: pp 261-277 (1977).
70   G. H. Emrich, W. W. Beck Jr and A. L. Tolman, Top sealing
      to minimise leachate generation. Case Study of the
      Windham, Connecticut landfill. Disposal of Hazardous
      Waste. Proc 6th Ann Res Symp, Illinois, US Environmental
      Protection Agency, EPA 600/9-80-010, pp 274-283 (1980).
71   D. E. Sanning, Remedial action technologies for uncontrolled
      hazardous waste sites - needs and solutions. Export
      Seminar on Hazardous Waste Problem Sites, OECD Paris,
      (1980).
72   W. W. Beck, A. L. Dunn and G. H Emrich, Leachate quality
      improvement after top sealing. 8th Ann Res Symp on Land
      Disposal of Hazardous Wastes. Ft Mitchell, Kentucky,
      1982, D. W. Schultz, ed, US Environmental Protection
      Agency, pp 464-474 (1982).

MANAGEMENT AND TREATMENT OF GROUNDWATER:

AN INTRODUCTION

K A Childs

Environment Canada

1   SCOPE

Chapters 7 to 9 are concerned with the management of the liquid phase associated with contaminated land sites. Preceding Chapters have described procedures for in-situ and on-site treatment of contaminated soils. When carrying out such treatments it will often be necessary to employ barrier systems or hydraulic measures as essential components of the overall remediation of the site. Covering systems, as described in Chapter 5, will be required as a component of macroencapsulation systems employing in-ground barriers and hydraulic measures. It is thus difficult to discuss treatment options in isolation from one another. This is particularly the case with liquid phase management which will invariably involve the utilization of more than one remdial or control measure. For convenience, however, the subject has been dealt with as follows:

Chapter 7 – Bottom sealing, vertical barriers and hydraulic
            systems.
Chapter 8 – Treatment of contaminated groundwater.
Chapter 9 – Mathematical modelling of contaminant transport.

The interdependence of the topics within and between Chapter is clear.

2   BACKGROUND

Chapters 7-9 deal with measures that might be taken when the contaminated soils remain in the ground at, or close to the point of deposition.

Measures described in Chapter 7 include:

(a) what is commonly referred to as macroencapsulation when the entire body of contaminated material is enclosed (including provision of an adequate cover) and thus isolated from the surrounding environment;

(b) the use of barriers, other than surface covering, to isolate the waste from groundwater by deflection or by discontinuous isolation;

(c) water table adjustment usually by pumping to separate the groundwater from the waste;

Chapter 8 includes descriptions of some of the treatment processes and systems that might be used for the treatment of water that has been contaminated as a result of contact with wastes (or contaminated soils) or as a result of contaminants migrating from the contaminant source to the groundwater. Chapter 8 does not address the treatment of relatively concentrated contaminated liquids that might be extracted from within the site or collected as a leachate at or within the site. The Study Group believes that these situations and technologies are adequately dealt with elsewhere.

Chapter 9 addresses the use of mathematical modelling to predict both contaminant migration and the effect of remedial actions.

Chapter 5 examined the potential for partially isolating the contaminated materials by placing, or constructing, a barrier above the contaminated soil. This form of isolation provides two main services:

- infiltration can be significantly reduced;
- a separation between the contamination and the land users can be achieved.

The measures described in Chapter 5 are frequently an important adjunct to the works described in Chapters 7 and 8. Reducing or totally preventing infiltration from above can significantly affect the need, extent and nature of, vertical barriers, horizontal underside barriers, pumping systems and treatment facilities.

3   DEFINITIONS

Terms common to Chapters 7, 8 and 9 are:

Liquid phase   Both the liquid within the contaminated land site and liquid migrating from the site together with the liquids remote to the site whose qualities have been influenced by the liquids migrating from the contaminated land site.

Groundwater   All water present in the saturated voids between the soil particles or rock fissures.

Groundwater table   The elevation below which all voids are filled with water - the saturated zone (ignoring the capillary fringe).

Zone of aeration   The zone between the ground surface and water table in which voids are filled with water and/or air (gas) - the unsaturated zone.

Leachate   All liquids emanating from a contaminated land site. These liquids, at the point of discharge, may be solely, or a combination of, the following:

(a)   the liquid product of decomposition
(b)   liquid or semi-liquid wastes
(c)   percolated precipitation
(d)   groundwater that has percolated through the site.

Contaminated groundwater   Groundwater in the vicinity of the contaminated land site that has changed in quality as a result of contact and mixing with leachate. This is a dynamic process with the degree of contamination changing as the mixing action continues and as reactions occur between the contaminated groundwater, the soils or rock, the leachate and the uncontaminated groundwater.

Contaminant plume (Plume of contamination)   The pattern of migration of contaminants from a contaminated land site. The extent (outer limit) of the plume will vary in both the horizontal and vertical planes depending upon a variety of factors - soil or rock conditions (permeability, sorption capacity, etc), the groundwater flow pattern and the nature of the contaminant, particularly its density. Different contaminants 'migrate' or are transported at different rates through the same environment. Therefore the plume of contamination for different contaminants will have different boundaries or limits.

Plume limits are directly related to concentration parameters. Therefore, in addition to the

physical factors, plume limits are also delineated
by 'political' considera-tions, ie legislated
acceptable limits of contamination.

It has been stated above that the water to be treated by the
methods described in Chapter 8 is the water that would be extracted
or intercepted at some point in the plume of contamination outside
the limits (ie hydraulically down gradient) of the contaminant
source. Collection systems that are either integral with the site or
constructed so as to restrict the flow of contaminated liquids beyond
the limits of the contaminated soil, are considered to be leachate
collection systems and are not addressed. This premise also applies
to pumping systems installed into that contaminated mass and likely
to extract concentrated leachate or liquid contaminants.

IN-GROUND BARRIERS AND HYDRAULIC MEASURES

K A Childs

Environment Canada

1   THE CONCEPT OF IN-GROUND MACROENCAPSULATION

The construction of bottom seals and walls at contaminated sites
will be undertaken for one of two main reasons:

 (i)   to encapsulate the contaminant source (soil or waste)
 (ii) to modify the local environment to create a diversion or
       separation of the contaminant plume.

In-ground encapsulation and isolation systems can be a means
of alleviating a problem or modifying unacceptable conditions.
However, the materials, techniques and procedures currently available
cannot provide either total and/or permanent isolation of the
contaminants from the environment. Encapsulation measures do not
influence the nature of the contaminants and in addition the
undertaking of additional activities in the milieu can provide
further opportunities for the environment to be adversely affected.
These additional factors must be accommodated at the time of
selecting both the encapsulating or isolating method and materials.
System failure due to reactions between the contaminants and the
encapsulating material may result in the release to the groundwater
of either the original target contaminants or additional contaminants
released as a result of these reactions. Care must also be taken to
ensure that contaminants are not released to the groundwater as a
consequence of the construction activities. Barrier materials may
possess significantly lower coefficients of permeability compared
with either natural soils or the materials that they replace.
However they must be considered as permeable and, therefore,
encapsulation is acting merely to retard the ingress and egress of
liquids.

Macroencapsulation systems or isolating features will probably be employed as long-term or, ostensibly, permanent features. Consideration must be given to what might occur to the encapsulated mass over the long term and what precautions must be taken to protect the integrity of the system against injurious external forces. In particular it is essential that the lateral and vertical extent of the system be recorded to protect against damage being caused by subsequent activities – construction, drilling, changes in land use, etc.

There is extended discussion with respect to long-term effectiveness of remedial measures in Chapter 2. Macro-encapsulation measures usually require significant initial investment and should be designed for the longest effective, predictable, or reasonable, life expectancy. Consideration must be given to the changes that may occur to the contaminants either naturally or by other processes during the life-span of the system and to the consequences of either sudden or gradual failure of the system.

Monitoring of encapsulation or isolation systems should be included as an integral part of the total monitoring program. It is essential that the effectiveness of the system be recorded during its entire lifetime, particularly registering the changes in the rate of release of contaminants (initially, during its lifetime and at the time of breakdown should it occur) which should be correlated to the 'load' on the system. Monitoring parameters should include measurements of changes in permeability and variations in the contaminant load. Monitoring of the system may assist in sensing if any reactions are occurring and therefore allow corrective actions to be initiated before the system breaks down entirely.

Throughout the life of any encapsulation system decisions will have to be made regarding the cost/benefits of the system. This exercise will be particularly important in the determina-tion of the nature and scope of corrective measures and maintenance activities.

2   THE CONCEPT OF REMEDY BY HYDRAULIC MEANS

Liquid phase management using hydraulic means has great potential for a number of reasons, principally the fact that the science of hydrogeology, including well design, has been practised for many years. However when hydralic measures are used as control mechanisms or aspects of a remedial technique, certain caveats should be observed.

The impact and effectiveness of hydraulic systems, when employed as control or remedial measures, are dependent on many external, influences. Frequently the source of these influences is remote from the site and control over them may be either impossible or limited.

For example – power sources and mechanical systems may fail, the groundwater conditions may change owing to a number factors (climate, remote extraction, etc) and the rate of release of the contaminant(s) may fluctuate. All of the elements in a hydraulic system change either naturally with time or by the imposition of other factors – the system is dynamic. It is imperative that remedial systems that have a hydraulic component are extensively monitored to ensure that the stated or requisite objectives are achieved. Those responsible for either maintenance of the system or achieving the objectives should be prepared to modify the system as conditions or objectives change.

In the discussions which follow later in this Chapter the systems and measures identified are described in isolation (ie not as part of larger systems) and it is assumed that the conditions are static. It is assumed that the practitioner will acknowledge that the system is dynamic necessitating system flexibility, that objectives and effectiveness will be monitored and the system adjusted accordingly.

3   BOTTOM SEALING – HORIZONTAL UNDERSIDE BARRIERS

3.1 General

This Section addresses the construction or development of a discrete impermeable layer, at depth, in contaminated land. Some of the techniques referred to are not limited solely to use in macroencapsulation but can be used as adjuncts to other forms of remedial action. Specifically, in-situ treatment (see Chapter 4) can employ grouting and ground injection as a means of stabilising or isolating contaminated soils and surface covering systems to isolate the wastes from the targets and/or to reduce infiltration, as discussed in Chapter 5.

The principal objectives of the procedures described here are to seal the underside either to reduce ingress/egress of liquids or to form a component in a macroencapsulation system (in combination with capping and vertical barriers) to achieve total isolation of the contaminant source.

The bottom seal could take basically four forms:

-    Injection into either (i) the contaminated soil or waste or (ii) into the native soil (including rock) if it is sufficiently permeable to allow the formation of a solid impermeable layer.

-    Creation of 'slab' or 'floor' in either (iii) the waste or contaminated soil or (iv) the native soils, to create a solid,

continuous isolating layer by displacing the in-place
materials.

The material used for sealing or for the floor must be selected
and 'designed' to withstand the pressures which might be exerted on
it and to protect against any adverse reaction due to contact between
the slab material and liquids or contaminant source. Throughout this
Chapter the materials referred to for use for bottom sealing or
vertical walls are required to provide sealing properties only and
any stabilising or neutralising effects are considered to be
incidental.

3.2   Grouting and Ground Injection

For reasons of clarity it is worthwhile to reiterate what
generally is understood[1] by these two terms.

'Grouting is used solely in the 'normal' engineering sense,
namely, the injection of appropriate materials – usually a
viscous fluid under pressure – into pores and cracks of another
material so as to decrease permeability or compressibility, or
increase strength, or a combination of the two. The term
'ground injection' is used to encompass injection of fluids
where the treatment objectives do not necessarily coincide with
the normal engineering objectives just mentioned; for example,
the use of chemical neutralising agents or forced leaching. The
techniques and equipment involved are identical to those
employed in grouting for engineering purposes. Thus it is
treatment objectives, and hence treatment agents, that
distinguish them.'

3.3   Bottom Sealing by Grouting of In-place Material

There are two methods of grouting: (i) claquage (hydraulic
fracturing) and (ii) permeation.

Claquage is frequently used by the petroleum industry to improve
the permeability of an oil bearing-formation. The primary concern is
that accidental fracturing will occur if grouting pressures are too
great. Normally 'designed' fracturing occurs when a borehole is
pressurised causing a failure of the borehole wall. Fracturing of
this nature can extend for considerable distances as a result of
pressure alone. Controlled fracturing in the horizontal plane can be
achieved by cutting horizontal slits in the bottom of the borehole.
By placing the boreholes judiciously, fractured sections can be made
to intersect. Grout is then introduced, under pressure, into adjacent
packed boreholes. The grout moves laterally outwards either through

the fractures in the borehole wall or, initially, along the horizontal slits until it meets the grout from the adjacent hole. In theory, an impermeable barrier is created when a continuous grout layer is created between all the injection points.

Permeation grouting is designed to achieve a uniform distribution of grout to fill the existing voids in the block of soil or waste (or a combination of both) immediately above, below or at the interface.

Whether permeation or claquage grouting is used consideration must be given to grout selection or design. It is important to match the grout to the conditions of the receiving media. Permeability, groundwater regime, chemistry and geochemistry, porosity and in-situ stresses must be considered. Particulate or suspension grouts (eg cements, clays, sand and fly ash) have relatively coarse particles resulting in the grout becoming trapped in the soil thus preventing the continuous flow of grout. Chemical or solution grouts being true solutions, have fewer practical transport problems. The rate at which grout is injected will depend on the relationship between grout viscosity, pumping gradient and soil or receiving media permeability.

3.4   Bottom Sealing by Means of a Slurry Floor

This method of bottom sealing requires that intersecting voids be created under the area of concern. The voids are subsequently filled with bentonite slurry. The voids are created by a cutting or 'kerfing system', essentially a high pressure fluid-jet cutting tool. At normal operating pressures it will drive nozzles about one millimetre in diameter and in most soils will cut a slit 1 to 3 metres long. Cutting rates up to several tens of centimetres per second can be achieved.

The high pressure kerfer can be used in two modes to achieve either vertical or horizontal cutting. If it is used as a means of creating a cavity for bottom sealing, the jet is oriented to cut horizontally from the bottom of a previously drilled borehole. If the cutter is rotated without raising a slit, a thin disc is produced. If the cutter is raised, a column section will be formed. The barrier material, usually in the form of a bentonite slurry, is then injected into the intersecting cavities to form a continuous floor (Figure 7.1).

There are obvious practical limitations with respect to the applications of this system. The soil or the waste to be removed (ie cut) must be of a type that will allow the floor to be constructed without the 'roof' collapsing.

3.5 Precautions

The normal design and installation precautions that must be observed when either of these systems is employed include:

- Selection of proper grout or slurry mix (noted earlier)
- Recognition of fire and explosion hazard – during installation
- Worker safety
- Protection against adverse environmental impact – eg release of toxic components from the grout, reactions between grout and surrounding environment.

3.6 Advantages and Disadvantages

The applicability of grouting and ground injection for in-situ treatment of contaminated land are discussed in Chapter 4. There is, however, merit to reviewing these processes as they apply to isolation and bottom sealing.

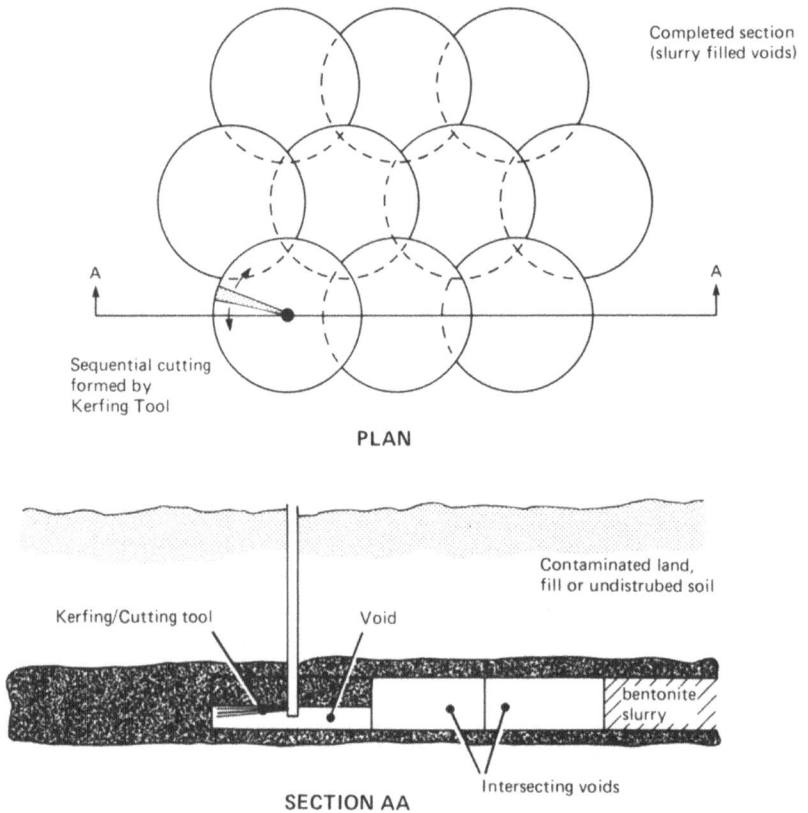

Figure 7.1 Undersealing by Kerfing (after Huck et al[7])

ADVANTAGES

Permeation and claquage grouting
have been standard practices for
many years and can be very effective
in selected soils and rock formations, ie they are proven techniques.

Construction is relatively easy
and can usually be performed
at any time of year, ie climate
is not a factor.

Environmental and aesthetic
impact on adjacent areas are
minimal during construction.

DISADVANTAGES

Drilling through
contaminated soil may
be difficult and even
dangerous because of
unknown materials.

The grout-take may be
erratic when uncharted
pockets of fine-grained
soils are encountered.

There are no totally
reliable methods of
determining that the
entire void between
boreholes or any faults
have been effectively
grouted. Infilling of
an ungrouted void or the
repair of a developing
void, subsequent to
initial constructions
would be difficult.

Structural overdesign of
the grout barrier or
slurry floor is almost
unavoidable because of
the uncertainties involved
in the attempt to create a
solidified mass beneath
the site. The sealing
layer could be thinner
and still achieve an
acceptable degree of
separation.

The potential deleterious
effects of the
contaminants on the
integrity of the grout or
slurry are not fully known
or assessed.

The effective use of
bottom sealing at
contaminated sites has
yet to be proven.

3.7 Applicability

The practice of bottom sealing can best be described as being in
the embryonic stage. For the present, given the lack of surety,
bottom sealing using grouts and slurries should remain as a supplementary line of protection only with principle protection being
afforded by other more reliable systems.

4   VERTICAL BARRIERS

4.1 General

Vertical barriers are considered to be trenches, walls or
membranes that are constructed or installed in an effort to isolate
contaminated soils. Such barriers may be either part of a macroencapsulation system or part of a control system designed either to
reduce or to eliminate interchange (contact transfer) between the
contaminated soil and the groundwater.

It was recently reported[2] that:

  'Cut-off wall techniques are well known in water and soil
  engineering and are therefore available for use on 'problem'
  sites. However the effectiveness and durability of these
  procedures have not been fully established in the context of
  hazardous waste (or contaminated land) sites. Consequently
  these countermeasures can only be considered as suitable for
  short- and medium-term protection of the environment. The long-
  term suitability cannot yet be assessed sufficiently.'

As noted earlier, in the case of bottom sealing, the 'relationship' between the soils, (both contaminated and uncontaminated) the
groundwater, the wastes and any construction materials must be
considered. In the case of a vertical barrier, in addition to the
foregoing considerations, the impact on the hydrogeology of the area
must be fully analysed.

Vertical barriers cannot be relied upon to prevent totally the
migration of hazardous substances even if they are keyed into
underlying impermeable layers. The construction of a vertical
barrier will normally influence the groundwater flow patterns and the
level of the groundwater table may be influenced in the vicinity of
the barrier. Hydraulic head differentials across a barrier
predictably create a need to pump the groundwater/leachate mixture.
This liquid will probably require treatment with the treated effluent
and any process residue requiring controlled discharge or disposal.
The provision of pumping facilities, which may have to be operated
and maintained for an unknown period of years, is discussed later in
this Chapter.

4.2   Types of Vertical Barriers

Vertical barriers are usually divided into two main categories and one lesser grouping that encompasses miscellaneous methods.

      Displacement – eg steel sheet piling
      Excavated   – eg slurry wall
      Low permeability barriers

Within these broad classifications there are many sub-categories, the differences usually being related to the type of barrier material or wall cross-section. Regardless of the type one of the controlling parameters must be compatibility between the barrier material and all the components in the local environment. A listing of the most common types of vertical barriers is given in Table 7.1.

4.3   Displacement Type Wall

This type of barrier wall can only be used when the in-situ material will allow penetration of the wall material or the form for the wall. The presence of boulders, rocks or bulky waste (eg demolition material) can easily preclude the use of these systems.

   4.3.1   Steel Sheet Piling.   Steel sheet piling is probably the least expensive, is relatively reliable and can be installed to a depth of 20–30 m for vibrated systems and 10 m for jetted systems. There are numerous sections and interlock systems available from numerous manufacturers.

Two common sections (see Figure 7.2) are heavy gauge steel wall and light gauge steel wall. Both are proven under construction conditions. The junction between sections is usually the main point of leakage. Infiltration through the joints may be reduced over a period of time as they become clogged with foreign materials or soil particles.

Where the objective is to isolate the contaminants, the piles are ideally driven, or injected, into an impermeable or less permeable stratum below the site. In acidic conditions (low pH) steel sheeting is subject to attack and may become less effective over a period of time.

   4.3.2   Vibrated Beam Slurry Wall.   In this process a beam having an 'I' section is driven to the required depth and then withdrawn. The void created is filled with a slurry mix. Each section is driven to overlap the preceding section to ensure complete coverage of the area. Figure 7.3 indicates the driving sequence, the extent of overlap, and the type of wall section achieved.

Table 7.1 Applicability and Limits of Vertical Barrier Systems

	Applicability			Watertight or impermeable	Width (m)	Feasible depth (m)	Proven
	Sand	Clay	Organic				
Steel sheet piling – heavy gauge – driven	Y	Y	Y	NW	0.1	20	Y
Steel sheet piling – light gauge – jetted	Y	N	Y	NY	0.01	10	Y
Slurry wall – vibrated beam	Y	Y	Y	1.0	0.1	25	Y
Jet grouting	Y	Y	Y	1.0	0.1 – 0.2	15	N
Panel wall	Y	Y	Y	0.1	0.3	35	N
Clay trench	Y	Y	Y	10–0.1	1.0 – 2.0	–	N*
Slurry trench	Y	Y	V	1.0	0.6 – 1.5	15	Y*
Deep wall	Y	Y	V	1.0	0.6 – 1.5	70	Y*
Deep wall with membrane	Y	Y	V	NR	0.6 – 1.5	–	N*
Continuous membrane	Y	Y	V	1.0	0.0001	5	Y*
Precast bentonite – cement	Y	Y	V	1.0	0.2 – 0.3	5	N*
Cutting pile	Y	Y	V	10.0	0.6 – 1.5	30	Y*
Cutting pile with injection	Y	Y	V	10.0	0.6 – 1.5	30	Y
Narrow trench and membrane	Y	Y	V	NR	–	5	Y*
Chemical grout curtain	Y	N	N	10.0	1.0 – 2.0	50	Y
Soil barrier	Y	N	N	V	1.0 – 2.0	–	Y
Freezing	Y	N	N	NR	1.0	–	Y

Y – applicable
V – variable
NW – not watertight
Permeabilities – metres/second x 10^{-8}

N – not applicable
* – can be used in the presence of obstacles
NR – not recorded

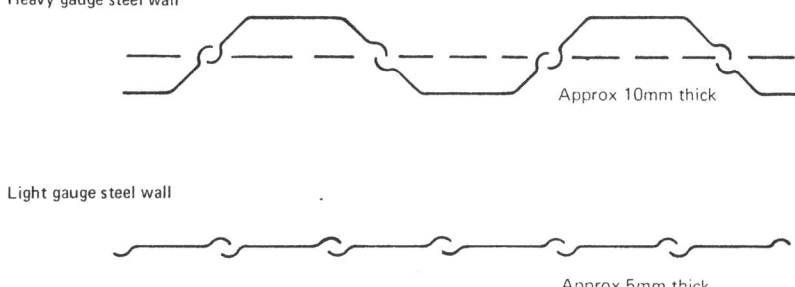

Figure 7.2   Steel Sections used in Diaphragm Walling

Traditionally slurry mixes have been a mixture of soil plus cement and bentonite. However, mixtures are now available using additives such as asphalt emulsions. Indications are that these mixtures will withstand the action of aggressive chemicals more effectively than the bentonite mixes used in the past. Tests of compatibility of soil and the wall material should always be made.

The average thickness of the vibrated beam slurry wall is approximately 0.1 m and construction to depths of 25 m are feasible. One identified advantage is that the 'key' with the impermeable layer can be more continuous than it might be with a slurry trench because penetration into the impermeable layer can be monitored at the time of driving the beam. Daily production rates can be 100% greater than those achieved with steel sheeting. If difficulties are encountered the vibrated beam slurry does have one potential 'escape path' in that the wall can be completed using a trenching system.

Like most barrier systems the continuing effectiveness should be monitored, particularly the condition (ie permeability) of the barrier.

4.3.3   Jet Grouting. This process has been employed in Japan, Germany and more recently the UK (see Section 3.3). To create a vertical wall holes are pre-drilled to the required depth and, as described in section 3.4, special equipment is used to create a cavity and simultaneously fill the void with grout. The resulting wall of intersecting columns provides a secure vertical barrier. Figure 7.4 shows a variation of this technique where the pre-drilled hole is filled prior to directional injection. The shape and width of the wall will depend upon the type of injection system used.

The maximum practical depth of installation using this system is approximately 15 m. The problems encountered are those that are common to all displacement systems - the soil conditions must be

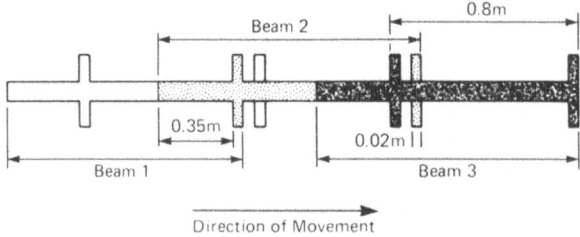

Figure 7.3   Vibrated Beam Slurry Wall

suitable and the presence of obstacles can frustrate attempts to
achieve a complete seal. Results indicate that a good seal can be
achieved at the time of initial placement but the integrity of the
system and the condition of the wall material need to be monitored.
Earlier remarks, with respect to grout selection, are equally
applicable to this system.

4.3.4 Panel wall. This system, reported by the Netherlands,
is a variation of the vibrated slurry wall. The series of sketches
in Figure 7.5 describes the sequence of actions. The maximum
feasible depth of construction in this system is approximately 35 m
with a wall thickness of 0.15 to 0.30 m. This system is in the final
development stages and information on effectiveness, durability and
costs are incomplete.

4.3.5 Membrane Wall. This method of isolation uses materials
such as H.D.P.E (High Density Polyethylene) or PVC (Polyvinyl
chloride). The material that forms the screen is injected into the
ground coupled to a special injection plank or form. The adjoining
screen plank is driven to the same depth with the two screen sections
now coupled together. The first screen and plank are uncoupled and
the injection plank withdrawn. The types of materials used in
membrane walls are generally resistant to contaminants such as
benzene and toluene which frequently have adverse effects on walls
made of bentonite, etc.

4.4 Open Trench - Excavated Systems

This type of construction can be more reliable for the
preparatory work can be undertaken with greater ease and certainty,
ie the excavation is continuous and not as adversely affected by
obstacles (rock, rubble, etc). The one facet which may not be as
'secure' is the key into the impermeable layer. Penetration into

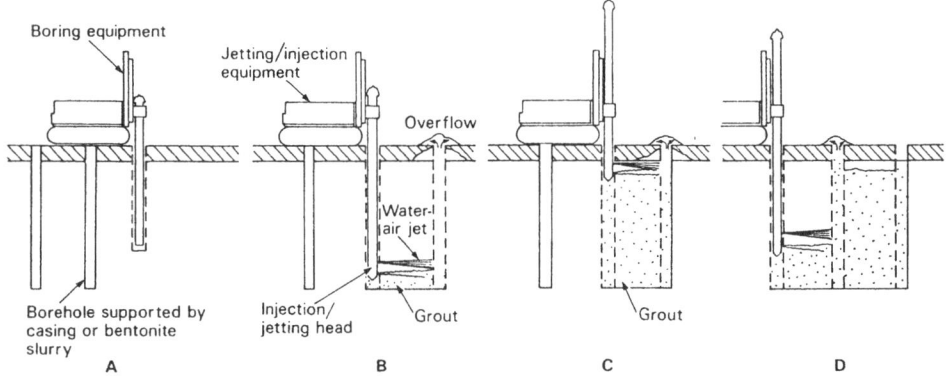

Figure 7.4   Construction of Vertical Barrier by
Means of Jet Grouting

some low permeability soils may be more difficult with the type of
equipment used in the construction of these systems. Various barrier
materials can be employed – clay, bentonite, synthetic membranes and
injectants.

4.4.1 Clay Barrier.   The simplest form of barrier is the clay
filled trench designed to achieve a wall having a lower permeability
than the in-situ materials.   Caution is required in the level of
reliance placed on this type of wall for there is increasing evidence
that permeability is increased when certain clay minerals are in
contact with organic substances.

4.4.2 Slurry Trench.   The following is an extract from the US
Environmental Protection Agency publication 'Remedial Action at Waste
Disposal Sites'[3].

'Slurry trenching involves excavating a trench through or
under a slurry of bentonite clay and water, and then backfilling
the trench with the original soil with or without slurry mixed
in.   Most commonly, the trench is excavated down to, and often
into, an impervious layer depending upon the application.   The
width of the trench can vary, but is typically from 0.6 to 1.5 m.
Depending on the depth of the trench, light or heavy equipment
is used for excavation.'

'Excavation of a trench under a bentonite slurry causes two
things to happen.   First, the slurry acts as shoring, supporting

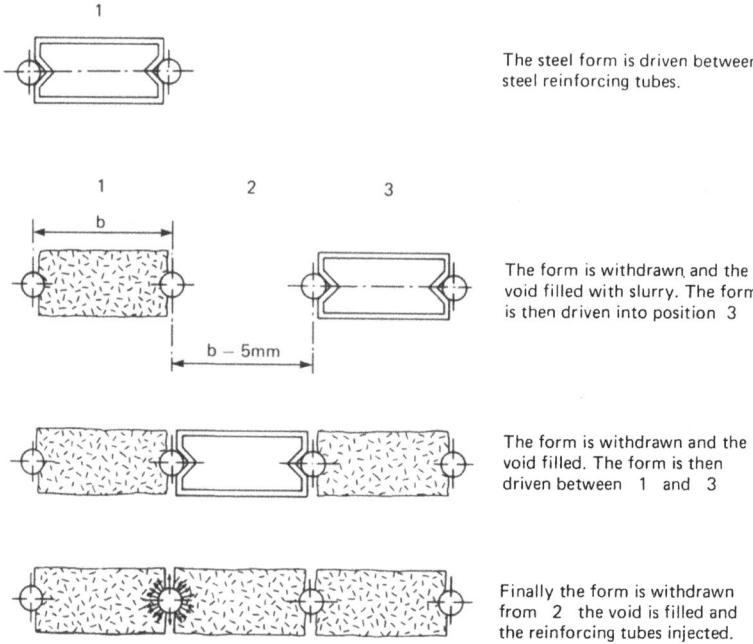

Figure 7.5   Panel Wall System

the trench walls to prevent cave-ins and slumping during further excavation. Secondly, and most importantly, the weight of the slurry forces bentonite into the soil matrix on the trench walls and bottom. As more and more bentonite is forced into the soil, a filter cake is formed, the thickness of which depends on the permeability of the soil and the weight of slurry. In essence, the trench becomes completely lined with a layer of soil and bentonite of extremely low permeability.'

Slurry trenches can be constructed to a depth of 15 m and a typical slurry mix will contain 4-7% bentonite. The addition of polymer compounds can prevent increase in permeability occurring due to the presence of an aggressive leachate. However, because of the method of construction slurry walls are not universally suitable. In situations where the natural soils have high permeability the use of slurry trenches may be totally impractical. A handbook[4] on slurry trench construction for control of pollution migration was published by US EPA in 1984. This discusses in detail the theory, potential application, site investigation requirements, design and construction, and monitoring and evaluation methods.

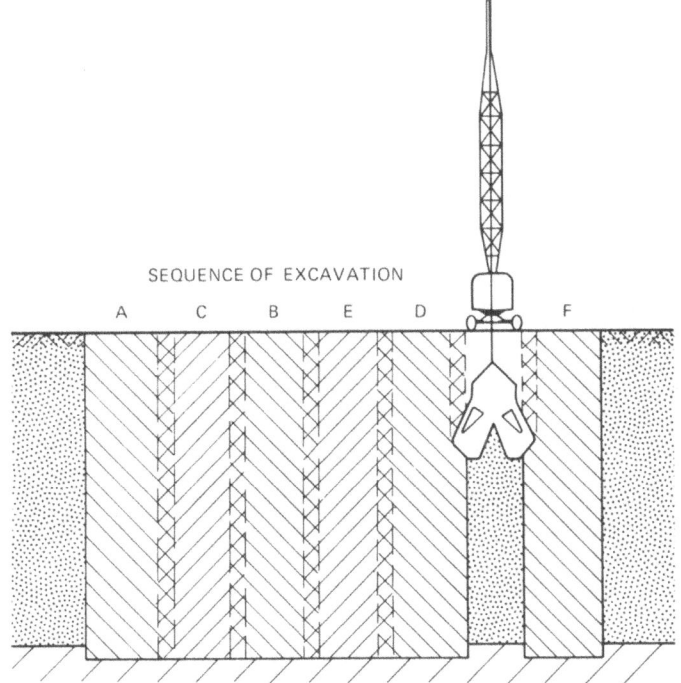

SEQUENCE OF EXCAVATION

A   C   B   E   D   F

Figure 7.6   'Deep Wall' Barrier (Bentonite-Cement)

4.4.3 Bentonite – Cement 'Deep Wall' or Barrier. This is
a variation of the slurry wall and will allow construction to a
greater depth; up to 80 m can be secured. No increase in wall
thickness over the conventional trench system (0.6 – 1.5 m) is
required.

Construction to these greater depths is achieved by use of a
clam shell excavator. The barrier is constructed of overlapping
'panels' or sections as shown in Figure 7.6. The wall material is
normally a concrete mix containing bentonite and cement. As noted
before, a wall constructed using a slurry mix of this type is not
recommended when the environment in which it must function is, or
might become, acidic or otherwise aggressive.

4.4.4 Trenches and Slurry Walls with Membrane. The
impermeability of clay trenches and slurry walls can be improved by
the addition of a synthetic membrane. The supplementary
membrane, usually plastic, would be placed integral with the wall

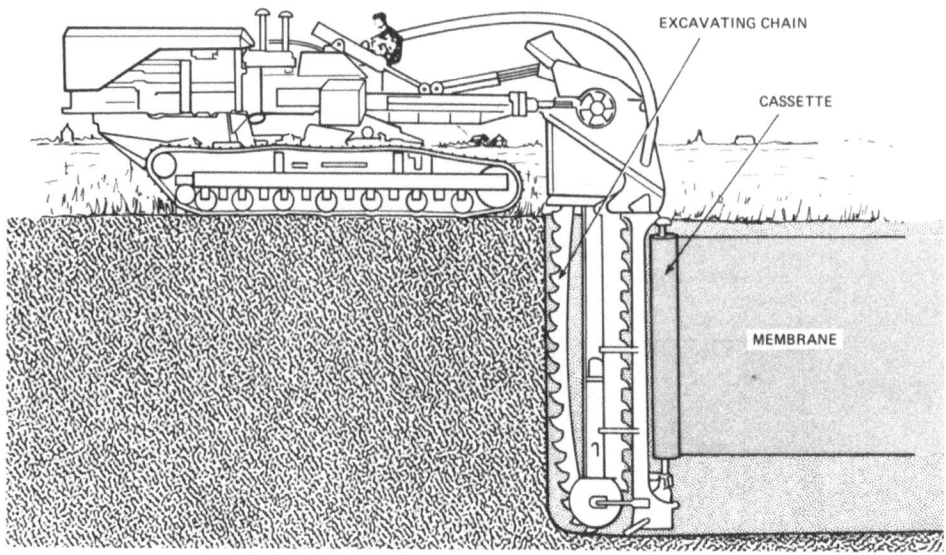

Figure 7.7   Synthetic Membrane Installed with Chain Excavator

or incorporated into the clay fill to provide an extra continuous barrier.

4.4.5 Narrow Trench with Membrane. A system has recently been developed that uses a chain excavator to form a narrow trench immediately followed by placement of a synthetic flexible membrane using a specially designed dispenser. Patents are currently being applied for. The concept is shown in Figure 7.7.

The membrane could be installed to depths of 5 m using this system. The membrane material would require sufficient strength to withstand the stresses imposed at the time of installation and would have to be compatible with the environment in which it will function. Linear production rates up to 800 m/day are projected.

Monitoring over the short- and long-term would be required for knowledge of the effects of chemical, biological and mechanical stresses is limited.

4.4.6 Precast Bentonite - Concrete Wall. Reference to a precast bentonite concrete is made in a report from the Netherlands. No working details are provided. However like similar concepts it would be limited in its application in terms of maximum depth and its resistance to chemical attack.

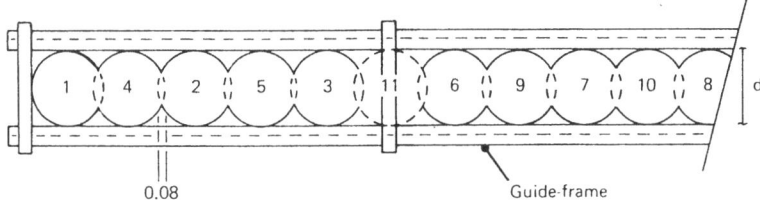

Figure 7.8   Construction Sequence for Pile Wall

4.4.7 Pile Wall. This barrier system is constructed by
drilling a series of overlapping holes to a specified depth and
filling the continuous void formed with a bentonite-cement or
concrete mix. To ensure that a regular pattern is maintained the
holes are drilled inside a frame in a planned sequence as shown in
Figure 7.8. The width of the wall can range from 0.35 to 1.50 m.
Construction to a depth of 30 m is feasible. Use of this method is
advocated for smaller applications because mobilisation and equipment
costs are usually lower. An optional step is to use a chemical
injectant (grout) between alternate posts or piles. This system is
subject to the same limitations (eg obstacles), potential adverse
reactions (eg chemical or leachate attack) and effects of hydraulic
head differential.

4.5 Vertical Walls of Low Permeability

Two methods of isolating contaminated lands by construction of a
wall or curtain of low permeability have been identified.

4.5.1 Chemical Grout Curtain. Grouts having a silicate base
are the most commonly used chemical grouts; they account for over 75%
of the grouting performed for water containment. Silicate grouts are
composed of a sodium silicate base, a reactant, an accelerator, and
water. The concentration of sodium silicate in the grout varies
between 20 and 60%, and when used for water containment would
normally be less than 30%. The concentration of materials in the
grout will depend on the receiving environment that is to be
injected. The lower limit of injectability has been identified as
soils composed of more than 20% sand. A chemical grout curtain can
be formed by a single line of holes but is more likely to be formed
using secondary or even tertiary curtains.

4.5.2 Natural material trench. In natural soils of high or
medium permeability a decrease in permeability may be achieved by
compacting (eg clay) in a trench. High in place density and low
permeability can be achieved by judicious selection of materials and
designed installation.

4.6 Applicability

The practice of installing vertical barriers to modify a local
environment, usually to facilitate construction by improving working
conditions, is a time tested activity. Selection of wall materials
has been influenced by a number of requirements or conditions - the
degree of water tightness (and impermeability), structural strength
requirements and the working environment.

In the case of the management of contaminated land the working
environment becomes a major consideration. In the short-term it is
desirable to undertake materials testing to determine the suitability
of currently available materials if they are required to perform as
ostensibly permanent barriers in a potentially aggressive environment. In the longer term it will be necessary to develop new
techniques, new materials and monitoring capabilities in order to
achieve and maintain objectives.

At the present time the materials and construction methods
available are, with a few exceptions, used in conventional
construction activities. In the future the imposition of additional
parameters may negate their suitability. Research and development is
required to provide barrier systems that are easy to install,
permanent and capable of being monitored.

5   UTILISATION OF HORIZONTAL, CONTINUOUS
        AND DISCONTINUOUS VERTICAL BARRIERS

5.1 General

This section addresses the use and placement of barriers to
isolate the contaminated land or to modify local groundwater flow
patterns in an effort either to reduce or to prevent contact between
the contaminated land and the groundwater.

To simplify and shorten the discussion certain assumptions are
made that apply to this section only. These are:

(i) That infiltration resulting from precipitation on the
     contaminated land will be minimised by the provision of a
     properly designed cover, ie infiltration vertically downwards
     is not a significant factor.

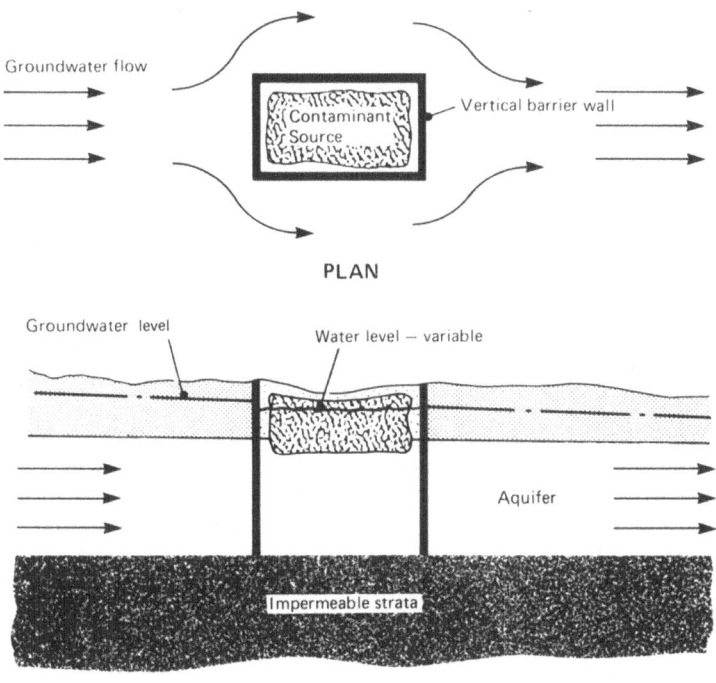

Figure 7.9 Total Perpheral Isolation (Vertical Barrier to
Impermeable Strata) after Rinnert and Kobus[6]

(ii)   That the vertical barriers are virtually 'impermeable', ie the
flow through the barrier, due either to faults or breakdown of
the integrity of the barrier, is minimal.

(iii)   That supplementary measures such as groundwater pumping are
not necessary.

5.2 Total Isolation

The likelihood of achieving total separation between source and
the surrounding environment is a function of the efficacy of each
component in place. The variation in level of efficacy is as
follows:

Potential efficacy of top sealing – (specially designed cover)	– very good (but see Chapter 5)
Potential efficacy of vertical barriers	– good to poor depending upon site conditions and barrier type
Potential efficacy of in-ground horizontal barriers	– still in the development stage and unproven for this purpose. They are likely to remain of limited value due to the fact that it will be difficult to ensure reliability.

The selection of the barrier system and the barrier material will be determined by the required depth of installation, the site conditions, the degree of integrity acceptable (ie initial effectiveness) and compatibility with the environment within which they will be required to function. Because of the current state of the art (ie developmental) the decision to use and, ultimately, the selection of the type of vertical barrier or bottom seal will depend primarily upon professional judgement. Selection of any system from an engineering point of view (eg structural strength) can be made by more precise means and objective assessments.

At present, there are no recorded examples of total isolation systems being installed and functioning over a long term at any contaminated land site. Even in the absence of practical operating experience it can be predicted that most isolation systems will require the support of a hydraulic system to maintain a balance between the water regimes inside and outside the barrier.

5.3   Total Peripheral Isolation

5.3.1   Total Peripheral Isolation down to Impermeable Strata.
In this case (see Figure 7.9) the original (natural) flow of the groundwater is deflected around the enclosed area. The groundwater level outside the barrier will stabilise in response to the intrusion of the barriers. The water level inside will fluctuate modestly in response to the influence of a number of continually changing variables including modified external groundwater level (due to intrusion of the barriers), the permeability of the vertical barriers, the degree of infiltration through the cover, and the permeability of lower strata. If it is presumed that the cover and vertical barriers are secure and retain their integrity and that a tight permanent bond is achieved between the impermeable strata and the barriers, then this system represents the closest to macro-encapsulation that is currently achievable.

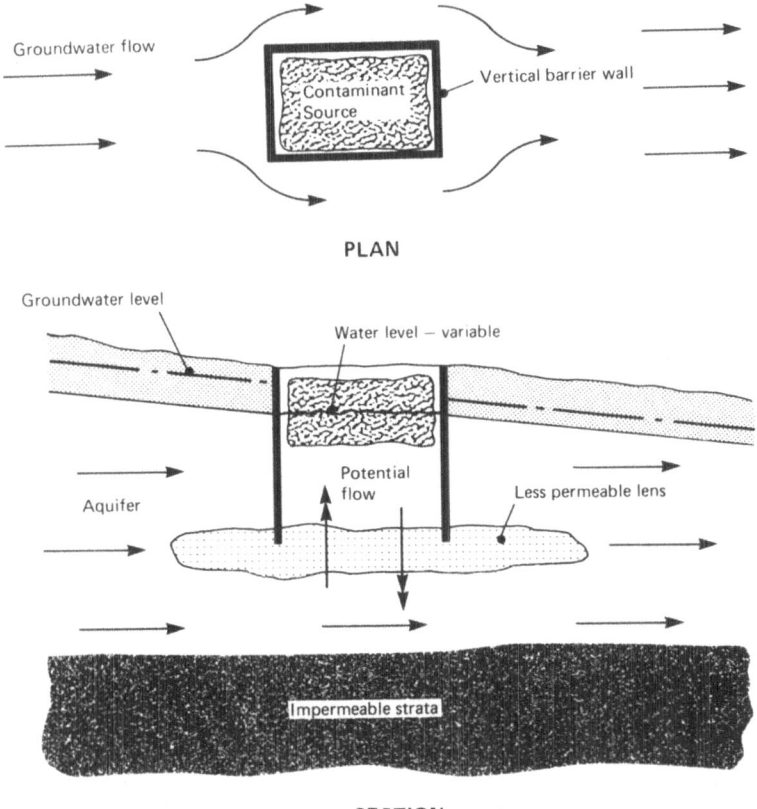

PLAN

SECTION

Figure 7.10   Vertical Barrier Walls to Discontinuous or Less
Permeable Strata (after Rinnert and Kobus[6])

     If bottom sealing is considered to be equivalent of an
impermeable layer the schematic presentation would need to be
modified only slightly. There would again be lateral deflection
around the barrier in the upper levels of the aquifer where the
barrier is intrusive. There could also be downward deflection of the
water table that would cause greater pressure on the underside of the
bottom seal. This could result in an increasing potential for upward
movement of water through the seal.

5.3.2 Total Peripheral Isolation down to Discontinuous or Less
Permeable Strata. In this case (see Figure 7.10) the groundwater

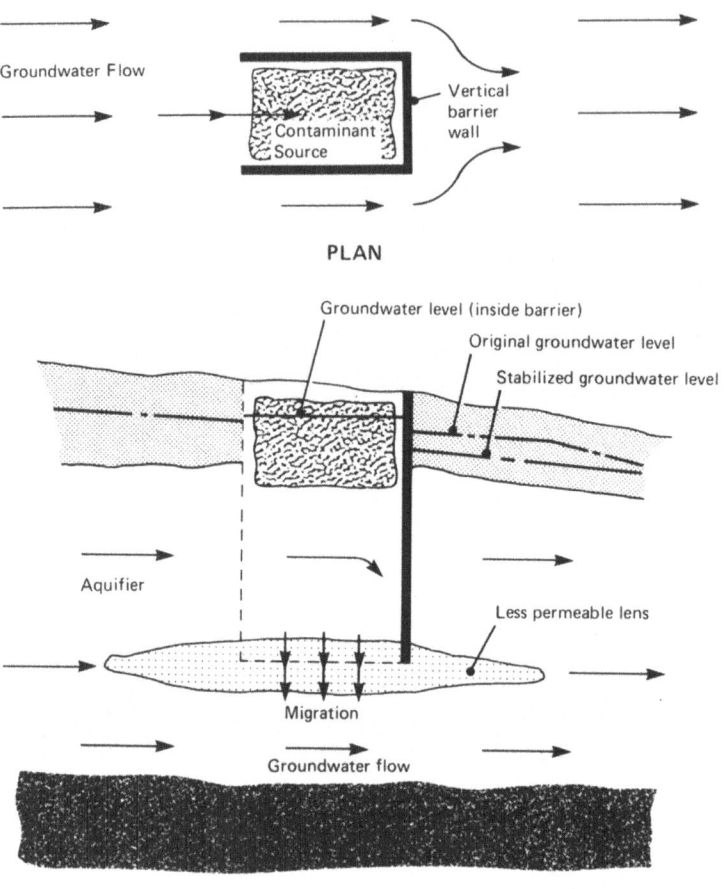

PLAN

SECTION

Figure 7.11   Barrier on Three Sides – Upstream Side
Open (after Rinnert and Kobus[6])

level outside the barriers will stabilise to a modified profile and
the groundwater inside the barriers will fluctuate depending upon (a)
ingress of water either due to infiltration because of cover or
barrier imperfections (increasing permeability or loss of integrity)
or due to the permeability of the material under the site; (b) water
loss due to egress through the walls or 'floor' or (c) the effect of
pumping inside and/or outside the barrier.

This case can also be compared to what is likely to be achieved when a combination of vertical barriers and bottom sealing is used and the mass to be isolated intrudes below the water table or into an aquifer. In the diagram (Figure 6.10) the discontinuous lens would be replaced by a constructed bottom seal.

5.4 Vertical Barriers on Three Sides

5.4.1 Upstream Side Open. In this case (Figure 7.11) the bottom component of the barrier system might be either a less

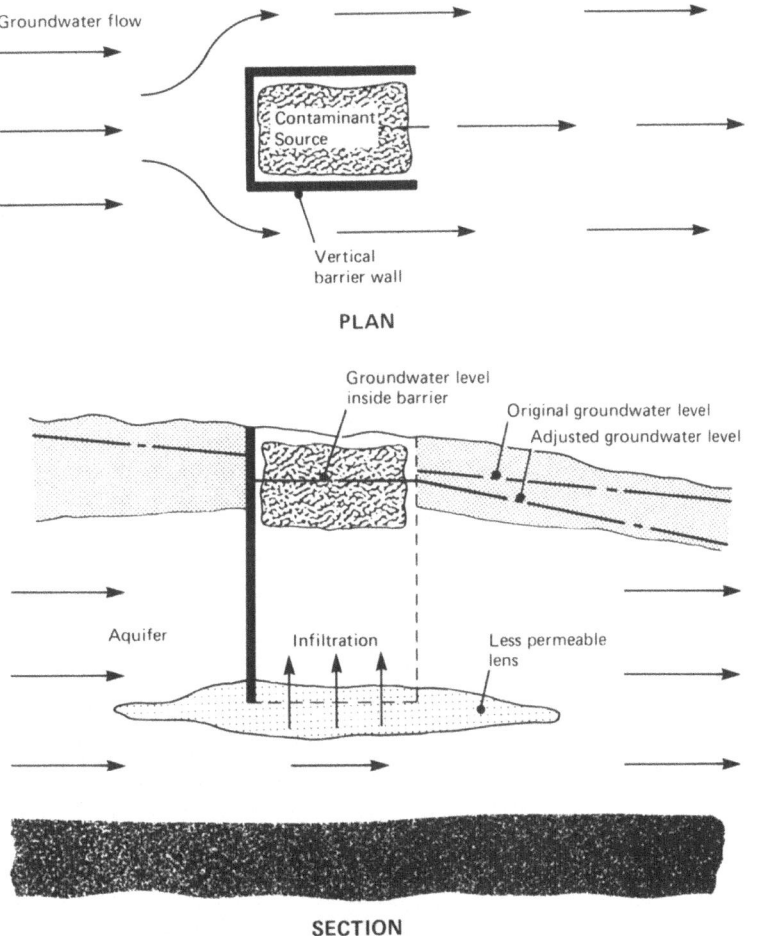

Figure 7.12 Barrier on Three Sides – Downstream Open
(after Rinnert and Kobus[6])

permeable lens or a constructed bottom seal. The groundwater
level outside will stabilise in response to the intrusion of the
vertical barriers. Inside the walls the water level will fluctuate
depending upon a number of variables (permeability of walls and base,
effectiveness of covers, variations in flow in the aquifer etc). If
the 'floor' is not totally impermeable there is a distinct
possibility that migration will take place downwards and under the
toe of the wall.

5.4.2 Downstream Side Open. In this case (see Figure 7.12)
the outside groundwater level will stabilise and the inside water
level will stabilise to the level of the groundwater at the downstream side. This balance may be achieved in part by migration
upwards through the 'floor' or a 'backing-up' of the groundwater
system.

It should be noted that in the last three cases the objective is
not to achieve absolute separation between the groundwater and the
contaminant source, but to achieve a lowering in the release rate of
the contaminant.

5.5 Applicability

Cut-off walls with no hydraulic support system are of limited
value. Even in the case of a continuous cut-off system in a
homogeneous aquifer the effectiveness of the measure could be
significantly impaired if the walls do not extend down to, and
preferably into, an impermeable layer. Cut-off systems for isolating
the contaminated soils that rely in part on either a constructed
bottom seal or a discontinuous natural lens of less permeable
material should be viewed with caution and probably limited to low
risk situations.

6   GROUNDWATER MANAGEMENT BY HYDRAULIC MEANS

6.1 General

Because the subject of hydrogeology is fundamentally complex and
because there is a paucity of documentary information relevant to
application of hydraulic means at contaminated sites the discussions
that follow are, perforce, superficial. Before embarking on any
programme of groundwater control, management, modification or
protection, reference should be made to the many learned texts and
that recognised experts should be consulted.

Groundwater modification systems can be employed to achieve a
number of objectives:

(a)   To lower the groundwater table in the vicinity of the contami-
      nated sites to create a separation between the contaminants and
      groundwater.

(b)   To contain or isolate a contaminant plume.

(c)   To supplement a barrier system.

(d)   To facilitate treatment of the contaminated water in a post-
      extraction system.

      The components in a hydraulic groundwater management system
include:

Extraction wells   – either singly or in groups to withdraw
                              groundwater from the aquifer

Infiltration or   – either singly or in groups to inject water
injection wells   into the ground after treatment or as part
                              of a hydraulic system for plume manage-
                              ment or water balance

Infiltration trenches   – used primarily as a route for recharging
                              the aquifer after inter-ception and/or
                              treatment.

6.2   Lowering the Water Table by Pumping

      6.2.1   Prevention of Discharge of Contaminated Water to the
Surface in an Adjacent Discharge Area.   The principal objective in
this application (Figure 7.13) is to change the location of the
discharge zone, moving it beyond bodies of surface water or otherwise
sensitive discharge areas.   Two supplementary effects can be
realised:

  (i)   creation of greater separation between the contaminant plume
        and the groundwater table and

 (ii)   the travel distance through the soil is greater thus affording
        more opportunity for natural processes to attenuate the
        contaminants thus possibly causing a greater degree of natural
        purification of the groundwater.

      6.2.2   Prevention of Direct Contact Between the Contaminants
and the Groundwater to Maintain the Quality of the Aquifer.   In this
case (Figure 7.14) the sole objective is to modify the level of the
groundwater table to creat a separation between the contaminant
source and the top of the saturated zone.   Monitoring is required to
ensure that a separation continues to exist even though conditions

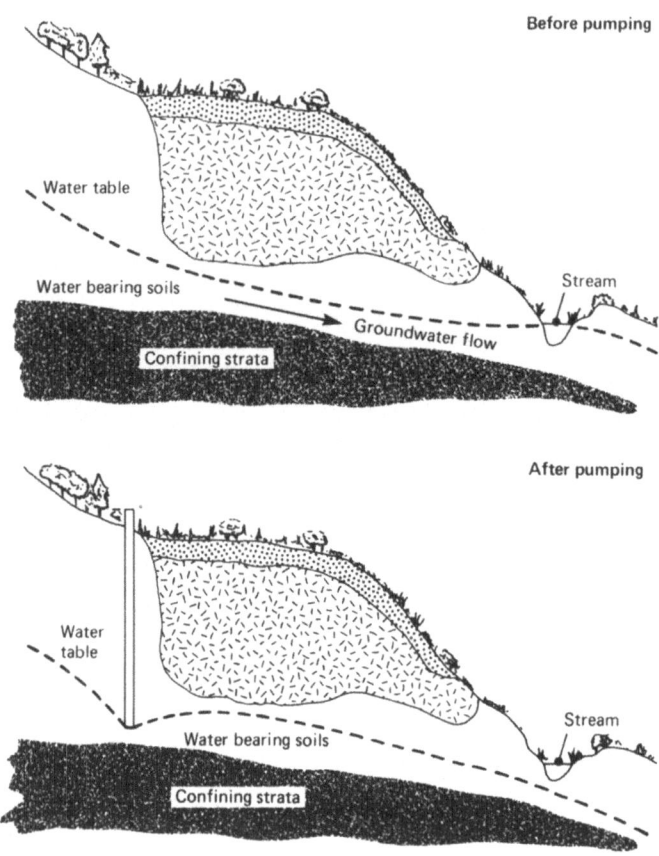

Figure 7.13 Discharge Zone Management (after Ref 3)

change. Pumping rates may have to be varied to accommodate
fluctuating conditions.

6.2.3 Prevention of Contamination of an Underlying Aquifer
by Creating a Localised Upward Hydraulic Gradient. The objective
in this instance is to create a localised change in the direction
of flow of the groundwater (Figure 7.15). Contaminants migrating
from a source at a higher elevation are transported laterally and
vertically downwards. Construction of extraction wells around
the source permits the development of a local upward hydraulic
gradient thus limiting or even eliminating the migration of
contaminants. Design, monitoring and maintenance of the system
are essential to ensure that the zones of influence 'cover' the
area of concern.

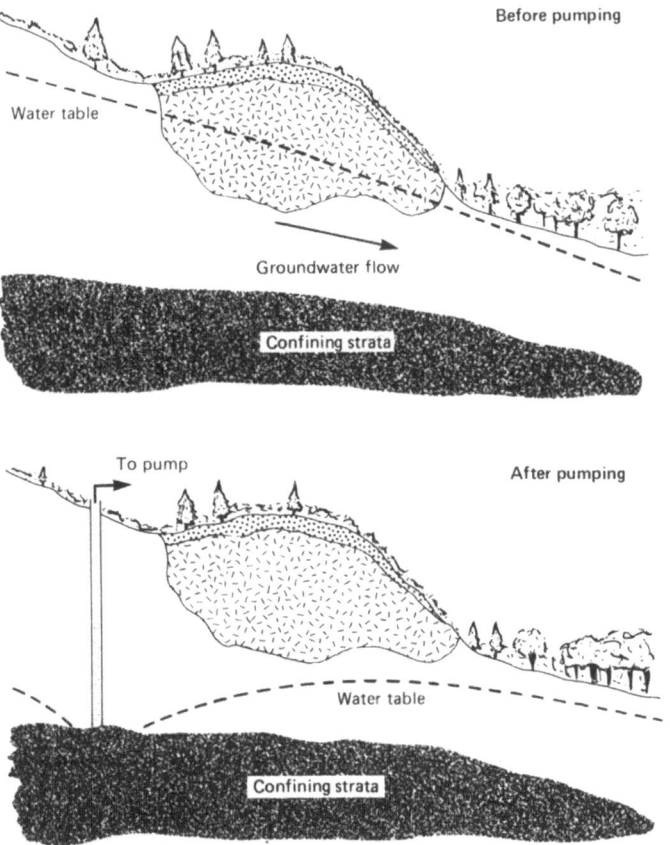

Figure 7.14   Separation of Groundwater and Contaminant Source
by Pumping (after Ref 3)

6.2.4 Water Table Adjustment by a Well Point System. The use
of a well point system to dewater (ie lower the water table) in a
defined area is a time-tested proven construction procedure and lends
itself readily to application at contaminated sites. The system (see
Figure 7.16) consists of a group of closely spaced wells, usually
connected by a header pipe. A pump may be connected to one well
point, or a central pump may be used for the entire well point
system.

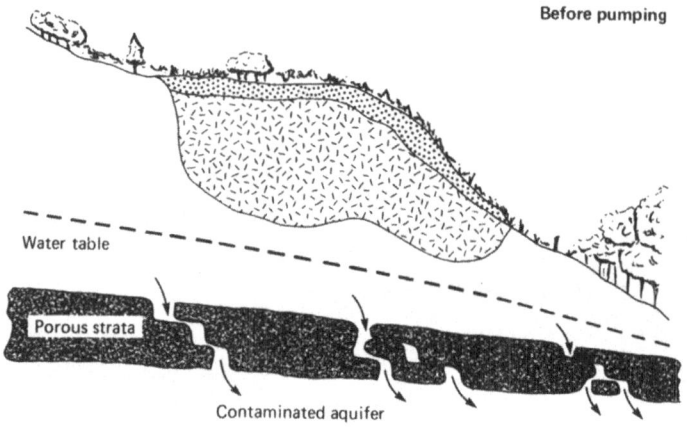

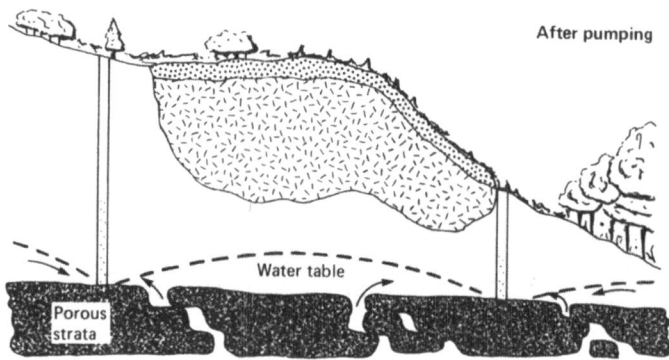

Figure 7.15   Creation of Local Upward Hydraulic Gradient
              (after Ref 3)

Lowering the groundwater level over the site involves creating a composite cone of depression by pumping from the well point system. The individual cones of depression must be close enough to overlap and consequently pull the water table down at intermediate points between pairs of wells. The design of any well (extraction, injection or well point system) requires specialised knowledge of well hydraulics and should only be carried out by qualified professionals.

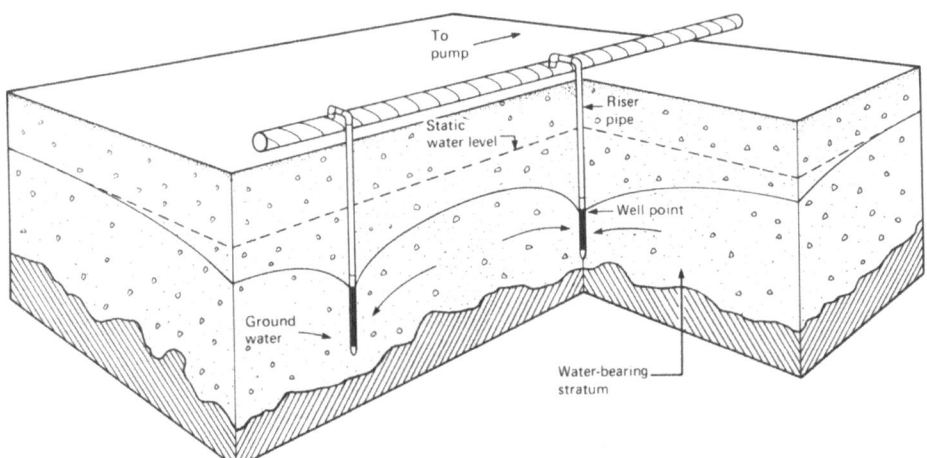

Figure 7.16   Schematic of a Well Point Dewatering System
(after Ref 3)

6.3   Plume Containment by Hydraulic Means

In some cases there might be a plume of contaminated groundwater
which cannot be eliminated in the short-term.   In cases of this
nature it is desirable to introduce measures that   will reduce or
limit the extent of the plume or arrest its further expansion.   There
are three basic cases (see Figures 7.17 to 7.19):

Extraction — treatment (optional) — injection
Extraction — discharge to surface (no treatment)
Extraction — recharge via basin or gallery (with or without
            treatment)

Plume containment by pumping, with or without subsequent
recharge, can be an effective means of preventing or limiting the
contamination of wells that are being used or could be used as
potable water supplies.   The concept can also be used to prevent the
pollution of streams or confined aquifers that are hydraulically
connected to the contaminated groundwater.   The technique could also
be applied to limiting the influence of migration from surface
impoundments.   Pumping without subsequent direct recharge (Figure
7.17) may be an acceptable approach when a small flow of groundwater
is involved.   However, when large groundwater flows are involved or
when residents are dependent on groundwater as a source of drinking

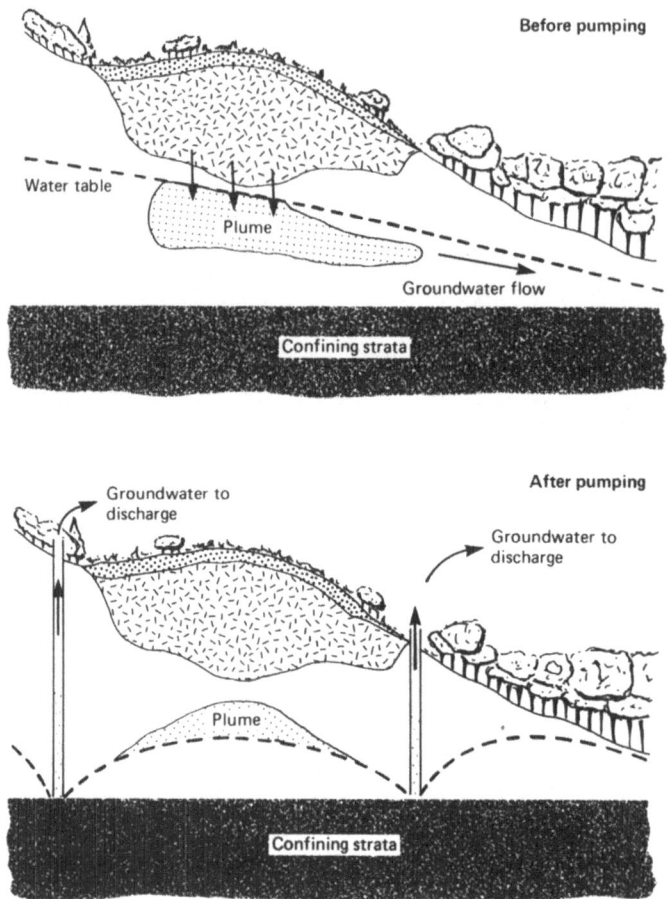

Figure 7.17   Groundwater Extraction to Contain a Plume
               - no recharge (after Ref 3)

water, recharge will probably be necessary. Pumping of large volumes
of water without subsequent recharge may lead to changes in the
potentiometric surface (level) or direction of flow within a confined
aquifer.

Containment of the plume is achieved by intercepting the plume
within the radius of influence of the well. Monitoring of is
essential to ensure interception in the first instance and to ensure

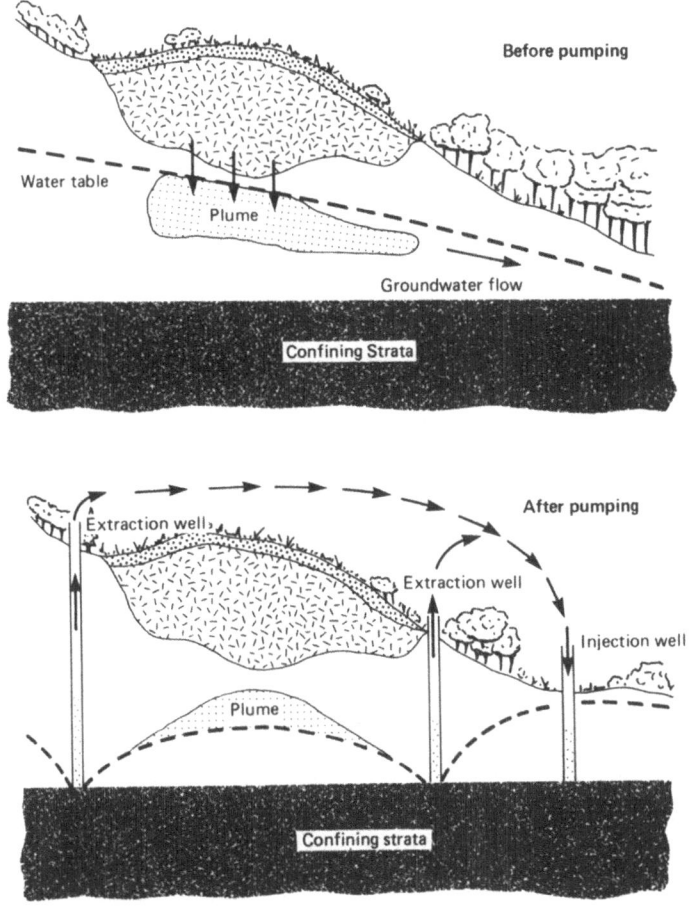

Figure 7.18   Use of Extraction and Injection Wells for
Plume Containment (after Ref 3)

continued effectiveness. Changes in the hydrological environment may require that pumping rates and pumping levels be modified.

The designer of a system using both extraction and injection wells (Figure 7.18) must have knowledge of the impact of withdrawal upon recharge, particularly, if there is any increase in the area of interference between two zones of influence, for, if this occurs, the behaviour of both components is slowed until hydraulic equilibrium is achieved.

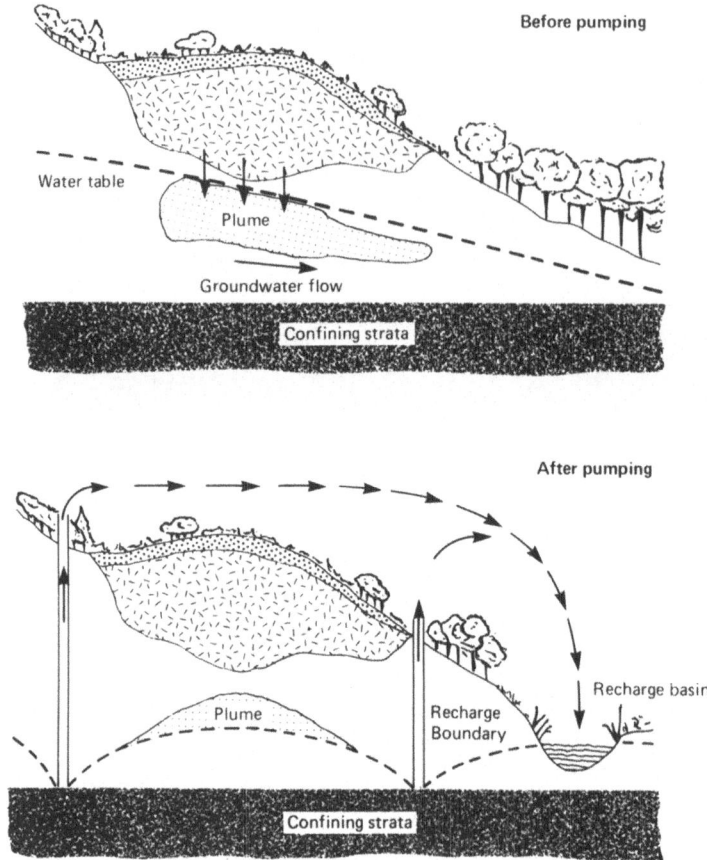

Figure 7.19   Groundwater Extraction Wells Followed by Subsequent
              Recharge Through Seepage Basins or Galleries
              (after Ref 3)

When hydraulic equilibrium is achieved the result is a net
reduction in the effectiveness of the extraction well which can mean
a change in the zone of influence of the system. Monitoring is
required to determine if the modified zone of influence is large
enough to contain the contaminant plume. To avoid potential
interference between wells and to allow maximum flexibility (ie
system modification) it is desirable if extraction and injection
wells are located beyond their respective zones of influence. Figure
7.20 demonstrates the relationship between injection and extraction
wells.

There are a number of advantages to using a system that employs only extraction wells followed by surface discharge. Advantages include simplicity of design and potentially lower maintenance costs. This type of system is ideally limited to low pumping rates when the aquifer is of less utility. Even though the system is more basic, monitoring is still essential.

Extraction followed by recharge through basins or galleries (Figure 7.19) requires more detailed design and potentially more maintenance. At the design stage the hydraulic impact of the recharge element on the extraction system must be recognised and accommodated in the hydraulic system. The rate of migration from the basin must be determined for it may be necessary to provide a series of basins. Ideally, to avoid interference between the components there should be hydraulic separation between the zone(s) of influence of extraction and the zone(s) of influence of recharge. Monitoring of basin performance and maintenance of the system are essential. In particular adequate seepage rates in the basins must be maintained. The main operating problems affecting recharge efficiency are clogging of the well or gallery, walls and filter beds.

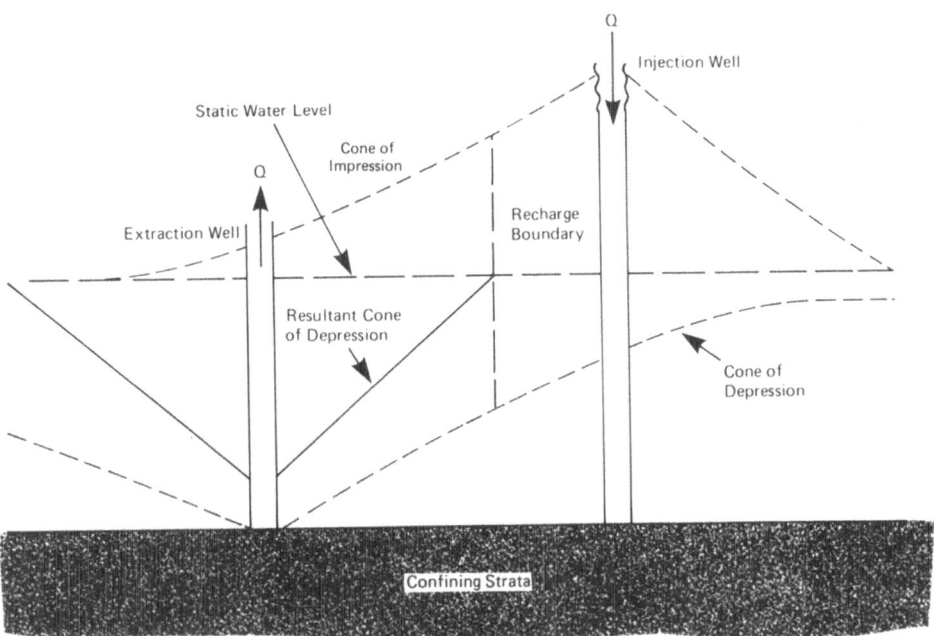

Figure 7.20   Effect of an Injection Well on the
Cone of Depression (after Ref 3)

Well point systems, described earlier, can also be used for plume containment and are more flexible in terms of being required to operate under changing conditions.

6.4 Advantages and disadvantages

There are advantages and disadvantages to all systems for groundwater management. In the case of containment of a contaminated plume using well systems the US Environmental Protection Agency identified them as follows:

ADVANTAGES	DISADVANTAGES
System may be less costly than construction of an impermeable barrier	The volume to be pumped and plume characteristics will vary with time, climatic conditions, and changes in the site. This can result in a need to provide a continuing monitoring program
High degree of design flexibility	System failures could lead to the plume expanding resulting in the contamination of drinking water
Moderate to high operational flexibility, which will allow the system to respond to varying pumping demands	Operating and maintenance costs are higher than for artificial barriers

7   GROUNDWATER MANAGEMENT USING HYDRAULIC MEANS IN COMBINATION WITH BARRIER SYSTEMS

7.1 General

As noted earlier there are a number of elements that can constitute an isolation system – top cover, bottom sealing, vertical barriers and hydraulic systems. This section deals with the combination of vertical barriers and hydraulic systems (wells and basins).

For ease of presentation the following assumptions are made:

(i) that there is no bottom sealing other than the natural material that was in place when the contaminant source was first deposited or migrated into the area.

(ii) that no exceptional measures have been taken to seal totally the top or any other part of the area influenced by the well system.

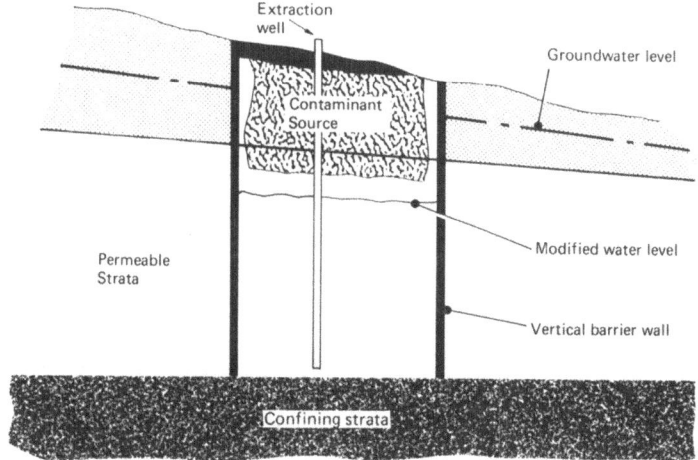

Figure 7.21   Continuous Vertical Barrier to Impermeable Strata
              in Combination with Internal Pumping (after Rinnert
              and Kobus[6])

        It is stressed that the science of using barriers, pumping
systems and combinations of both is relatively new and, as a
consequence, information on the long-term effectiveness and impacts
of this type of control action is limited.

7.2   Continuous Vertical Barrier to Impermeable
         Strata with Internal Pumping

        The objective (Figure 7.21) is to achieve and maintain a
separation between the contaminant source and an adjusted water table
inside the barrier (ie the water level inside the barrier resulting
from vertical and horizontal migration through the barrier(s) as
modified by pumping). Occasional withdrawal of water by pumping is
virtually inevitable if the separation between the wastes and the
water table is to be maintained. Assessment must be made on a
continuing basis as to the need for continued monitoring and
operation.

        It is probably easier to justify a system composed of pumps and
a shallow vertical barrier. In this case the vertical barrier would
be required to extend into the water table but not into the
impermeable strata. This combination, although not as secure, could

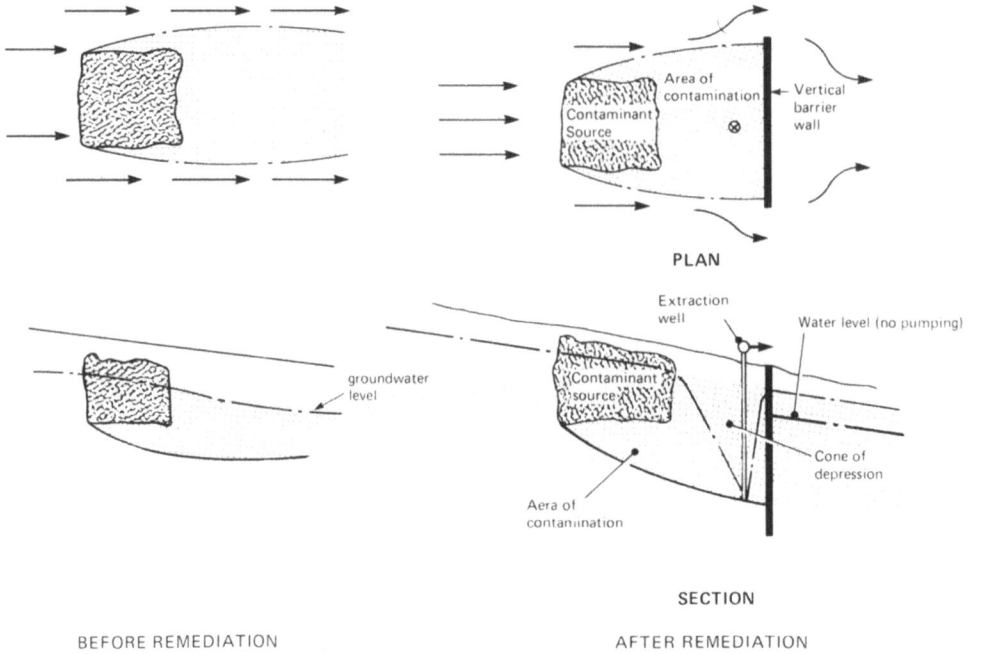

PLAN

SECTION

BEFORE REMEDIATION   AFTER REMEDIATION

Figure 7.22   Downstream Barrier with Extraction
(after Rinnert and Kobus[6])

be designed to achieve a separation of groundwater and contaminant
source. In addition to the potential for financial economies, the
systems, efficiency is not dependent upon achieving a key between the
wall and an impermeable stratum.

7.3   Discontinuous Barriers and Groundwater Pumping

Barrier systems that are either discontinuous (ie do not totally
encircle the contaminated land in the horizontal plane) or do not
extend to an impermeable stratum are of limited value. There are,
however, some instances where the combination of discontinuous
barrier and pumping system can be used to modify certain undesirable
groundwater/contaminant conditions.

7.3.1   Downstream Groundwater Withdrawal – No Barriers.   As
noted in 6.3 plume containment can be achieved without any barriers.
This option should be considered first.

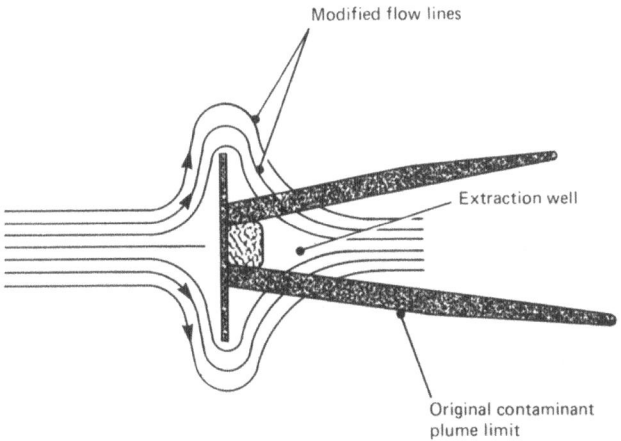

Figure 7.23   Upstream Barrier with Extraction
(after Rinnert and Kobus[6])

7.3.2   Downstream Groundwater Withdrawal - With Barriers.
A barrier is constructed close to the downstream limit of the
site.   Such a barrier would have to extend laterally to the outer
limits of the plume and, ideally, to a depth such that the bottom
of the barrier was below the bottom of the plume.   The lateral
limits of the plume and the water level upstream of the barrier
would be modified by the impact of the extraction well, ie the
well would be designed to create a zone of influence that would
contain the plume and inhibit any plume expansion (see Figure
7.22).

7.3.3   Upstream Deflection - With Barrier.   A barrier is
installed immediately upstream of the contaminant source to
modify the groundwater flow pattern in the vicinity of the site
(see Figure 7.23).   The length of the wall should be sufficient
to change the flow patterns to the point where the contact
between the groundwater and the contaminant source is either
eliminated or significantly reduced. The barrier should be of
sufficient depth to allow 'keying' into an impermeable layer, or
at least into a less permeable layer. Construction of a barrier
in this location will probably allow the extraction well to be
located closer to the site.   The volume of contaminated water
will be less and the pumping system can therefore be designed
with a lower capacity.

8   REFERENCES

1   D L Barry, 'Treatment Options for Contaminated Land', Atkins
       Research and Development, Epsom, UK (1982)
2   Organisation for Economic Co-operation and Development, Proc
       Experts Seminar, Hazardous Waste 'Problem' Sites, Paris 1980
       (OECD), Paris
3   'Handbook for Remedial Action at Waste Disposal Sites, US
       Environmental Protection Agency, Cincinnati, Ohio, EPA
       625/6-82-006 (1982)
4   Slurry Trench Construction for Pollution Migration Control,
       US Environmental Protection Agency, Cincinnati, Ohio, USA.
       EPA 540/2-84-001 (1984)
5   Huck, Walker and Shimondle, 'Innovative Geotechnical
       Approaches to the Remedial In-site Treatment of Hazardous
       Materials Disposal Sites, Earth Tech Research Corporation
6   B Rinnert and H Kobus, 'Groundwater Rehabilitation by
       Hydraulic Means in the Vicinity of Abandoned Disposal Sites
       (Study for Environmental Protection Agency, Baden,
       Workenburg, FRG)

TREATMENT OF CONTAMINATED GROUNDWATER

K A Childs

Environment Canada

1   INTRODUCTION

The terms and concepts employed in Chapters 7 to 9 have been
explained in Chapter 6. This Chapter deals with the situation when
water to be treated is extracted or intercepted at some point in the
plume of contamination beyond the limits (ie hydraulically down
gradient) of the contaminant source. The level of contamination may,
consequently, be low and comprised of a number of constituents.

2   TREATMENT TECHNOLOGIES – GENERAL

The science and practice of groundwater treatment, as part of a
remedial process, is in its infancy. With the exception of a few
innovative systems, available technology is limited to mainly
hypothetical applications of known leachate treatment processes.

There is an abundance of information readily available on
the leachate treatment and concentration technologies. The latter
are more applicable to groundwater treatment. Technologies in both
groups are similar. However, 'leachate treatment' is usually
discussed in the context of a collection system intercepting the
leachate before it can migrate from the site and become diluted.

An exhaustive listing of concentration technologies appeared
in a recent US Environmental Protection Agency publication and an
edited version is included as Appendix G. Three additional
technologies are also referred to in the context of leachate
treatment and groundwater rehabilitation. These are chemical

Table 8.1 Applicability of Treatment Systems (from Ref 2)

Contaminant classification	Biological	Activated carbon	Chemical precipitation	Chemical oxidation — Alkaline chlorosis	Chemical oxidation — Ozone	Chemical reduction	Reverse osmosis	Ion exchange	Stripping	Wet air oxidation
1 Alcohol	1	V	0	NA	E	NA	0	V	0	0
2 Aliphatic	V	V	0	NA	4	NA	0	V	0	0
3 Amine	V	V	0	NA	NA	NA	0	0	0	0
4 Aromatic	V	E	3	NA	4	NA	0	V	0	0
5 Ether	2	V	0	NA	0	NA	0	0	0	0
6 Halocarbon	4	2	0	NA	3	NA	1	0	0	0
7 Metal	4	V	1	NA	0	2	2	1	NA	0
8 Misc – Ammonia	1	NA	NA	E	0	NA	0	0	2	0
– Cyanide	2	NA	NA	E	E	NA	1	0	NA	0
– TDS	NA	NA	NA	NA	NA	NA	1	1	NA	NA
9 PCB	NA	1	0	NA	0	NA	0	0	0	0
10 Pesticide	NA	1	0	NA	1	NA	0	1	0	0
11 Phenol	2	1	0	NA	1	NA	0	V	0	0
12 Phthalates	2	1	2	NA	0	NA	0	0	0	0
13 Polynuclear Aromatic	NA	1	R	NA	2	NA	0	0	0	0

1 Excellent   3 Fair   NA Not applicable   0 Not reported
2 Good   4 Poor   V Variable   R Reported as removed

Table 8.2   Contamination of Water in the Proximity of Waste
            Disposal Sites

Classification	No of sites	No of contaminants in classification
1 Alcohol	2	5
2 Aliphatic	4	12
3 Amine	2	2
4 Aromatic	8	33
5 Ether	Not reported	–
6 Halocarbon	9	29
7 Metal	15	24
8 Miscellaneous	11	30
9 PCB	2	4
10 Pesticides	7	8
11 Phenol	7	13
12 Phthalates	2	2
13 Polynuclear	5	6

oxidation, chemical reduction and wet oxidation.   These are also
described in Appendix G.

To indicate in general terms the compatibility of a process with
a particular contaminant, a classification has been developed and is
cross referenced with the most commonly employed treatment processes
in Table 8.1

The magnitude of the problem that may be faced by a government
authority is indicated by the figures in Table 8.2 which refers to an
examination of water in the proximity of twenty-seven waste disposal
sites in the United States.

3   PROCESS TRAINS

When contamination occurs due to waste disposal operations,
spills, industrial activities and other community activity it is
usually a complex matter with the contamination consisting of a
number of compounds or elements. By the time the wastes have been
collected, randomly deposited with subsequent leaching and mixing
with the groundwater, it is not likely that one simple or standard
treatment process will be adequate to remove all the contaminants of
concern.   The need to provide more than one process results in the
development of a process train where a series of processes forms a
unique treatment system.   The sequence and treatment elements in the

system will be determined by the nature and concentration of the
contaminants requiring treatment or detoxification.

Currently available literature contains many examples of process
trains designed to treat leachates or contaminated water. Process
trains, commonly identified as being applicable to problem aqueous
wastes or contaminated water, are described below.

3.1 Biological/Carbon Sorption

This system (Figure 8.1) would be used if the waste could be
characterised as having high TOC, low toxic organics and heavy
metals. The essential process features are pH adjustment,
flocculation (coagulation) to remove metals to sub toxic concentrations, followed by biological treatment. Biological treatment
might be an activated sludge or biological plate process either of
which can realise a reduction in both BOD and bio-degradable toxic
organics. The subsequent filtration and settling (clarification)
steps reduce the suspended solids load prior to transfer to the
activated carbon sorption process. In the activated carbon stage
refractory and toxic organic residuals are removed.

Potential weaknesses have been identified as follows:

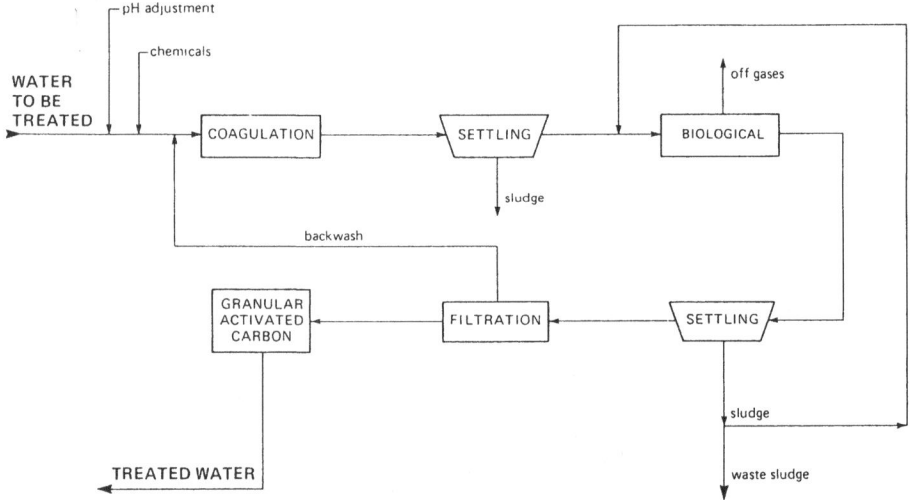

Figure 8.1   Process Scheme for Treatment of Water by
Biological/Carbon Sorption (from Ref 1)

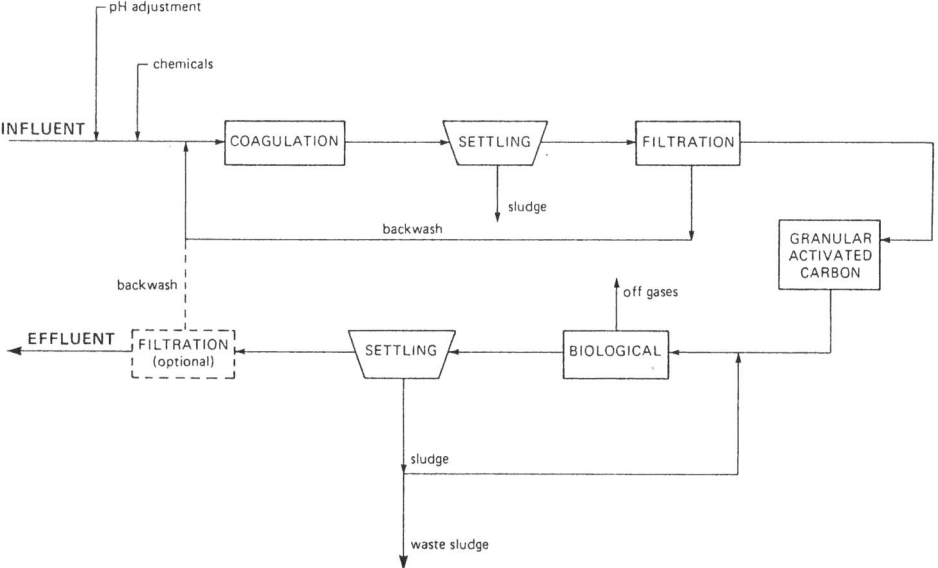

Figure 8.2   Scheme for Treatment of Water by Carbon
              Sorption/Biological Processes (from Ref 1)

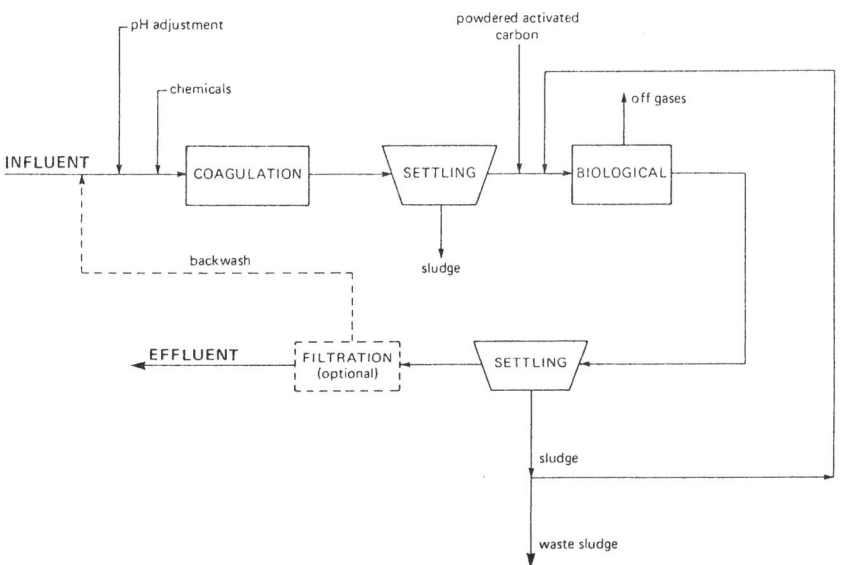

Figure 8.3   Biophysical Treatment Scheme (from Ref 1)

(i)   the overall system efficiency depends substantially on the efficiency of the biological treatment component;

(ii) the presence of volatile organics could cause air pollution problems;

(iii) disposal of process residues can be a problem.

3.2 Carbon Sorption/Biological

This system (Figure 8.2) is essentially the same as the preceding system except that the sequence of the process components is different to accommodate wastes having different characteristics. The activated carbon component precedes the biological component so that any organics that are potentially harmful to the biological system, are removed prior to transfer to the biological treatment. For reasons of both cost and system efficiency, the rate that toxic organics are 'allowed' to reach the biological system is controlled. If the biological system is capable of accepting and is allowed to accept a higher concentration of organics, the life expectancy of the carbon filter will increase. The coagulation, settling and filtration stages are designed to remove soluble inorganics and particulates thus minimising the head loss in the carbon filters.

3.3 Biophysical

This system (Figure 8.3) employs a combination of biological and carbon treatments and is less complex than the two preceding systems.

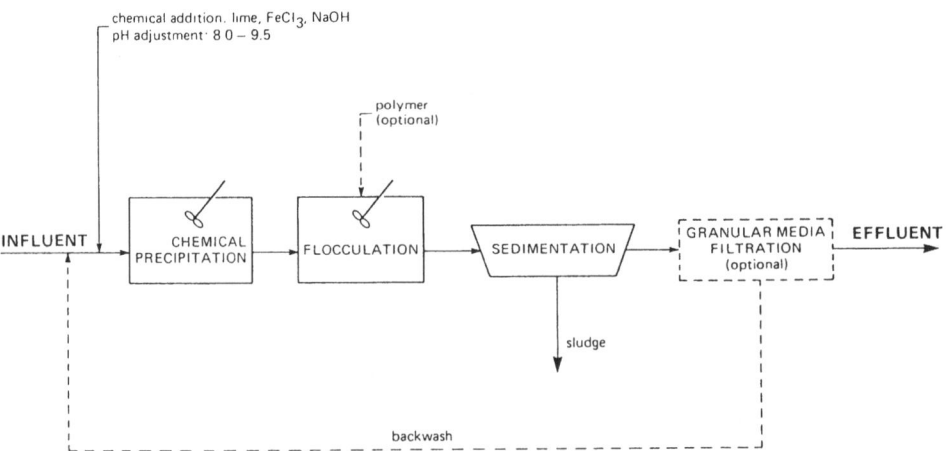

Figure 8.4 Process Scheme for Metals Removal (from Ref 2)

Powdered activated carbon is used and the resulting biological/carbon sludge, after dewatering, can be disposed of more readily. The need for physical facilities (mechanical plant, structures, etc) is also reduced. Solids retention times of 100-150 days are recommended with suspended solids concentrations up to 25 000 mg/l (60% PAC and 40% biomass).

3.4 Metals Removal

Heavy metals can be removed from the process waste streams and other aqueous streams by employing chemical precipitation using either $FeCl_3$ (ferric chloride) or NaOH with the pH adjusted to 8.0-10.0. Flocculation efficiency can be improved by the addition of a polymer with precipitation of the whole in a subsequent sedimentation phase. A higher percentage of solids can be removed and the effluent 'polished' by the introduction of a granular media filtration component (see Figure 8.4).

3.5 Hexavalent Chromium Removal

This is essentially the same as the preceding system with the addition of the chemical reduction stage to change the toxic hexavalent chrome to a trivalent state (see Figure 8.5). The chemical reduction occurs at a low pH (below 3.0). Sulphur dioxide is used as the reducing agent followed by pH adjustment

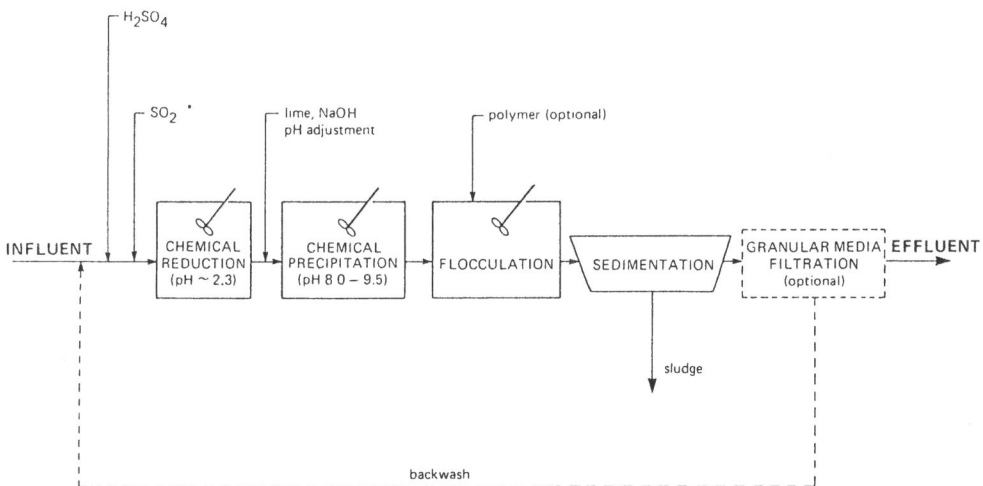

Figure 8.5   Process Scheme for Hexavalent Chromium Removal
              (from Ref 2)

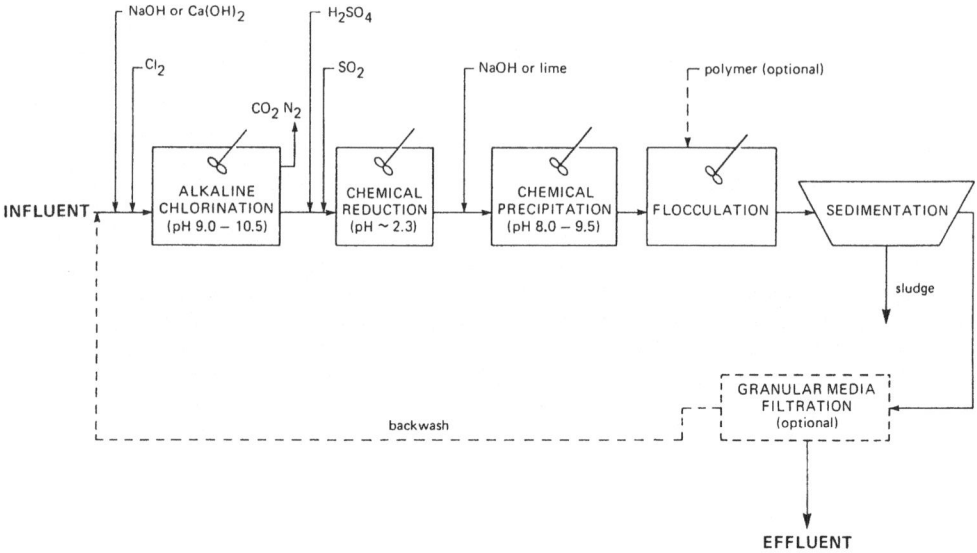

Figure 8.6   Process Scheme for Removal of Heavy Metals
and Cyanide (from Ref 2)

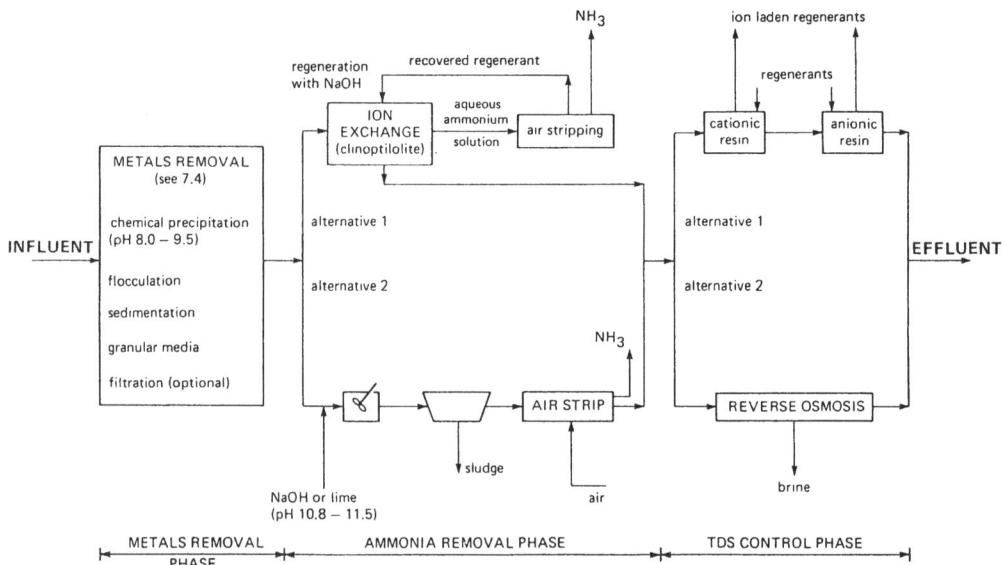

Figure 8.7   Process Scheme for Removal of Metals and Ammonia
and for TDS Control (from Ref 2)

(to 8.0 +) using lime or sodium hydroxide. Trivalent chromium and other metals are precipitated out at this stage. The balance of the system is as described in the metals removal system (3.4).

3.6   Heavy Metals and Cyanide Removal

This system (Figure 8.6) has been developed to treat water containing heavy metals (including hexavalent chromium) and cyanide. In the first stage the objective is to oxidise the cyanide using alkaline chlorination by the addition of sodium hypochlorite or chlorine to adjust the pH to 9.0 or higher. Oxidation of the cyanide requires an excess of chlorine under strictly controlled pH conditions. All the cyanide must be oxidised at this stage. If cyanide passes through the process to the reduction stage there is a risk of generating hydrogen cyanide gas. Oxidation using ozone represents an attractive alternative particularly when the wastes have an organic component which may be converted to a chlorinated compound. In the reduction stage hexavalent chromium is converted to its trivalent form by the addition of sulphuric acid. Subsequent to reduction the pH is adjusted to 8.0 or greater and the trivalent chromium and the other heavy metals removed in a precipitate. The balance of the system is similar to that described in 3.5.

3.7   Removal of Metals, Ammonia and TDS Control

This process train represents alternative systems for aqueous waste streams containing heavy metals, ammonia and a level of total dissolved solids (TDS) requiring some control. The three phases shown in Figure 8.7 are designed to function as follows:

Phase 1 – Heavy metals removal – discussed earlier (3.4)

Phase 2 – Ammonia removal – achieved by selecting one of the following options

Option 1	Option 2
Ion exchange using zeolite. Ammonia in the regenerate is removed by air stripping and the lime slurry reused.	Sodium hydroxide or lime is added to the waste to adjust the pH to 10 +, sludge is removed and the ammonia removal from the liquid phase is accomplished by air stripping.

Phase 3 – TDS removal. Using ion exchange, cationic or anionic resins, or reverse osmosis. Any of these processes could be used for final treatment prior to discharge to the environment.

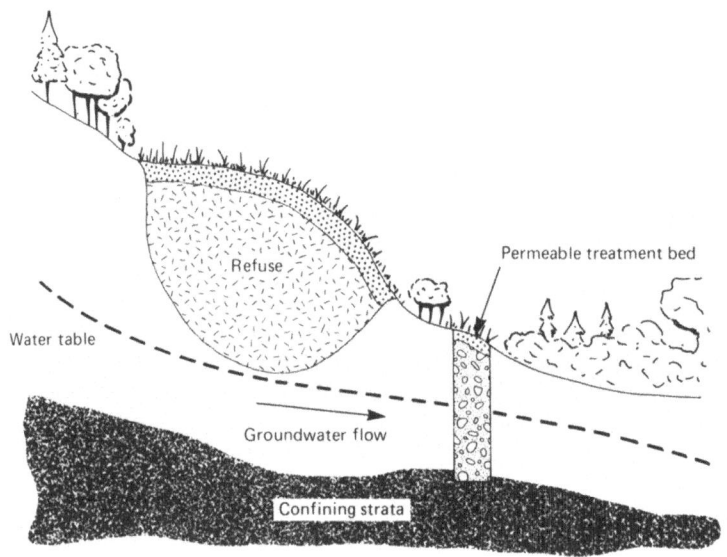

Figure 8.8   Installation of a Permeable Treatment
             Bed (from Ref 3)

4   IN-SITU TREATMENT

In certain cases it may be possible to treat groundwater without
withdrawing it from the aquifer. Most in-situ treatment processes
must at present be considered as develop-mental requiring further
research before they can be employed with confidence on a widespread
basis. It appears that these systems will be selective in their
effectivness meaning that only a specific contaminant or contaminants
will be affected whilst others remain unaffected.

4.1   Permeable Treatment Beds

If the aquifer is close to the surface and of limited depth the
contaminated groundwater can be 'treated' by allowing it to pass
through a permeable bed (Figure 8.8). If a particular contaminant is
targeted and the medium materials selected appropriately, physical
and chemical reactions between the contaminated groundwater and the
bed medium can improve the quality of the water.

In addition to the fact that the concept is still substantially unproven obvious limitations include:

- depth limitation

- potential for clogging

- continually changing hydraulic conditions.

Possible media materials identified include limestone (for neutralising acidic water), activated carbon (for removing organic compounds) and, theoretically, zoelite and synthetic ion exchange resins (for heavy metal removal). The last are probably not practical due to high cost, short life, etc.

The material most likely to be used is crushed limestone as this has been subjected to the most scrutiny. Even so, information on contact time, design parameters, efficiency, etc, is sparse. Before this material or any other medium is used as a treatment bed more research is required so that pH adjustment mechanisms and metal removal efficiency are known and the systems designed and operated accordingly. Activated carbon could be used for removal of organics and some metals. However costs could be extremely high particularly if regeneration was not possible. Disposal of exhausted medium material could present a problem.

The US Environmental Protection Agency has undertaken treatment tests using glauconitic greensand as a filter medium. It has been

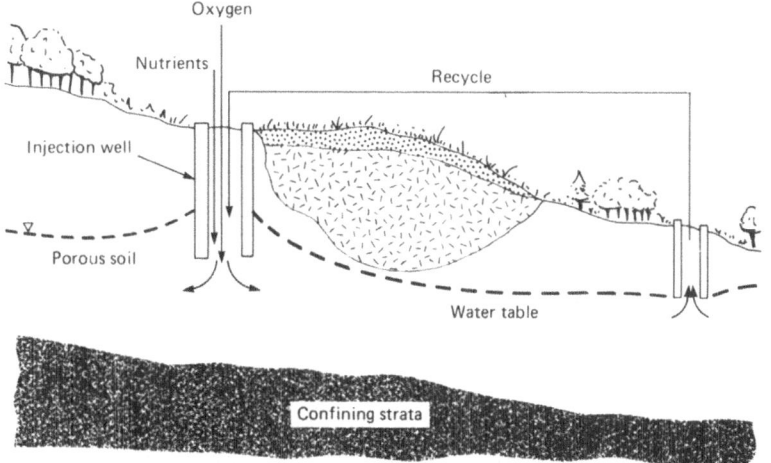

Figure 8.9   Scheme for Microbial Reclamation of Contaminated Groundwater In-situ (from Ref 3)

reported that this material has a capability to reduce concentrations
of metals including copper, mercury and nickel significantly. Odour
reduction occurs suggesting that this material may also adsorb some
organics. This commonly occurring (North American and Europe)
material contains glauconite, a clay mineral containing ferric iron
and potassium. Additional work is required before this material can
be judged as having widespread potential.

Design considerations include:

Length – must extend laterally beyond the limits of the plume
Depth – sufficient to prevent 'escape' under the bed
Width – depending on groundwater velocity, permeability
          of the media and retention time needed.

Advantages and disadvantages have been identified as follows:

ADVANTAGES   DISADVANTAGES

Limestone:

Reduces acidity   Tendency to plug
Heavy metal removal achieved   Organics are generally
Potentially controls chromate anion   unaffected
Cost effective   Channelling through the
                                        medium may occur

Activated carbon:

Removes non-polar organic compounds   Tendency to plug
Readily available   Limited effect on polar
Easily installed   organic compounds
                                        Influenced by other
                                        chemicals
                                        May allow desorption
                                        with subsequent
                                        recontamination
                                        Spent medium material
                                        requires proper disposal
                                        Potentially short life

Glauconite:

Heavy metal removal   Capacity unproven
Good working characteristics   Not universally available
(short resistence time)   Tendency to plug
                                        Changes in pH

4.2 Bioreclamation

In-situ treatment of groundwater can also be achieved by introducing selected microbial organisms into the groundwater. The process is sometimes assisted or accelerated by the addition of nutrients in an aerated hydrological environment (Figure 8.9).

The microbial organisms are selected on the basis of their proven affinity with the leachate or contaminant, in particular their known capability to degrade the target contaminant and, as a consequence, they must be selected on a case by case basis. Matching the bacteria to the target is of paramount importance and this can only be done after the water has been analysed and all the constituents identified.

Contaminants such as hydrocarbons are well suited for treatment using this process whereas chlorinated solvents are less universally suited to this treatment.

Bioreclamation with no external stimulation is a slow process. However, as noted earlier, the addition of oxygen and nutrients may accelerate the process. Gradual acclimatisation of the bacteria to the environment is also essential to ensure no shocks and to achieve effective biodegradation of the contaminant.

A variation of this concept is reported by the Federal Republic of Germany. In this case groundwater from an aquifer contaminated by hydrocarbons was extracted and treated by aeration and nutrient addition (nitrate and phosphate) followed by re-injection. Degradation of the hydrocarbons was accelerated by increasing the temperature of the re-injected water.

The most advantageous uses of bioreclamation have been identified as treatment of water containing hydrocarbons and some other organics. It is accepted as environmentally secure, is relatively inexpensive and effective in the short term. Disadvantages include the selectivity of the culture and the fact that nutrients used to stimulate reactions may have adverse effects on surface waters. Maintaining the reactions may be a problem and the long-term effectiveness is not yet known.

4.3 Arsenic Reduction Using Potassium Permanganate

The use of potassium permangante to oxidise arsenic is reported in detail in Chapter 4. However, the essential facts warrant repetition. Arsenic levels were monitored as follows:

Trivalent arsenic   1 mg/l

Pentavalent arsenic   0.1 mg/l

Maximum arsenic concentrations in groundwater   56 mg/l

Background arsenic concentration   0.01 mg/l

The source of the arsenic was the residue from a zinc ore smelter located near Cologne, FRG. It was established that because of the nature of the arsenic it would be precipitated out if the oxygen supply was improved. Over a 6 month period 29,000 kilos of potassium permanganate was injected into the site using 17 injection wells. During the year subsequent to commencing the injection process the arsenic concentration in the groundwater was lowered to 0.06 mg/l. However, over the following two year period the concentration increased to 0.4 mg/l.

It was concluded that the addition of potassium permanganate had accelerated the natural oxidation process. Natural oxidation had caused a reduction in arsenic concentrations in the preceding years.

4.4   Oxidation of Groundwater

Reference is again made to Chapter 4 wherein a FRG report[5] on a project where oxygenated water was injected into an aquifer is discussed. Theoretically the addition of oxygen increases the redox potential. The increased oxygen level promotes beneficial biological activity thus improving the quality of the water.

A number of commercial systems have been developed and are available for purifying groundwater in-situ by enriching water with oxygen. It is claimed that these systems are effective for the removal of iron and manganese. It is conceiveable that these systems could be readily adapted for aquifer remediation.

4.5   Treatment of Groundwater with Ozone

The Federal Republic of Germany reported that since 1980 the City of Karlsruhr has used ozone to treat groundwater contaminated with mineral oil. Water from four deep wells is passed through a reaction vessel where the contaminated water is reacted with ozone (O_3) and subsequently reinjected into the ground. The impact on the groundwater table, due to reinjection, is to create a hydraulic barrier reducing the migration of the contaminants into the zone of influence of the extraction wells. The effect of ozonation has been beneficial in a number of ways including an increase in biological activity, reduced DOC and a reduction of cyanide (one of the original contaminants).

5   REFERENCES

1   Touhill (Shuckraw and Associates Inc), 'Concentration
     Technologies for Hazardous Aqueous Waste Treatment'. US
     Environmental Protection Agency Contract 68-03-2766.
2   Touhill (Shuckraw and Associates Inc), 'Management of
     Hazardous Waste Leachate, US Environmental Protection Agency
     Contract 68-03-2766.
3   'Handbook for Remedial Action at Waste Disposal Sites', US
     Environmental Protection Agency, Cincinnati, EPA 625/6-82-
     006, (1982).
4   G Matthess, 'In-situ Treatment of Arsenic Contaminated
     Groundwater', Sci Total Environment 21, 99-104 (1981).
5   U Rott, 'Protection and Improvement of Groundwater Quality
     by Oxidation Processes in the Aquifer', in: Proc Symp
     Quality of Groundwater, Noordwijkerjout, Netherlands 1981.
6   G Nagel, W Kuhn, P Werner and H Sontheimer,
     'Grundwassersanierung durch Infiltration von ozontem
     Wasser', GWF-Wasser'Abwasser 123 (8), 1982.

9

MATHEMATICAL MODELLING OF POLLUTANT TRANSPORT

BY GROUNDWATER AT CONTAMINATED SITES

K A Childs

Environment Canada

1   INTRODUCTION

A responsible designer of a remedial action plan will use every valid analytical resource in order to optimise the design and the effect of the system. Mathematical modelling is a potentially effective tool and can assist the analyst or designer in a number of ways including:

   (i)   Providing indications of what will occur if nothing is done.

  (ii)   Providing indications of the likely impact of different remedial actions.

 (iii)   Providing indications of what constitutes the optimal remedial system.

Mathematical modelling is discussed below with the following important caveats in mind:

   (i)   Modelling cannot replace field information.

  (ii)   Not all models are the same and 'off the shelf' selection of a model is frequently an unwise action.

 (iii)   Interpretation or analysis of outputs must be undertaken by professionals having specialised knowledge.

2   THE NATURE OF MATHEMATICAL MODELS

Fried[1] has noted that the modelling of groundwater pollution
consists of describing, by mathematical expressions, the following
processes:

(i)   The advection of the contaminant, ie its transport by ground-
       water flowing at the mean velocity;

(ii)   The dispersion of the groundwater within pore spaces about
       this mean value;

(iii) The chemical interaction of the contaminant with the solid
       matrix of the aquifer system, eg absorption;

(iv)   The decay of the contaminant due to biodegradation or radio-
       active disintegration.

These last terms describe the attenuation of the contaminant.
Mercer and Faust[2] have pointed out that hydro-geologists have two
types of mathematical model available to them to assist in the
analysis of groundwater flow and contaminant transport:

(i)   Analytical models, in which the transport equation is solved
       by an exact solution for the initial and boundary values
       stipulated.

(ii)   Numerical models, in which the two dimensional transport
       equations describing the behaviour of a continuous variable
       are approximated numerically by finite-difference or finite-
       element methods resulting in a finite number of algebraic
       equations which may be solved by matrix techniques.

Since 1960 a great deal of interest has been shown in the
development of numerical models of groundwater flow (ie the advective
flux) and contaminant transport (ie the advective and dispersive
fluxes with various attenuation terms).

In the following pages a brief review of mathematical modelling
for studying both flow and transport problems at contaminated sites
is presented. However, it is desirable first to issue a warning
concerning the necessity of acquiring adequate, reliable data for use
in the modelling operation. A most eloquent caution is given by
Wang and Anderson[3] in their book 'Introduction to Groundwater
Modelling'.

'The answers generated using a mathematical model are dependent
on the quality and quantity of the field data available to
define the input parameters and boundary conditions. Modelling
can never be a substitute for field work. Used in conjunction

with good field data, a model can provide insight into the
dynamics of the flow system and also serve as an invaluable
predictive tool.'

3   GROUNDWATER FLOW MODELS

Following the acquisition of a complete set of data defining the
hydrogeologic conditions and properties of the groundwater flow
system within which the contaminated site is situated (eg hydraulic
heads and conductivities, geometries of the permeable and impermeable
units comprising the flow system), it is possible to begin to model
the pattern of groundwater flow in the aquifer. The equation takes
account of the permeabilities of the aquifer soils or rocks, the
hydraulic heads (ie water levels) and the sources and/or sinks
(recharge and/or discharge points) within the flow system.

A best match of simulated water levels with observed water
levels is obtained by adjusting the geometries and the permeabilities
in a trial and error fashion. The first attempts at modelling will
probably result in a desire to collect further field data so that a
better match between simulated and observed values may be obtained.
Consequently there will be cycles of field and modelling studies
resulting in the prediction of the hydraulic head pattern within the
flow system from which groundwater and contaminant flow patterns may
be deduced. Furthermore by employing Darcy's law, which relates the
specific discharge of groundwater to the hydraulic head gradient and
permeability, a map of the groundwater velocities may be drawn.
Knowledge of the velocities, so determined, are essential for the
next step of groundwater modelling.

4   CONTAMINANT TRANSPORT MODELS

During the 1960s it was customary to study contaminant transport
in groundwater systems either by the use of analytical solutions (eg
Ogata and and Banks[4]) or by groundwater flow models in the method
described above (eg Freeze[5]). Since 1970 considerable progress has
been made in developing numerical solutions to the advective-
dispersion equation describing contaminant transport. This equation
describes the mass balance of a contaminant during groundwater flow-
through whilst it is undergoing dispersion and is subjected to
attenuation.

The dispersion coefficients used in the equation are a measure
of the rate at which the concentration gradients are dissipated. The
longitudinal and transverse dispersion coefficients used may be
considered to be measures of the dispersive capacity of the hydrogeological system being modelled.

All early numerical modelling of contaminant transport treated the dispersion coefficients as constants which could be used as a 'fitting' parameter in the same way as permeability is used in groundwater flow models. Consequently a wide range of dispersion coefficient values have been reported, from centimetres in sand aquifers to tens of metres in fractured basalt rocks. Pickens and Grisak[6] have shown that this wide variation is partly due to the heterogeneous nature of hydrogeologic systems and the dilution of groundwater samples when large-diameter wells rather than small diameter sampling systems are used. Since dispersion causes dilution of the contaminant, any dilution of the contaminant during sampling implies a greater rate of dispersivity than may really exist.

In recent years groundwater modellers have begun to treat the dispersion coefficient as a function of either the mean travel distance or travel time of the contaminant. Stochastic modelling techniques used by Gelhar and De Marsily[7] suggest that the dispersivity is dependent on the time of travel of the contaminant from the site and should reach an asymptotic value at large travel times where hydrogeological boundaries are encountered or a significant transverse velocity (ie transverse to the direction of flow) develops. Many contaminant transport problems require a chemical reaction term to describe the sorption of heavy metals, radionuclides, organic compounds or other contaminants during their transport through the groundwater flow system and/or the biode-gradation of organics. This can be accommodated in the invective dispersion equation.

Much research work in the NATO/CCMS countries is currently directed to finding improved ways of measuring sorption biode-gradation and other chemical reaction terms in groundwater models. This includes the development of in-situ distribution coefficient measures which describe the retardation of the contaminants relative to the groundwater flow because of sorption[8].

5   APPLICATION TO THE DESIGN OF REMEDIAL MEASURES

5.1 General

The nature of groundwater flow and contaminant transport models has been outlined. Opportunities for their application in the development of systems for the decontamination of aquifers and/or minimising or preventing contamination in aquifer systems are apparent. In fact the ability of models to predict system responses to remedial actions make them virtually essential if any program of site restoration is to be optimised. Cole and McKown[9] have summarised their uses in such situations as:

(i) To aid in the design of the site investigation program;

(ii)   To help assess whether remedial measures are required and
       which ones would be the most effective;

(iii) To assist in the design of a site monitoring program.

5.2   Model Selection

There are four groups of models; analytical and numerical, flow
and transport. Flow models are readily available whereas transport
models without a flow model component are now rare if not totally
unavailable.

For complete and accurate predictions, the use of a transport
(ie transport plus flow) model is desirable. However, one of the
governing factors will be the level of funding and accessibility to
the program - many programs are retained 'in-house' requiring that
the client purchase the total package without having access to the
program. Selection will be influenced by the degree of sophistication
required and it should be realised that even a simple analytical
model will give meaningful results and assist in problem solving.
Numerical models using digital computers will, however, always
provide more detailed information.

A person or agency charged with developing a remedial action
plan will probably choose modelling as a tool because the situation
to be addressed obviously or readily lends itself to a modelling
analysis. Having chosen to use modelling, the choice of a modelling
package will be based on a number of reasons including:

Proven model   - it has been shown that there is good correlation
                 between numerical and exact analytical solutions.

Familiarity   - if the client is familiar with a system an element
                 of self-confidence is introduced.

Flexibility   - the model may be required to respond to many
                 different situations.

Documentation - this is particularly important for the first-time
                 user. The program must be adequately documented to
                 allow optimal utilisation.

Availability   - as noted above some models remain 'in-house'
                 requiring that a total package be purchased.

5.3   Current Availability and Applications

There are many models currently available which have been

employed in site remediation activities, for example:

(a)   The Battelle model - seepage from uranium mill tailings.

(b)   The Geotrans model - Love Canal remedial plan assessment.

(c)   The GTC model - Gloucester, Ontario, Canada - contaminant
             migration prediction.

(d)   The USGS model - Rocky Mountain Arsenal remedial plan; and
             in the Federal Republic of Germany, nutrient
             addition to accelerate biological action.

5.4 Future Applications

In the future hydrogeologists will have available both effective proven groundwater-flow and contaminant transport models together with combined transport and optimisation codes[10]. These can be employed for the assessment of contaminated sites and for optimising the locations for decontamination well fields. All models will require input of reliable hydrogeologic (field) information. If the input of field information is limited the value of the modelling exercise will be reduced and, in the worst case, totally ineffective.

Because of difficulties in measuring and interpreting values of the dispersion and distribution coefficients, contaminant transport modelling is still in an evolutionary state and is likely to remain so for some time. At the same time modellers and field scientists continue to develop better approximations of basic phenomena.

6   CONCLUSIONS

In summary, groundwater models provide an unequalled aid in:

(i)   Evaluating hydrogeologic conditions at contaminated sites and
       determining what additional field activities (eg drilling,
       piezometer installation) should be conducted;

(ii)   Predicting the migration of groundwater contamination (from a
       calibrated model);

(iii) Assessing the likely effects, both beneficial and adverse, of
       various remedial measures (eg purge wells, grout curtains,
       etc) as they relate to migration and change in concentration
       of contaminants.

It cannot be stressed too strongly that the model results depend upon the quality and quantity of the hydrogeologic data used.

Modelling is a tool and its function could be demonstrated by the term 'model-supported investigation' ie it should have a support role in a more broadly based investigatory or monitoring process. Even allowing for the uncertainties in the data, groundwater modelling permits rational development of a program of remedial measures at landfills and contaminated land sites.

7   REFERENCES

1   J J Fried, 'Groundwater Pollution Mathematical Modelling: Improvement or Stagnation?' In: Quality of Groundwater, pp 807-822, Elsevier Scientific Publishing Co. The Netherlands, (1981).

2   J W Mercer and C R Faust, 'Groundwater Modelling', National Water Well Association, (1981).

3   H F Wang and M P Anderson, 'Introduction to Groundwater Modelling', Freeman, San Franscisco, (1982).

4   A Ogata and R B Banks, 'A Solution of the Differential Equation of Longitudinal Dispersion in Porous Media', US Geol Survey, Professional Paper 411a, (1961).

5   R A Freeze, 'Subsurface Hydrology at Work Disposal Sites,' IBM Journal of Research and Development, 16 (2), (1982).

6   J F Pickens and G E Grisak, 'Scale-dependent Dispersion in a Stratified Granular Aquifer,' Water Resources Research, 17, (4) (1981).

7   G De Marsily, A Dieulin, E Ledoux and P Goblet, 'Are we able to Measure the Parameters Governing Transport of Solute in Porous Media, and thus, to Predict Long-term Migration?' International Workshop on the Comparison and Application of Mathematical Models for the Assessment of Changes in River Basins, Both Surface Water and Groundwater, UNESCO, IHP, la Corunna, Spain, April 1982, Preprint.

8   C R Cole and G L McKnown, 'The use of Mathematical Models to Assess and Design Remedial Action for Chemical Waste Sites.'

9   J F Pickens, R E Jackson, K J Inch and W F Merritt, 'Measurement of Distribution Coefficients using a Radial Injection Dual-tracer Test'. Water Resources Research, 17 (3) (1981).

10   S M Gorelick and I Remson, 'Optimal Dynamic Management of Groundwater Pollution Courses', Water Resources Research, 18, (1) (1982).

10

TOXIC AND FLAMMABLE GASES

S C James*, R N Kinman** and D L Nutini**

* US Environmental Protection Agency
** RNK Environmental Inc, USA

1   INTRODUCTION

Emissions of toxic and flammable gases may be an environmental hazard whenever uncontrolled hazardous waste sites or other contaminated sites are assessed or cleaned up or hazardous waste treatment, storage or disposal facilities are operated. This Chapter is concerned specifically with problems associated with toxic and flammable gas at uncontrolled hazardous waste and other contaminated sites. However, this information is also applicable to operating hazardous waste treatment, storage and disposal facilities. The emphasis is on volatile organic compounds (VOC's) and less commonly occurring gases rather than on methane and 'landfill gases' about which there is already comprehensive literature available.

2   SCOPE

'Superfund' legislation (US Comprehensive Environmental Response, Compensation and Liability Act, PL 96-510) requires the cleanup of hazardous materials at many remedial action sites. This may cause the release of toxic and flammable gases to the environment. Even the opening of containers for sampling purposes may cause such releases. Site surveys, monitoring programs for air or water contamination, soil or geotechnical surveys, cleanup activities etc, may also lead to gas release.

Man may be harmed due either to the specific toxicity of gases or because of a lack of the necessary oxygen required for respiration. Gases may emanate either from biological activity for example hydrogen sulphide or from planned or accidental activity of man.

Gases including volatile substances recently separated from either
either solids or liquids at the site may emanate from disposal sites,
surface impoundments, storage tanks, storage containers, treatment
facilities, etc.

Flammable gases are those that will burn. They may be toxic
as well as flammable. The most prevalent flammable gas from natural
biological activity is methane. However, the purpose here is not to
discuss methane, but rather other gases such as benzene, toluene,
ethylbenzene, tetrahydrofuran, vinyl chloride, etc.

3   VOLATILISATION AND MIGRATION OF TOXIC AND FLAMMABLE GASES

Volatilisation is the transfer of a substance from a liquid or
solid phase to a vapour phase. It occurs readily for most organic
compounds at a relatively low temperature. The rate of volatilisation of an organic compound depends mainly on its vapour pressure,
molecular weight, solubility, diffusion coefficient, mass transfer
coefficient, concentration, exposed surface area of the specific
compound, and the surrounding environment eg temperature, pressure,
and wind speed. Volatilisation also involves desorption of vapours
from sorbents such as soils or other materials. Generally, low
molecular weight organic compounds are more volatile than high
molecular weight organic compounds. Appendix K lists the major toxic
organic substances identified at sites that have caused problems due
to volatilisation into the air from landfills, waste lagoons and
chemical storage areas.

Many organic molecules of environmental concern have relatively
low vapour pressures owing to relatively large enthalpies of
vaporisation which depends upon the molecular weight of the compound
of interest and other variables. Thermodynamics predict that
compounds which have a high enthalpy of vaporisation will be more
sensitive to the effects of temperature changes than materials with
low enthalpies of vaporisation. Therefore, materials that have
relatively low vapour pressures are expected to be proportionally
more sensitive to ambient temperature fluctuations than are materials
that are more volatile.

Recent studies[1,2] have shown that seasonal temperatures significantly affect even very stable chlorinated hydrocarbons with a
very low vapour pressure such as PCBs; volatilisation rates are
higher during summer than winter months. Shen[3], found that each
$10^{\circ}C$ rise in soil temperature increased PCB vapour pressure by a
factor of two. Farmer et al,[4] found similar results with
hexachlorobenzene (HCB) and HCB waste reporting that increasing the
temperature $10^{\circ}C$ increased the vapour density about three and one-
half times. Certainly, volatilisation is influenced by temperature

changes through the day,[5] since high temperature strongly increases
vapour pressures of liquids and/or solids.

Wind speed will affect the volatilisation rate of organic
materials from surface impoundments. The gas phase mass transfer
coefficient is increased at higher windspeeds. There is also a
dilution effect at higher windspeeds so that the gas phase concentration of volatile organic materials can be lower than at the slower
windspeeds, even though the total amount of organics that is emitted
into the air is greater. The overall mass transfer coefficient is
proportional to the velocity raised to a power which is somewhat less
than one, thus permitting a dilution of the emitted vapours.
Thibodeaux et al,[6] concluded that 'no-wind conditions' result in
smaller mass transfer coefficients than when wind is present. The
gas transfer coefficient measured for methanol in surface impoundments averaged 2680 cm/hr, while the turbulent gas transfer
coefficient averaged 13,900 cm/hr. Thibodeaux et al,[7] also found
similar results with benzene and 1,1-dichloroethane (DCE). Here, the
gas transfer coefficients measured 741 and 909 cm/hr for benzene and
1,1-DCE, respectively, while turbulent gas coefficients increased to
792 and 981 cm/hr for benzene and 1,1-DCE, respectively. Wind can
also increase the dilution factor; so it may compensate, depending on
the location of the receptor, for the ambient air concentration.[3,8,9]

Volatilisation of an organic chemical is affected by absorption/
adsorption onto a solid adsorbent or soil. Shen and Tofflemire,[1]
showed that the rate of volatilisation of Aroclor 1242 was less for
organic topsoil than for coarse sand. The organic topsoil was more
adsorbent than the sand because of the greater surface area and the
nature of the surface. Materials such as humic acids in the organic
topsoil are expected to adsorb organic chemicals. The adsorption of
organic chemicals onto the soil reduces the concentration of organic
chemicals in the vapour phase, and thus slows the diffusion process.

Since many combinations of wastes and conditions may occur at
contaminated sites, other factors must be considered to influence
volatilisation. The solid and/or liquid waste types, waste concentration, landfill waste and soil density, soil permeability, moisture
content, gas pressure, etc, all influence diffusion of volatile
substances. The compaction of soil cover, soil permeability,
moisture content and the vapour pressures of the organic substance
may determine how fast these materials might diffuse. McCord and
Zeigler[10] made theoretical calculations for diffusion of methanol
through hypothetical soil columns of varying permeabilities and
predicted gas pressure buildup at the bottom of the columns at 25°C.
There was almost no buildup for sand (1 atm), but some pressure
buildup (1.132 atms) for clay. Farmer et al,[4] also demonstrated
the effect of soil layers and materials on volatilisation of HCB
waste (see Table 10.1).

Table 10.1   Effect of Various Covers on Volatilisation
of HCB Waste (from Farmer et al[4])

Cover	HCB vapour flux (kg/ha/yr)
None	317
1.9 cm topsoil	4.56
0.15 mm polyethylene film	201
1.43 cm water	0.38
120 cm topsoil	0.066

Another pathway of gas migration besides volatilisation
(diffusion) is the adsorbtion/absorbtion of hazardous toxic organic
compounds onto fine particles (particulate matter, soil dust, coal
fines[11]) which then become air-borne. Natural or man induced
activities like wind erosion of the waste materials or soil cover,
excavation, or vehiclular traffic around a site, may create movement
of these particles producing an air pollution problem.[11,12] This
may be localised or may affect locations far from the site.[3] These
particulates may be deposited back onto land or water in the form of
fallout or may precipitate from the air during periods of rain, snow,
etc. This could be a cyclic process as shown in Figure 10.1.

Consideration must also be given to migration of toxic and
flammable gases associated with movement of leachate and/or groundwater containing organic compounds.[13] Groundwater is not only
contaminated, but also organic compounds volatilise into the air.
This is a problem in several case studies reviewed, where contaminated leachate migrated into areas surrounding the landfill site.
Such sites as the Sylvester Site in New Hampshire[11,14,15] had a
groundwater plume of contaminants extending over an area in excess of
8 hectares. Contaminants in the groundwater included chloroform,
toluene, tetrahydrofuran, benzene, and methylene chloride. At Kin
Buc Landfill in New Jersey[16,17] the main problem centred around
controlling leachate that contained contaminants such as benzene,
PCBs, and 50 to 60 different organics (mostly chlorinated).
Concentration ranges in air at Kin Buc are reported in Appendix K.

4   AIR EMISSION PROBLEMS BEFORE REMEDIAL ACTION

One of the most salient signs of an air emission problem is
complaints by residents about odours.[18-27] Some of the main
chemical constituents identified that may have caused these odour
problems are toluene, benzene, phenols, ethylbenzene, naphthalene,
vinyl chloride, methylene chloride, chloroethane, trichloro-

ethylene, chlorotoluene, tetrachloroethane, chlorobenzene, xylene, trichloroethane, chloroform, coal tar residues, tetrachloroethene, vinylidene chloride, 1,2-dichloroethane, trans-1,2-dichloroethene, pesticides, and hydrogen sulphide gas. Appendix K reports peak concentrations found at various sites. Some are above the Threshold Odour Concentrations (TOC) values for the compounds in question.

In some cases odours were detected at substantial distance from the waste site. In one case,[22] odours from pesticides emanating from the soil were noted five miles away. Other odour problems were were noticed by workers performing maintenance construction over or near abandoned sites when the soil cover was disturbed.[20,22,28]

A common visual effect indicating a gas problem is damage to vegetation.[19,21,25,26,29,30] Gas buildup results in an anaerobic environment in the soil which affects the root systems of the plants. In the cases cited, the main chemical constituents noted where vegetation deaths occurred were toluene, phenols, benzene, ethyl-

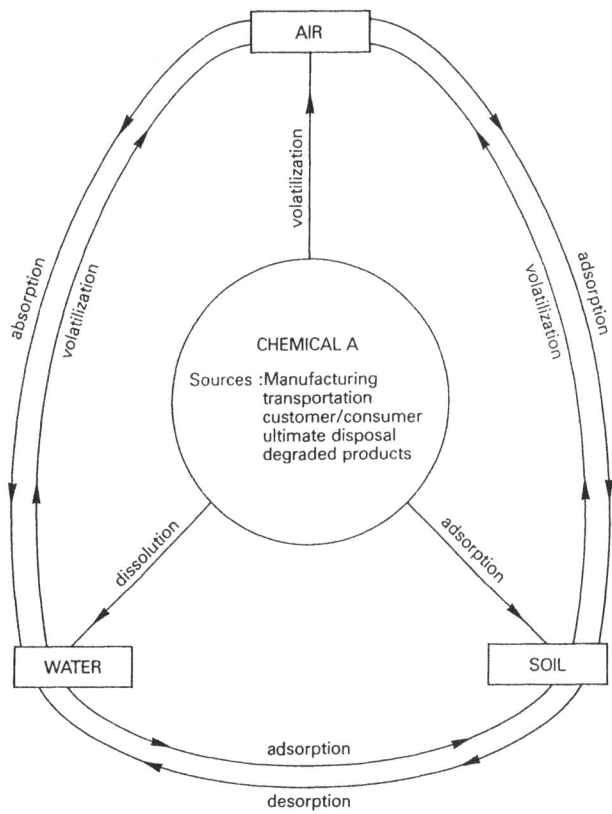

Figure 10.1 Movement of Chemicals in the Environment

benzene, naphthalene, vinyl chloride, methylene chloride, chloroethane, trichloroethylene, chlorotoluene, tetrachloroethane, chlorobenzene, xylene, trichloroethane, chloroform, and methane gas. It has
been observed that in many cases, gases damaged vegetation not only
on the site, but also at farms and homes adjacent to landfills.[25]
This was due mainly to the migration of methane gas.

Explosions and fires have provided[29,30,31,32,33,34] direct
evidence of air pollution problems. Chemical constituents that did
or may have contributed to these problems are highly flammable and/or
explosive industrial solvents, methane, acetone, benzene, hexane, and
the comingling of wastes with acid wastes from leaking drums.

Other clues to gas problems included health effects to
animals[2,4,19,30] and man.[30,34] One case in Southern
Louisiana[4] involved contamination of beef cattle, which were
found to have high levels of HCB (hexachlorobenzene) in their fat
tissues. The cattle were exposed when uncovered trucks spilled and
contaminated areas adjacent to the roads while hauling the HCB waste
from the industrial source to the landfill. At the site the HCB
waste was spread over the landfill to help control the flies. The
levels of HCB in the blood plasma of residents exposed along the
truck route was as high as 23 ppb. The peak concentration of HCB in
the plasma of landfill workers was 345 ppb. A control group recorded
a high level of only 1.8 ppb.[4]

The health of individuals near the MIDCO I[34] site in Gary,
Indiana were affected as a result of heavy rains flooding the
site and causing chemical runoff containing liquid industrial
wastes to flow from the site. Residents complained of organic
vapours emanating from their basement sumps causing illness and
injury due to chemical contact or odour. Some of the chemicals
later identified at this site that could have caused these
problems were dichloromethane, benzene, 1,1-trichloroethane,
tetrachloromethane, trichloroethane, tetrachloroethylene,
methylbenzene, ethylbenzene, and dimethylbenzenes.[35]

Other consequences of gas problems include discolouration
of the soil. This was noted in two case histories.[21,28] The
chemical types identified as possibly causing this were benzene,
toluene, xylenes, trimethylbenzene, dichlorobenzene, trichlorobenzene, naphthalene, phenols, parathion, carbon tetrachloride,
tetrachloroethylene, ethylbenzene, oil, waste solvents, chemical
process waste, plating waste, and heavy metals containing sludges.

One other type of effect mentioned in a case history on the
Hyde Park Landfill, Niagara County, New York[27] is the premature
deterioration of metal surfaces. This was attributed to the
landfill gases, among which were many chlorinated hydrocarbons and
hydrogen sulphide although no proof for this occurrence was given.

5   AIR EMISSIONS ENCOUNTERED DURING REMEDIAL ACTION

Some investigators, workers, and hygienists have encountered
several air contamination problems while working at remedial action
sites. Leaking drums are one of the causes.[9,30] Removal by
excavation, redrumming aged and leaking containers, and opening drums
for sampling have also caused emissions of organic solvents to the
air.[30,37,36] Exposed Workers can be protected by wearing US
Environmental Protection Agency Level C protection, which includes an
approved air purifying mask with an organic vapour cartridge as well
as Tynek/Saran hooded disposable coveralls. Other equipment and
procedures to minimise emissions during drum sampling, include the
use of a pneumatic impact wrench and a hydraulic penetrating
device, as described by Blackman et al.[38]

Since waste removal is frequently part of the remedial action,
excavation procedures are necessary in many cases. Cases[21,25,28,32]
have been cited where this has resulted in emissions because of
the disturbance of the ground cover and opening of drums (mentioned
above), etc. When excavating a site, safety procedures can be taken
to prevent worker exposure. Protective clothing, respirators, heavy
equipment (backhoes) with enclosed cabs, etc can be used. Air
monitoring can be performed to determine levels of gas concentrations, as well as to determine the level of protection needed.

While workers are usually clothed and equipped for problems on-
site, nearby residents have no such protection.[5,34] In one
case,[34] persons working near the site, but not involved with site
cleanup, experienced adverse health effects (irritations to skin,
eyes, nose, throat; nausea; dizziness; etc). This occurred during a
spring warming trend, which increased the volatilisation of organic
chemicals such as benzene, tetrachloroethylene, trichloroethylene,
and methylbenzene. Since these people were not involved in the
cleanup and had no protection, they had to move their work operations
to a different location until conditions improved. It might be noted
here that air monitoring performed in this area by the on-scene co-
ordinator showed that ambient air levels of organic compounds did not
exceed 3 ppm, yet these people experienced the symptoms mentioned
above.

There has been concern for health, even when workers wear
protective clothing.[5] One concern is chemical substances
permeating clothing materials, especially gloves. Costello and
Kominsky[5] reported that breakthrough times for some substances were
very short (as little as 3 minutes) for many gloves when wetted with
carcinogenic substances such as chlorinated ethanes. This resulted
in dermatitis. They also pointed out that skin exposure is as
important a concern as inhalation at hazardous waste sites
particularly for substances like cyclopentadiene, which is common to
many landfills. The other concern they emphasised, as a result of

workers wearing protective clothing, is heat stress. The protective
suits are designed to prevent the exposure of chemicals to the skin;
however, the suit will not permit the natural evaporation of sweat.
This can cause the worker to become overheated and his heart rate may
increase. To compensate for this effect the worker should take
periodic breaks.

Drilling has caused releases of sulphur dioxide (SO_2) gas as
well as other noxious gases by breaking the surface of the
landfill.[39] If present, prevailing winds may carry these gases
directly into the adjoining homes, potentially creating further
problems. Water can be used to control these emissions. Blackman et
al[38] point out that methane and other explosive gases. could be
ignited by drilling rigs or other ignition sources. Drilling into
sites may also cause disposed incompatibles to be mixed. It is
therefore advisable to drill only at the periphery of the site in
order to characterise leachate or gas moving from the site. If
subsurface sampling is necessary on site, then excavation, should be
performed by hand and with spark-free tools. Further recommendations[38] for drilling at sites to help avoid problems include: have
a well-qualified person (geohydrologist) present; sweep the area with
metal detectors before drilling; and seal the test holes when the
investigation is completed.

Topographical features and extreme weather conditions have
caused problems.[30] A landfill site was located in a valley
setting which intensified the air problem by increasing gas and
volatile organic concentrations because of insufficient natural
ventilation. Another gas problem existed at the same site during
liquid-pumping operations, when foggy conditions caused a heavy
organic vapour buildup in the valley, affecting residents in the
area.[30]

Another concern is the chance of fire or explosion when working
at hazardous waste site. This could occur during activities already
mentioned above such as drum handling, excavation, and drilling. To
avoid these situations, it is recommended that the construction of
enclosed structures should be avoided; non-sparking tools used;
explositivity readings taken to determine levels of gases (if greater
than 20% Lower Explosive Limit (LEL) is detected a careful survey of
the area must be made – if 50% LEL levels are detected, evacuate the
area and notify the fire department);[38] incompatible waste types
not mixed as shown in Appendix L; and smoking by personnel on-site
prohibited.

Fugitive dust emissions have caused problems during work
site activity both on and off site.[11,12,40,42] In such cases,
trucks and equipment working on-site or travelling to and from the
site cause the light soil particles with absorbed chemicals to enter
the air, creating an increased level of exposure to workers and

residents. Dust emissions can be controlled by wetting the ground
surface with water, treating the surface with penetrating chemicals,
controlling vehicle speed, or paving roads (although this is highly
unlikely because of expense). Shen and Sewell,[12] point out the
efficiency of these fugitive dust emission controls. Of the most
practical and inexpensive methods, watering has 50% efficiency, while
vehicle speed controlled at 15 miles/hour has an 80% efficiency.
Rosburg[128] thoroughly addresses dust control measures.

Certain odourless contaminants may act as olfactory
anaesthetics[40] so that the ability to detect other gases that might
cause ill effects is impaired. Similar hazards can arise from
olfactory fatigue such as can occur with hydrogen sulphide (after a
period of exposure the substance can no longer be smelt).

Another gas problem that is difficult to quantify involves the
alteration of mental processes of workers by air pollution. Workers
with fatigue, mental confusion, and a lack of co-ordination may be
more prone to injure themselves and others.

6   RESPONSE TO EMISSION PROBLEMS AT UNCONTROLLED
       HAZARDOUS WASTE SITES

To properly respond to a gas emission problem at an uncontrolled
hazardous waste or other contaminated site, many factors must be
considered during site investigations, mitigation, and closure.
Procedures employed at a number of uncontrolled hazardous waste sites
are summarised below. They are generally found to be useful in
response to site problems, but not all are necessarily applied at
every site reviewed.

A comprehensive assessment of the site should be made before any
plan of action is prepared.[11,38,40,42,43] Elements of such an
assessment include:

   (i)   Assessment by an experienced environmental specialist;
   (ii) A review of site records and other previous studies such as
         citizen information, federal, state, and local files, or
         aerial imagery;
   (iii) Mapping the site;
   (iv) Determination of the extent of remedial action needed;
   (v)   A determination of the extent of the contamination and hazard
         which should include:

         (a) Noting container conditions, types and amounts of
              wastes, and their locations; a preliminary inventory of
              the suspected contaminants, their composition and any
              byproducts associated with them.

Table 10.2   Types of Equipment Used in Air Monitoring
             For Specific Hazards[44]

Hazard	Equipment

Explosive atmosphere
Oxygen-deficient atmosphere
Toxic atmosphere
	− combustible gas indicator
	− oxygen meter
	− organic vapour analyser −
	  (OVA−GC)[30,45]
	− photoionisation detector (PID)
	− flame ionisation detector (FID)
	  with gas chromatograph (GC)
	  option
	− colourimetric tubes
	− collection systems using
	  sampling pumps in conjunction
	  with absorption tubes, detector
	  tubes, filters, and impingers.
	  After collection, lab analysis
	  would include use of gas
	  chromatograph/mass spectro-
	  meter (GC/MS), GC/electron
	  capture, atomic absorption (AA),
	  or wet chemistry methods.
	  Mathamel,[44] lists the specific
	  applications for collection media
	  including the lab analysis required.
	  Montgomery et al,[45] describe
	  some of these instruments and cite
	  case histories where they have been
	  applied.

Radioactivity	− radiation survey meters (alpha, beta, gamma) − passive monitors (alarms) − personal dosimeters (film badges)

Note: Melvold et al[46] and Montgomery et al[45] evaluate several
      of these devices.

(b) Noting contaminated soils (check the potential sources
    of contaminated dusts).
(c) Noting existing and the potential for fires.
(d) Noting odours and vapours by using instruments such
    as gas detector tubes and combustible gas indicators on
    site, and flame ionisation detectors and GC/MS in the
    laboratory.

(e)   Noting visual evidence of contamination, such as
      vegetative or animal deaths, or soil discolouration,
      on-site or nearby.[21]

(vi)   A determination of access control;
(vii)   Determination of safety steps needed to work the site;
(viii) Determination of the resources needed to respond to the
        site problems;
(ix)   A focus on specific areas of the site, if necessary.[38,44]

Other procedures used to characterise gas problems are air
monitoring[11,43,44,45,46] and modelling (emission estimates).[1,12]
As mentioned above, air monitoring on site, at the perimeter of the
site, and off site, is used in assessment of a site before work
begins. Useful common air monitoring equipment for characterising
hazards in the air are given in Table 10.2 (see also Chapter 11).
Trained personnel are required when using flammable and toxic gas
detectors. Air monitoring is also used to determine types and
quantities of gases, to monitor exposure of workers or residents,
and to record residual contaminant levels after remedial action
is completed.[11] This is an extra precautionary measure to
demonstrate that air risks have been reduced and also to reduce
the chances of liability suits. A final consideration when
employing air monitoring would be the merits of using a qualified
public agency or contractor independent of the cleanup to avoid
conflict of interest[33] and to assure worker safety.

Once the need for remedial action is established, a
comprehensive site plan called a Remedial Action Master Plan (RAMP),
integrating health and safety procedures, should be established. The
plan involves many considerations including:

(i)   Establishing a base of operations.[42,43] This command
       post should be located upwind and central to all areas of the
       work on-site. It should be manned by experienced personnel,
       serve as the communications centre, and include such things
       as emergency shower and eye wash, first aid and other
       emergency equipment and supplies, and indices of operations,
       procedural and safety manuals for hazardous waste cleanup.
       Some good reference manuals are given by Melvold et al[46]
       and Townsend,[40]. Other relevent literature includes
       references 47-52.
(ii)   Responsible and experienced personnel required on site,
       including:

       (a)   on-scene coordinator
       (b)   environmental specialists

(iii) Coordinating activities with local, state, and federal
       officials.[42]

(iv)   Determination of levels of protection,[33,38,42,46] including
       protective clothing or equipment to be worn or used. (Note:
       overprotection may in some cases cause worker stress and/or
       community stress. Guidance on what level is required under
       specific conditions is given in Table 10.3.)
(v)   Examination of site characteristics:[38,42,54]

       (a)   topography of site
       (b)   meteorological data
       (c)   geological and hydrological data
       (d)   ground cover
       (e)   adjacent tenants (residential and/or industrial).

(vi)   Meetings to describe situation at site before work commences
       and daily while work proceeds.[33,43]
(vii) Pretesting equipment before site visit to insure proper use
       by personnel and proper functioning of the equipment
       itself.[42]
(viii) Providing medical surveillance before, during and after
       working the site.[40,43,46,54] This should include an
       awareness of exposure limits for gases encountered on site.
       Appendix K gives information for the most frequently reported
       toxic and flammable organics. This type of information
       should be kept on site.
(ix)   Provision of a decontamination station for personnel and
       equipment.[33,38,42,44,46,54] Buecker and Bradford,[33] give
       a description of personnel decontamination procedures used
       while cleaning up the General Disposal Co facility in
       California. This involved a sequence of 3 stations to remove
       clothing and equipment, for showering, and for dressing.
(x)   Use of proper equipment for safe operation. In several
       cases[30,32,38,55] this included such equipment as backhoes
       and drum grapplers with enclosed cabs with fresh air for
       working at a 'safe' distance, a hydraulic penetrating device
       for opening drums at a 'safe' distance, SCBA and protective
       clothing (already mentioned), and the use of non-sparking
       tools.
(xi)   Detection of ignition sources (possibly metal drums) by using
       a metal detector[42] or other remote sensing technique. This
       is particularly useful if the contractor is contemplating
       core drilling on the site for sampling or for building vents.
(xii) Determination of the type of containment and/or
       control[1,14,15,16,21,25,46,56,57,58] required including:

       (a)   gas barriers
       (b)   trench vents
       (c)   pipe vents
       (d)   wells (extraction)
       (e)   collection systems
       (f)   treatment systems

Table 10.3   Conditions Associated with EPA Defined Levels of
             Personnel Protection[53,54]

Level	Condition
A	requires full encapsulation, including a self-contained breathing apparatus (SCBA), and protection from any body contact or exposure to materials that are toxic by inhalation and skin absorption.
B	requires SCBA and skin contact protection where air concentrations may be immediately dangerous to life and health (IDLH).
C	hazardous constituents known; protection required for low level concentrations in air by using air-purifying cartridge respiratory protection and skin contact protection (gloves and disposable overalls); exposure of unprotected body areas (head, face, and neck) is not harmful.
D	no identified hazard present, but conditions are monitored and minimal safety equipment is available including mechanical filter, air purifying respiratory protection, and skin contact protection.

(xiii)   Perform community odour surveys to detect odours and their
         extent under various conditions.[59]   These are performed by
         experienced odour judges going through areas (particularly
         resident neighbourhoods) adjacent to a site noting
         meteorological conditions, time of day, location, quality,
         intensity and detectability of any odours.   Intensity
         measurements are made by butanol referencing. Detectability
         is measured using a scentometer (field instrument). This
         information can later be used to calibrate an odour
         dispersion model for a site.
(xiv)   Availability of proper safety equipment.[42]
(xv)   Provision for safe handling and sampling procedures,
         including:[13,14,30,32,38]

         (a)   excavation of drums
         (b)   drum sampling
         (c)   drilling wells and sampling
         (d)   taking core samples from solids
         (e)   sampling of test pits and lagoons.

(xvi) Establishment of an emergency plan with backup equipment.
This includes a 'hot line'.[60] Alcott et al[42] list
anticipated emergencies including: weather extremes, fire and
explostion, work accidents, medical emergencies, civil
disobedience or unauthorised entry to site. They also give
established procedures and responsibilities for proper
attention to emergencies. Many of these require consideration in site planning eg communications, fire control/
suppression, and medical care. Other elements of the
emergency plan are specific eg rescue of injured or ill, site
evacuation, evacuation of adjacent properties, and site
security.

(xvii) Use of safe conduct at all times must be practiced.[43]

(xviii) Develop an information service to keep the public aware of
daily operations.

(xix) Provide training and instruction for all on-site personnel.

The above listing is not an exhaustive compilation of what must
be considered when working a site, but includes the more important
aspects. Reference should be made to the sources used in compiling
this list[33,38,42,46,54] for further details. Reference should also
be made to Kletz,[55] who discusses several myths in working with
hazardous materials including some related to flammable and toxic
gases, use of non-sparking tools (as an unnecessary expense), sources
of ignition, and pressure effects. For example, Myth No 1 says that
flammable mixtures are safe and will not catch fire or explode if
everything possible has been done to remove known sources of
ignition. Kletz points out that this can be misleading, because
cases have been cited where flammable gases or vapours mixed with air
in flammable concentrations have been present, then sources of
ignition turned up which ignited these gases. In many fires and
explosions, a source of ignition is not found. So a safe rule to

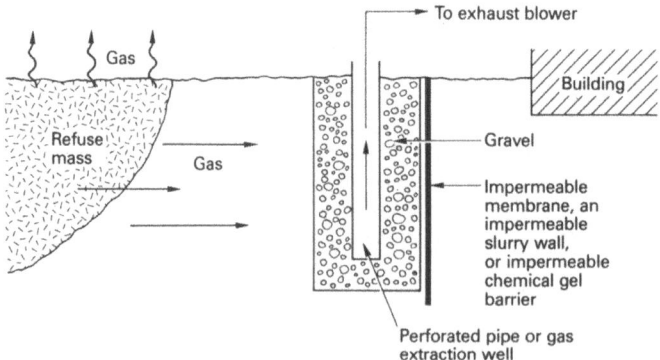

Figure 10.2   Permeable Trench System with Induced Draft
              Header System

follow is to assume that mixtures of flammable vapours in air will
catch fire or explode.

7   EMISSION CONTROL METHODS

Table 10.4 lists the type of gas control methods presently used
to limit the concentrations of pollutants and their lateral movement
and emission. Because the controls applied to remedial action sites
have been site specific they are listed as separate unit operations.
Several of the controls listed have been used in combination with
one another.

8   STRUCTURE PROTECTION METHODS

8.1 Existing Structures

Buildings on or adjacent to a remedial action site with no
built-in gas control facilities must be protected from gas migration
into the structure. Combinations of gas collection and diversion
techniques, such as a permeable trench, impermeable membranes, or an
induced exhaust system and an induced gas extraction system, may be
used. A possible protection scheme (see Figure 10.2) would be a
permeable trench system, complete with gas extraction wells, header
system, and exhaust fan system. A negative head of 250 mm of water
on the header system with sufficient venting, either by pipe vents in
the trench or by extraction wells, should be adequate. This system
would require maintenance and operation for the period that
protection was needed for the building.

A permeable trench with pipe vents may work well and be somewhat
less expensive and less costly to operate than the system with the
exhaust fans and header system. Site conditions for optimal use
include porous soils for gas movement to trench, space for
construction, and no leachate problems. The particular design chosen
would, of necessity, be site specific and should be designed by an
experienced specialist.

8.2 New Structures

Buildings newly constructed on or near remedial action sites
where gas migration can occur must be designed to prevent gas
migration into the building. One method is to raise the structure
off the ground and leave space for wind action beneath the structure,
but this may not entirely prevent some flammable and/or toxic gas
buildup beneath the building and, hence, into it during no-wind
movement periods.

Table 10.4   Gas Control Methods

Method	Remarks	Reference
A NATURAL		
1 Dilution & dispersion into the atmosphere	Used for less reactive hazardous air contaminants; use caution to meet acceptable ambient air guidelines.	1,12
2 Barriers	Consist of nearly saturated fine-grained clay soils (if not fractured); the water table if shallow; or combination of soil and water; limits the extent of gas migration.	56,60
B ARTIFICIAL		
1 Gas treatment systems	Include such systems as thermal oxidation systems (flares and afterburners), carbon absorption units, air and steam stripping units, to remove and destroy toxic organic substances; high dependence on method used; site specifications will designate the method; generally expensive.	1,62,63,64
2 Covers (capping)	Used to lower gas diffusion rates and provide control of gas volatilisation and transport at remedial action sites. This proved useful at the Caputo Site to control increased PCB volatilisation rates during the summer;[13] with a 'gas tight' cover, a collection or vent system should be included (see Chapter 5).	65
(a) Soils	Clay covers effective but temporary. Problems include cracking, plants penetrating cover, animals burrowing holes in cover; with soils, one can lower the diffusion rate by increasing depth of soil cover and reducing the soil porosity; porosity of soil is decreased by increasing soil (density) compaction.	1,4,12,13 15,21,65
(b) Water & Soil	Water, in combination with soil, can aid in reducing the rate of diffusion short term; also decreases porosity of soil.	4,65
(c) Membrane	Not as effective as above (a and b) because soils have adsorbing capacity, reducing volatilisation, while plastic liners are of little help in providing this reduction; there are differences in the gas transmission rates of liners such as Butyl and Neoprene, but these are only effective in reducing volatilisation over a very short term; a study by Farmer et al[4] (see Table 10.1) demonstrates the effect of soils, water and soil, and polyethylene liner (4-6 mm) on volatilisation.	1,4,65
3 Chemical	(Watering) used for control of dust emissions; (Oils) used over lakes and lagoons; retards evaporation; controls odours; to further decrease emissions, one can increase the lagoon depth, construct wind barrier, and pretreat waste to the lagoon by removing volatiles from water before entering the lagoon.	1,12,62
4 Foams	State-of-the-art technique; used to isolate spills from ignition sources and radiant energy; reduces release of some volatile chemicals (for example, time for significant vapour breakthrough for benzene when using National 6% Regular Protein Foam Agent is 60 minutes); inexpensive; effective method of initial vapour control	66

Method	Remarks	Reference
	during an accidental spill, provided the correct foam made from a quality agent is used; lab and field tested for hazardous spills; could be applied to remedial action sites to control volatile releases of compounds such as benzene, toluene, acetone, etc; the problem is some foams are more effective for some volatile organic compounds than others; that is, time for significant vapour breakthrough may vary from compound to compound for one particular foaming agent; therefore, this technique would seem to be most useful at remedial action sites for protection from quick releases of volatile organic chemicals, for instance, when excavating or drilling, to aid in the protection of nearby workers.	
5 Predisposal treatment		
(a) Stabilisation	Chemically binds material to prevent or reduce volatilisation; heat may be transferred or released in the chemical reaction which could result in vaporisation of unreacted parts of the contaminant.	1,8,12
(b) Solidification	Prevents volatilisation by sealing the waste in a hard, stable mass; types include cement-based, lime-based, thermoplastics, organic polymers, self-cementation, and glassification; limited in application to remedial action sites because of costs and waste-specificity of the process.	63,64
6 Pipe vents	Moderately expensive; low upkeep; easy to place, but safety precautions must be taken; use closed, forced ventilation to prevent any toxic vapours from migrating; used to relieve localised accumulation of gases at sites; commercially available.	12,15,56 57,58,63
7 Trench vents	Expensive; moderate upkeep; easy to place with proper safety precautions taken; induced draught is more effective than atmospheric dissipation, especially when used in combination with barriers; passive open trenches may be applicable to the control of toxic vapours in an emergency situation where immediate relief is required; commercially available.	12,25,56,57, 58,63,67
8 Gas barrier	Controls lateral gas migration only if it extends downwards to another impermeable material (includes the use of low gas permeable material such as clay, concrete slurry walls, gunite, and flexible membrane liners); expensive, limited use, not effective generally over long term; commercially available; must be used with venting or collection systems to be effective.	58,63,64
9 Gas collection	Single fan/vent (forced ventilation) collection systems are cheap, effective, and easy to upkeep; could be applied to 2+ hectare grids; manifold collection system is more complicated, costly, and requires a great deal of upkeep; use with gas treatment systems, such as a flare or carbon unit; commercially available.	62,63,64,67
10 Air injection system	State-of-the-art technique; used to control methane migration and production consists of a series of injection wells around a landfill and as theory suggests, a ring of high pressure air which counteracts the positive pressure of the landfill gas, it may create an aerobic environment so the methane gas is no longer produced.	68

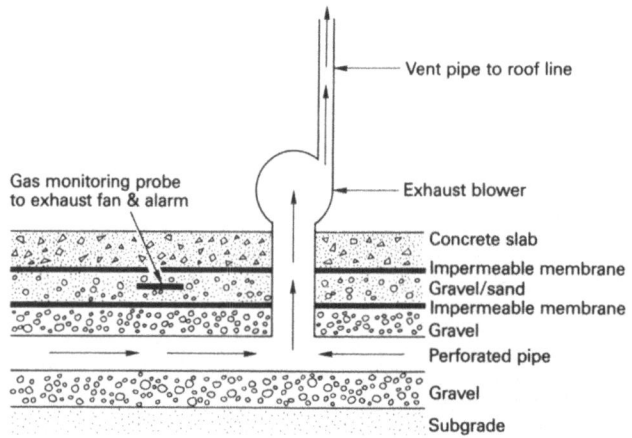

Figure 10.3   Gas Vents for New Buildings

One way to protect the building and the people in it, is to
provide an impermeable barrier to gas flow around the complete
bottom of the structure. Figure 10.3 shows how a hypothetical system
of two impermeable membranes may be placed beneath the floor slab to
stop the flow of gas into the building. The vent system with
exhaust blower to carry the gas out from under the structure
offers a positive protection from gas buildup beneath the
structure. A monitoring system can be used to activate the
blower system or the blower could run continuously depending on the
situation. All vents should be extended to the roof line of the
building. Again, the design of this system would be site specific to
include the size of the building to be protected and the various
other soil and site characteristics.

9   FEASIBILITY OF RECOVERY AND BENEFICIAL USE
        OF GASES AT THE SITE

There seems to be no feasible recovery methods or uses of toxic
and flammable gases at waste sites except for methane. The mixture
and quantities of waste types do not lend themselves to recovery of
other gases. The only way resource recovery could probably become
feasible would be to have stronger controls on waste disposal
operations and/or segregate the waste to produce a more 'pure' waste
stream.

Only one case of resource recovery other than methane gas
recovery systems for landfills has been identified. This was at the
Stroudsburg site in Pennsylvania[20] and was not, in the strict
sence, a gas recovery system. It was related to a gas odour problem

caused by coal tar residuals seeping out of the ground. Approximately 7,500 US gallons of coal tar were recovered from the site through a well system and sold to a chemical company, where it was used as fuel.

One reported problem associated with methane gas resource recovery at landfill sites is the presence of toxic and flammable materials such as benzene, toluene, xylene, hexanol, dichloromethane, and trichloroethylene. Ting et al[70] studied the potential problems associated with the removal of these organic materials from landfill gas by the selexol solvent process which employs air stripping, absorption, and flash stripping. It removes and disposes of at least 90% of the organic materials with molecular weights of 44 or greater. Areas of concern from this process include: (i) emissions from air stripping of absorbent from the flash process will contain solvents absorbed, but not removed by flashing, and (ii) the waste gas from flashing may contain solvents not removed by the purification process. They determined that modelled concentrations of pollutants were well below acceptable ambient air levels except for benezene. For this compound, the maximum ground level concentrations from the selexol solvent plant was modelled at 0.5 $\mu g/m^3$, compared to an acceptable ambient level was 0.14 $\mu g/m^3$ (the derivation of this standard level is given by Ting et al).[70] The conclusion arising from these findings is that there is a potential risk to public health, especially from exposure to carcinogenic compounds; therefore, hazard assessments should be conducted when considering landfill gas recovery, although it may be difficult to make the hazard assessment with any degree of certainty.

10   LONG- AND SHORT-TERM PROBLEMS ASSOCIATED WITH SITE
      RESTORATION TO MITIGATE GAS GENERATION OR MIGRATION

10.1 Short-Term Problems

In attempting to mitigate gas generation or migration at waste sites, several short-term problems have been encountered or raised by the investigator/worker. One problem involved attempts to estimate gases or volatile organics in the site area[5,11,13,22] before actually entering the site to evaluate risks associated with these materials. Other articles brought out the fact that published gas monitoring data on rate and amount of gas production from industrial waste landfills is limited.[1,13] This is discussed in a later Section. Also, because of the wide range of contaminants[1,22] and the interference posed by other gas sources in the site area,[5,28] measuring each contaminant is difficult.

McEnery,[22] points our that these problems are being overcome by gathering background data either from federal, state and local officials, manufacturers, and others who live or work near the site

or by performing field investigations using environmental air
monitoring equipment such as combustible gas indicators, oxygen
meters, gas detector tubes, radiation detectors, photoionisation
detectors and flame ionisation detectors. The uses and limitations
of this monitoring equipment is described by McEnery.[22]

In addition to this, Costello and Kominsky,[5] list the use of
other direct-reading instruments such as colorimetric detector tubes,
various electronic devices designed to quantify specific airborne
contaminants, and portable broad spectrum analytical devices (gas
chromatographs, infra-red spectrographs, and photoionisation
detectors). An experienced analyst is recommended when using these
instruments with knowledge of pertinent interferences and specific
calibrations of the device for the substances to be measured.

Other investigators[71,78] have turned to the use of air
emission modelling for estimating rates and amounts of gas or
volatile organics from landfills. The reliability of these models
depends on the field data gathered such as quality and composition of
wastes, soil properties, and its surrounding environmental
conditions.[13] These are also being developed to be able to predict
long-term problems from gas generation and migration. Emission
modelling is discussed later in Section 14.

Another short-term problem is the effect of variable meteorological conditions, on gas generation and migration rates.[2,5,11]
There are increases in the movement of gases and volatiles when
temperatures increase, winds increase, atmospheric pressure changes,
etc. These changes in physical conditions at sites could cause acute
or chronic chemical exposure to workers or may, suppress exposure
levels to workers depending on how the conditions influence the
volatiles.

Construction of dykes and fences were used at the MOTCO
disposal site to eliminate the immediate threat of migration of vinyl
chloride bottoms and other pollutants from unlined lagoons into
navigable waters.[79] However, this is only a short-term remedy. An
example of a short-term remedy that may result in problems at sites,
is 'capping' the landfill to reduce or prevent gas migration.
Problems arise from this remedial action as reported in Shen and
Tofflemire,[1] when the seal is disturbed as a result of animals
burrowing holes, plants rooting, or cracks developing in it.

Other remedial action options can increase short-term risks with
regard to exposure to chemical substances including physical removal
operations and pre-treatment techniques such as air or steam
stripping. Either may cause an increase or sudden release of a
volatile chemical or gas. For example, at Lehigh Electric in
Pennsylvania,[11] off-site transport of PCB-contaminated coal fines

posed a problem. The suggested remedy was a comprehensive dust suppression programme.

An additional short-term problem is toxic vapours and gases absorbed/adsorbed on small soil or dust particles. These particles will not remain in the air long. Once emitted into the air, they will eventually return to land or in waters by fallout or in precipitation. This will permit the substances to re-emit to the atmosphere by wind or volatilisation. This cyclic process[12,13] (Figure 10.1) may lead to a long-term effect by accumulating over time and/or providing a pathway for long distance migration causing more than a localised problem. For example, Shen and Sewell,[12] in a study of physical and chemical removal processes, revealed that for volatile chemicals, such as vinylidene chloride and perchloro-ethylene, chemical removal residence times in air were estimated to range between 3 to 70 days providing the time for long distance migration.

10.2   Long-Term Problems

Long-term contamination effects from toxic and flammable gases are not yet well known.[21] However, it is believed that long-term airborne effects from predicted concentrations of organic vapours could cause increased health risks. This has been pointed out, for example, in studies on the Gilson Road Hazardous Waste Dump.[11,15] The authors suggest that if no remedial action was to take place here, predicted airborne concentrations of the organic vapours cumulatively would exceed the recommended long-term exposure limits for a cancer risk of 1 in 1 000 000 by a factor of 100 for the residents of the mobile home park near the site.

As mentioned above, many short-term accumulations of substances could lead to long-term problems. However, the risks from eventual long-term accumulations can be abated or decreased with sufficient controls.

11   COSTS ASSOCIATED WITH CONTROL TECHNIQUES

Paige et al,[80] describes how cost estimates were derived for various remedial actions at uncontrolled hazardous waste sites. They specifically give the costs of several gas migration control techniques. These sample costs are for hypothetical sites and were obtained from published guides and other literature sources. The gas migration control system, system details, and total costs are given in Table 10.5.

Other gas migration control costs have been reported by Welsh et al,[67] and Lippitt et al.[53,129] Both sources give typical

Table 10.5   Cost of Specific Control Techniques

Technique	Details	Total cost (US $ - 1982)
Trench vents	As constructed; dimensions are 6.1 m deep, 1.2 m wide, 152 m long. Cost covers excavation, spreading of backfill material, well point dewatering, sheet piling, laterals, risers. Cost of treatment of water withdrawn during dewatering is not included.	273,510
Pipe vent	For a single forced ventilation pipe vent 100 mm in diameter and 9.1 m deep. Includes cost for installation of well, PVC casing, mushroom top, elbow piping, and a small fan applicable to one vent.	1,110
Gas collection system	This consists of a surface gas collection network connected to the heads of 4 pipe vents. The distance between each vent is 61 m and an additional 6.1 m of PVC pipe is needed to connect the last vent to the fan. Pipes are designed to maintain the required head throughout the system. Also, PVC elbows, PVC tees, butterfly valves and flowmeters are used as needed. Cost includes installation and materials.	14,230

remedial action unit operation costs for several remedial actions at three cost levels (i) upper US average, (ii) lower US average, and (iii) specific costs encountered in 1980 at Newark, NJ and how they arrived at these costs. As in the study by Paige et al, sample costs are for hypothetical disposal sites and were obtained from published guides (Dodge and Means Construction Cost Guides - 1980 Edition). The range (using the US high and low average) of life cycle costs, including initial capital and first year operation and maintenance, for the gas migration control techniques are given in Table 10.6.

Lippitt et al[53,129] also give cost information for health and safety equipment and materials when working a site. The cost which they give to equip a worker for four levels of personal protection as

Table 10.6   Range of Costs of Specific Gas Migration
             Control Techniques - Constant Units*[53]

Technique	Range of life cycle cost (US $ - 1980)
Passive vents	17,300 - 25,220
Passive trench barriers	29,015 - 44,180
Active gas extraction wells	28,900 - 48,200
Surface seal with clay	96,300 - 163,300
Surface seal with asphalt	67,300 - 92,700
Surface seal with fly ash	95,900 - 165,700

*   All units in dollars per hectare of site surface area.
    ha = 0.41 acres.

Table 10.7   Costs for Equipping a Worker at Four
             Levels of Protection

Level	Equipment	Total cost (US $ - 1980)
A	Fully-encapsulating chemical resistance suit; pressure demand, self-contained breathing apparatus (SCBA); gloves, outer, chemical resistant; boots, chemical resistance, steel toe and shank; hard hat; boots, outer, chemical resistant, disposable.	1,854
B	Chemicals resistant clothing - jacket, bib overalls; pressure demand, SCBA; gloves, as in A; boots, chemical resistant, leggin; boots, outer, chemical resistant, disposable. Boots, chemical resistant, steel toe and shank; hard hat.	1,136
C	Saran-coated disposable suit; full face air purifying cannister respirator: front belt mounted, back belt mounted; gloves, as in A; boots, as in B; escape mask, hard hat.	736
D	Disposable coveralls; gloves, outer, chemical resistant; boots, chemical resistant, steel toe and shank; escape mask; hard hat and face shield.	260

defined by the Interim Standard Operation Safety Procedures developed
by the USEPA Emergency Response Team is given in Table 10.7. These
authors also give some preliminary estimates for health and safety
costs for individual cost components.

12   REPORTED EMISSION LEVELS AT UNCONTROLLED
         HAZARDOUS WASTE SITES

Appendix K lists some of the more common chemicals found at
uncontrolled hazardous waste sites that have resulted in toxic and
flammable gas problems with some of the relevant chemical properties
such as molecular weight, boiling point, vapour pressure, flash
point, and solubility. Established limits are reported in the form
of Provisional Limits in Air (PLA), the Threshold Value (TLV), and
the Short Term Exposure Limit (STEL). These, and the other
characteristics cited, are defined in Appendix K.

Appendix K also lists the concentrations that have been
found at specific uncontrolled hazardous waste sites. These are air
concentration ranges measured at the site unless indicated otherwise
in the notes and are taken from current literature. These sources
do not generally state the conditions under which the measurements
were made. Sampling and analytical techniques must be included with
reported data if the information is to be useful for health
protection and other purposes.

13   CHEMICAL GROUP COMPATIBILITY

Appendix L lists the combination of chemicals that will result
in fire, explosion, heat generation, toxic gas generation, or
flammable gas generation.[81] This is not intended to be an all
inclusive list because of the infinite number of combinations that
may occur at uncontrolled hazardous waste and other contaminated
sites. It does, however, provide a good indication list of
combinations that should be avoided and can be used in determining
potential gas problems when assessing a site.

14   AIR EMISSION RELEASE RATE MODELS

14.1 Introduction

Atmospheric transport is the major mechanism for the movement
and emissions from the contaminated site. These emissions are
transported either as a vapour or sorbed onto fine particles and
eventually deposited onto the land or water. Since air quality
monitoring of toxic emissions is costly, emission models have been
developed and are being refined to estimate the volatilisation of

chemicals. This section deals with empirical equations used to
calculate emission rates of volatile organic chemicals.[127] Air
emission models for the following disposal/treatment/storage options
are discussed:

- landfills
- land treatment
- surface impoundments
- storage tanks
- drum handling and storage
- treatment processes
- waste piles (particulate emission)

Diffusion of volatile organic emissions from sites can be discussed in terms of diffusion through multiple layers, the last being
the lower atmosphere. The rate of diffusion through all the combined
layers determines the rate of release to the atmosphere. This mass
transfer rate is expressed in terms of the mass transfer coefficient.
A compound's volatility, characterised by its solubility, partial
pressure, Henry's law constant and diffusivity, provides an indication of the quantity of material in the gas phase which will be
subjected to the transfer process. Thibodeaux[90] provides a
thorough documentation of the environmental movement of chemicals in
air, water, and soil. Physical and chemical data can be obtained
from various handbooks.[91]

The intention of this section is to provide a discussion
of air emission release rate models. A thorough documentation and
description of these models is beyond the scope of this report.
However, the cited references will enable the reader to obtain the
further information on this subject.

14.2 Landfills

Six air emission models are reported in the literature[127] (see
Table 10.8). The models represent covered landfills with and without
internal gas generation, covered landfills with barometric
fluctuations and open dumps.

The model developed by Farmer was used for designing a landfill
cover that minimised the emission of heaxachlorobenzene (HCB) or
other volatile organic vapours.[92] The model could be used to
assess the effectiveness of an existing landfill cover for controlling
other emissions. However, the model was only verified experimentally
for HCB-containing waste in a laboratory-simulated landfill.

Using Fick's First Law for steady-state diffusion, Farmer
describes the volatilisation or vapour loss of HCB, or other
compounds, as a diffusion controlled process. With the assumptions

Table 10.8 Air Emission Release Rate (AERR) Models for Landfills

---

Model	Applicability

Farmer et al[92]	Covered landfills
Shen[94]	Covered landfills
Thibodeaux[95]	Covered landfills with no gas generation
Thibodeaux[95]	Covered landfills with biogas generation
Thibodeaux[96]	Covered landfills with barometric fluctuations
Shen[98]	Open dump

---

of no degradation from biological activity, no adsorption of the
compound, no transport in moving water and no landfill gas
production, the rate at which a compound will volatilise from the
soil surface to the atmosphere will be controlled by the diffusion
rate through the soil cover. To describe molecular diffusion through
a soil surface, Farmer adopted the effective diffusion coefficient
suggested by Millington and Quirk.[93]

Shen took the simplified version of Farmer's equation for vapour
flux from a soil surface and converted it to an emission rate by
multiplying by the exposed area.[94] To determine the emission rate
of a specific waste component, Shen multiplied by the weight percent
of component in the bulk waste.

Thibodeaux[95,96] presented three equations for estimating air
emissions from covered landfills; (i) without internal gas
generation, (ii) with internal gas generation, and (iii) with
internal gas generation including the effect of barometric pumping
pressure.

Thibodeaux's model for covered landfills without gas generation
appears similar to the two resistance theory of mass transfer: ie
Flux = (overall transfer coefficient) (concentration gradient). The
model is the algebraic sum of the two individual phase equations: the
rate of vapour movement within the soil phase, and rate of vapour
movement from air-soil interface to overlying air.

The rate controlling portion of Thibodeaux's flux rate equation
is the rate of vapour movement within the soil phase. The mass
transfer coefficient for vapour diffusion in soil is described for
vapour movement through a porous media. The second portion of
Thibodeaux's model describes the vapour movement through air, based
on the work of Sutton.[97]

In addition to molecular diffusion of landfill vapours through soil, there is a 'convective sweep' of chemical vapours toward the surface created by the formation of landfill gases (CO_2, H_2, CH_4), especially applicable to co-disposal sites and municipal waste sites. Therefore, the model with internal gas generation contains both a diffusive and convective term.

Additional work by Thibodeaux et al[96] addressed a third transport mechanism in addition to the diffusive and convective mechanisms previously discussed; the barometric pumping pressure. Atmospheric pressure fluctuations develop pressure gradients that pump vapours and gases from landfill cells to the air above. This pumping enhances vapour phase mass transfer.

The emission rate equation for open dumps presented by Shen[98] is derived from Fick's Law and Arnold's equation[99] for a surface exposed to open air. A major limitation of this model is that it only considers air-phase diffusion and it does not accurately describe the effect of boundary layer formation caused by ambient wind, as shown by Thibodeaux in describing the gas phase resistance of vapour movement.

14.3 Land Treatment Farming

The Thibodeaux-Hwang[100] model most accurately defines the physical situation existing in a landfarm, namely the mass transfer of a chemical species from a soil/waste mixture. One apparent limitation of this model is that it describes vapour diffusion as being soil phase controlled, based on the assumption that the oil-layer diffusion length is of the order of a soil particle size. This assumption may only be valid after a certain exposure time. For petroleum waste, the sludge material generally subjected to land farming is a multicomponent viscous material. Molecular diffusion coefficients for this material will be several orders of magnitude less than that for soil phase diffusion. Thus, the diffusion rate determining factor will be the application thickness of the waste material. It may be possible that both phases (soil and oil) will contribute equally to the diffusion rate of a specific component. This model is applicable to either surface application or subsurface injection methods.

14.4 Surface Impoundments

14.4.1 General. Surface impoundments (SIs) are defined as a facility or part of a facility which is a natural topographic depression, man-made excavation, or dyked area formed primarily of earthen materials (although it may be lined with man-made materials) designed to hold an accumulation of liquid wastes or wastes

containing free liquids. Examples of surface impoundments are
holding, storage, settling, and aeration pits, ponds, and lagoons.

Eight models or variations thereof appear in the literature
for estimating the air emission release rates (AERR) from SIs
(non-aerated and aerated) as follows:[127]

- Mackay and Wolkoff unsteady-state, non-aerated;[101]
- Mackay and Leinonen unsteady-state, two-film theory, non-aerated model;[102]
- Thibodeaux, Parker and Heck steady-state, two-film theory, one non-aerated and one aerated model;[103]
- Shen[104] modification of Thibodeaux, Parker and Heck non-aerated model;
- Smith etal[105,106] steady-state, first-order kinetic, non-aerated model;
- McCord[107] steady-state, modified Nusselt equation, one non-aereated and one aerated model.

The eight predictive models can be categorised according to
their theoretical basis. Five of the eight models incorporate the
two-film resistance theory and are currently considered more reliable
than those based on other concepts. The other three predictive
models have been criticised by more recent research. The two
predictive models presented by McCord are based on the Nusselt
equation but the theoretical basis of water evaporation implied by
the Nusselt equation is considered an unrealistic approach to the
volatilisation process of most slightly soluble organics. The Mackay
and Wolkoff predictive model was essentially revised by Mackay and
Leinonen's research.

Although models which incorporate the two-film resistance theory
are considered the most accurate for describing the actual volatilisation process, and subsequent flux rate of an individual compound
into the atmosphere; these models require determination of the
individual liquid and gas phase mass transfer coefficients for each
compound of interest. The availability of liquid and gas phase mass
transfer coefficients is limited. They are primarily based on
laboratory experiments or on field measurements of lakes, rivers, and
the ocean. Consequently, the accuracy of all available state-of-the-
art models is limited owing to problems involved with precisely
determining mass transfer coefficients for specific situations.
Table 10.9 summarises the models.

14.4.2 Non-aerated Impoundments. Mackay and Wolkoff[101]
proposed a model to quantify the volatilisation of low solubility
compounds (hydrocarbons and chlorinated hydrocarbons) from rivers,
lakes and oceans. Their approach was based on equilibrium
thermodynamic principles of water evaporation, in contrast with more
recent research which is based on mass transfer principles of a

Table 10.9   Air Emission Release Rate (AERR) Models
             for Surface Impoundments

Model	Applicability
Mackay and Wolkoff	Non-aerated Unsteady state
Mackay and Leinonen	Non-aerated Unsteady state
McCord	Aerated Non-aerated Steady state
Smith, et al	Non-aerated Steady state
Thibodeaux, Parker and Heck	Aerated Non-aerated Steady state
Shen	Non-aerated Steady state

concentration gradient across an interface. Mackay and Wolkoff
assumed that the AERR could be calculated on the basis of water
evaporation and the ratio of the contaminant to water in the vapour.
This approach assumes that the diffusion or mixing in the water phase
is sufficiently fast so that the concentration of the contaminant at
the water-air interface is close to that in the bulk of the water
body. However, recent research on mass transfer rates suggests that
diffusion in the water phase is the rate controlling variable for
most low-solubility compounds. In other words, current mass transfer
theory shows that the basic assumption in the Mackay and Wolkoff
model is inappropriate for modelling AERR from hazardous waste
surface impoundments. Other stated assumptions in the Mackay and
Wolkoff model include:

-   The contaminant is truly in solution, not in suspended,
    collodial, ionic, complexed, or absorbed form.
-   The vapour formed is in equilibrium with the liquid at the
    interface.
-   The water evaporation rate is negligibly affected by the
    presence of the contaminant.

Dilling's[108] preliminary work indicated that the Mackay and
Wolkoff model was inadequate since neither absolute nor relative
predicted rates were in agreement with experimental data.

Mackay and Leinonen extended Mackay's prior research in order to develop a more realistic AERR estimate for low solubility compounds from the entire water body, and not just from the water surface. They incorporated the work of Liss and Slater[109] to develop a model to determine the AERR of a single compound (Liss and Slater applied the two-film resistance theory to estimate the flux of gases across the air-ocean interface. Contrary to Mackay's earlier thermodynamic model theory, Liss and Slater's work suggested that for most low-solubility gases, the water (liquid) phase controls, ie the liquid phase offers more resitance to contaminant transport than the gas phase).. The model is based on unsteady state conditions, ie the chemical contaminant of interest enters the water body in a discrete slug as a pulse injection. All other SI models are based on steady state conditions, ie a fairly constant influx of the contaminant into the SI. As in the case of all other predictive models for non-aerated SIs, this model simplifies the actual situation by assuming well-mixed air and water phases separated by an interface with near stagnant films of air and water on either side. Thermoclines, ie other rate-limiting diffusion processes at depths in the water body, are not considered in this model.

Thibodeaux et al[103] proposed a steady-state model based on the two-film resistance theory that assumes a constant influx of contaminant. Empirical relationships for mass transfer coefficients developed from laboratory and field experiments at lakes, rivers and oceans were incorporated.[110,111,112]

Shen.[104] using the rate expression proposed by Thibodeaux et al,[103] described a simplifying method for estimating the volatilisation rate of volatile organic compounds from waste lagoons. Shen's modifications included the use of Owen's correlation and Matsugu's correlation for determination of mass transfer coefficients.

Smith et al,[105,106] proposed a technique for determining an overall volatilisation rate constant based on laboratory measurements. The volatilisation rate constant is essentially the overall mass transfer coefficient that represents the individual liquid and gas phase values combined.

A predictive model for estimating the emission rate of volatile compounds from non-aerated lagoons was presented by McCord.[107] This steady-state model, was based upon Nusselt's equation which describes water evaporation rates from a lagoon. Consequently, McCord's model is limited to situations where volatilisation is controlled by the gas phase mass transfer. This limitation is similar to that of Mackay and Wolkoff's model. It has been shown by more recent research that most volatilisation is liquid phase controlled. Additionally, McCord assumes that equilibrium at the air-water interface follows Raoult's law. However, for dilute

aqueous solutions, the equilibrium should be determined on the basis
of Henry's law.

14.4.3 Aerated Impoundments. Thibodeaux et al[103] proposed
use of their steady state model for non-aerated surface impoundments
for use with aerated impoundments with some modifications. The
theory of two-film resistance is applicable to an aerated impoundment
if one considers that turbulence, caused by aeration, creates two
distinct zones at the impoundment surface, ie (i) turbulent, and (ii)
convective or natural. The overall liquid phase mass transfer
coefficient for the entire system must be modified to account for
each distinct zone, proportional to the affected area of each zone.

One additional predictive model was proposed by McCord[107] to
estimate the emission rate of volatile compounds from aerated
lagoons. This steady-state model is based upon Arnold's studies on
the diffusion of volatile compounds from a liquid surface into air.

Discussions in the preceeding section concerning McCord's[107]
steady-state predictive model for non-aerated impoundments showed
that the basis for McCord's model is evaporation of water from the
impoundment surface, ie gas phase control. Therefore, this model
does not accurately represent the volatilisation rate of most
slightly soluble volatile organic compounds.

14.5 Storage Tanks

14.5.1 General. Air emissions of volatile compounds from
storage tanks are a function of several factors including:

  (i)   physical and chemical characteristics of the stored liquid;
 (ii)   tank design;
(iii)   operational characteristics, especially turnover frequency.

For a given liquid, tank design influences the emission rate
potential. The five types of storage tanks are:

  (i)   fixed roof;
 (ii)   external floating roof;
(iii)   internal floating roof;
 (iv)   variable vapour space;
  (v)   pressure tanks.

Of the five designs, vapour space and pressure tanks generally
produce the least air emissions. These will not be further
discussed. In addition to tank design, the true vapour pressure of
the material stored is one of the most significant parameters
affecting emissions.

Physical actions on the tank such as changes in temperature or pressure affect the volatilisation rate. Temperature increases from direct solar radiation and contact with warm ambient air increase volatilisation and emission potential. Danielson[113] notes that for a free vented tank, winds may entrain or educt some of the saturated vapours into the ambient air.

Operating conditions also affect tank emissions, ie frequency of filling (turnover rate), vapour tightness of the tank, and volume of the vapour space. When the turnover rate is long, ie extensive time periods between filling/emptying cycles, the free space in a tank becomes saturated with vapour from the liquid. Thus, during filling of the tank or during breathing cycles, a larger concentration of vapours exists in the air-vapour mixture vented to the atmosphere. Danielson states that vapour tightness of the tank can influence the evaporation rate, and a lack of tight vapour seal allows increased emissions. Proper seal maintenance for floating tanks is necessary to limit vapour losses.

The American Petroleum Institute (API),[114,115,116] EPA and others have developed empirical equations for fixed and floating roof tank emissions based on field test data. Masser[117] notes that emissions from pressure tanks occur only when the design pressure is exceeded, when the tank is filled improperly, or when abnormal vapour expansion occurs. Because these are not regularly occurring events, and pressure tanks are not a significant source of emissions under normal operating conditions, no equations were found available for estimating air emissions from pressure tanks.

14.5.2 Fixed Roof Tanks. Air emissions from fixed roof tanks occur from breathing losses and working losses. Masser[117] defines breathing loss as vapour expulsion due to vapour expansion and contraction from changes in tank temperature and ambient barometric pressure. Breathing losses occur in the absence of any liquid level change in the tank.

The combined loss from periodic filling and emptying is called working loss. When a tank is filled, vapours are expelled from the tank when the pressure inside the tank causes opening of the relief valve. Emptying loss occurs when air drawn into the tank during liquid removal becomes saturated with organic vapour, expands, and exceeds the capacity of the vapour space.

14.5.3 External Floating Roof Tanks. Standing storage loss, the major element of evaporative loss, results from wind induced effects acting on the top of an external floating roof tank. The types of seals used to close the annular vapour space between the floating roof and the tank wall dictate the nature of wind effects.

Standing storage loss emissions from external floating roof tanks are controlled by either a primary seal, alone, or a primary and a secondary seal. Three basic types of primary seals used on external floating roofs are: (i) mechanical; (ii) resilient (non-metallic); and (iii) flexible wiper. Although there are other seal system designs, Masser[117] indicates that the systems described here comprise the majority in use today.

Withdrawal loss is another source of emissions from external floating roof tanks. This loss is the vaporisation of liquids that cling to the tank wall and are exposed to the atmosphere when a floating roof is lowered by withdrawal of liquid.

14.5.4 Internal Floating Roof Tanks. Internal floating roof tanks generally have the same sources of emissions as external floating roof tanks, ie standing storage and working losses. Fitting losses through deck fittings in the roof, roof column supports, or other openings, can also account for emissions from internal floating roof tanks.

Typical internal floating roofs incorporate two types of primary seals, resilient foam filled and wiper. These are similar to those employed in external floating roof tanks and close the annular vapour space between the edge of the floating roof and the tank wall.

14.5.5 Open Storage Tanks. It is unusual to find pure volatile organic liquids stored in open tanks. However, certain wastes such as sludges may be stored in open tanks prior to treatment or disposal. The mechanisms of molecular and turbulent diffusion will control the process of evaporation. Thus, the methods for determining an emission rate for this type of storage becomes similar to that developed for surface impoundments.

The major difference between a surface impoundment and an open storage tank is the potential for a greater freeboard effect. For a substantial freeboard which restricts the motion of the wind across the surface of the waste. The mass transfer coefficients can be substantially lower in an open storage tank. When the gas phase mass transfer coefficients are lower, the gas phase has a proportionally greater share of the overall resistance and thus the emission rate controlling step is more likely to be the resistance of the gas phase.

14.6 Drum Storage and Handling

Drum storage and handling facilities encompass diverse operations and sources of air emissions. Sources of routine air emissions include the storage of volatiles in lagoons or storage tanks. Air emission release rate estimation techniques for these activities are

described in other sections of this report. The other important air
emissions category is thought to be accidental spills.

The purpose of this section is to identify data necessary for
estimating air emissions attributed to accidental spills at drum
storage and handling facilities. It provides a general description
of a drum storage and handling facility and also some spill rate data
for hazardous waste treatment facilities and petroleum handling
facilities. No data specific to air emissions from drum handling and
storage facilities have been found.

At drum storage and handling facilities, waste material, most
commonly solvents, may arrive by tank car or tank truck, as well as
in drums. Material may then be pumped to storage tanks, lagoons, or
other drums for storage. Materials are segregated by type until
enough are collected for reprocessing or disposal either on site or
off site with removal again either by tank car, or tank truck, and
less commonly, in drums.

Drums are removed from a tractor/lorry trailer on arrival with a
fork lift and are conveyed to and from the drum storage building
where storage occurs on concrete pads. Drums that are stored
directly on the ground may freeze to the surface, and may thus be
subject to rupture during removal.

A number of other operations may also occur. Damaged drums will
usually be replaced either upon arrival or upon later inspection.
Material may have to be resampled to check compatibility parameters
such as flash point, acidity or chloride content. Drums may be
intentionally burst if this is the most effective way of removing the
contents of damaged barrels.

The amount spilled is not the same as the amount of volatile
released to the atmosphere. Absorbent material will be added, when
spills are discovered during inspection. Spills on a concrete
surface, as opposed to ground, will be more amenable to cleanup.
Many handling operations occur in outdoor areas. The rate at which
material volatilises before it can be cleaned up will then depend on
temperature and wind conditions. Similarly, for indoor spills,
volatilisation rates will depend primarily on temperature and
ventilation conditions. Finally, human factors such as management
attitudes and worker training are important in determining the
accidental spill rate.

14.7 Treatment Processes

14.7.1 General. The purpose of this section is to identify
AERR models which can be used for quantifying air emissions of
volatile compounds from hazardous waste treatment systems. The

approach used for identifying the appropriate AERR model was to
categorise each treatment system into one of these main categories:

-    open tanks with no mixing;
-    open tanks with mixing;
-    closed systems.

       In all the cases where volatile species in the waste stream have
been identified, and their effluent concentrations measured, AERR
models for aerated or non-aerated surface impoundments could be
applied to open tank processes depending upon system dynamics (ie
aerated or non-aerated). For closed system wastewater treatment
processes, it can be assumed that aside from system leakage or
operational abnormalities, there are no air emissions. The
concentration of effluent contaminants can be calculated based on
process unit efficiency. Limited data are available regarding the
removal efficiency and effluent concentrations for organic pollutants
subjected to various biological treatment systems.

       14.7.2 Open Tank System - No Mixing. Typical wastewater
treatment systems that fall into the category of open tank-no mixing
are:

-    sedimentation;
-    chlorination;
-    equalisation.

       No AERR models are available that specifically describe air
emissions from these treatment systems. However, it can be assumed
that these systems fall into a broader category of non-aerated
surface impoundments. Treatment systems that fall into the open
tank-no mixing category represent plug flow systems. Therefore,
applying the non-aerated surface impoundment model one needs to
accurately define the in-tank concentration of the specific compound
in question.

       14.7.3 Open Tank-Mixing: Biological Treatment Systems.
Evaluation of open tank treatment processes with mixing focused
mainly on the activated sludge biological treatment process (high
rate mixing). Other processes, which might fall into the low rate
mixing sub-category, include neutralisation or precipitation
involving the addition and rapid mixing of a chemical reagent. The
emissions from these processes are best described by application of
Thibodeaux's Aerated Surface Impoundment model.

       Hwang's[118] Activated Sludge treatment AERR model employs
techniques presented earlier for predicting emissions from aerated
surface impoundments under steady state conditions (Thibodeaux et
al).[103] The model by Thibodeaux et al predicts the air emission
rate of a compound based on the concentration of the compound within

the impoundment. Hwang notes that for AS treatment systems, AERR models should consider removal of the compound by: (i) biological oxidation; (ii) air stripping; and (iii) adsorption to wasted sludge. Thus, Hwang's model predicts the effluent concentrations that result when the mass balance for the compound across the AS process is satisfied. The emission rate from the process is then predicted by the aerated surface impoundment equation. Note that if the pollutant's effluent concentration for an AS process is known, the emission release rate for that pollutant can be calculated directly by the aerated surface impoundment equation. This is based on the assumption that the effluent concentration adequately represents the concentration of the compound in the aeration tank.

Freeman's[119] original AS treatment AERR model appears very similar to the work of Thibodeaux[103] and Hwang.[118] The major difference between the Freeman and Hwang AS models are the biodegradation kinetics used. In addition, Freeman does not address adsorption of substrate onto the waste sludge, assuming that adsorption is not a major removal mechanism for organic pollutants.

In later work Freeman[120] developed a significant modification to the initial AS model to more adequately describe the mass transfer phenomena taking place in diffused air AS systems. Diffused air systems are commonly employed in laboratory bench testing and occasionally used in actual field applications. Freeman's Diffused Air Activated Sludge (DAAS) model, also called the subsurface aeration model, focuses on the mass transfer of the compound into the air bubbles released by the spargers as the bubbles rise to the surface. This is quite different from the models discussed earlier, since the area across which mass transfer takes place is the surface area of the bubbles, not the basin surface area. Note that this model is better suited to the activated sludge diffused air application than other models discussed. Aside from the variation in mass transfer expressions the DAAS model should be considered identical to Freeman's earlier AS model.

14.8 Waste Piles

Particulate air emissions from waste piles occur at several points in the storage cycle:

-   transfer of material to and from the pile;
-   wind erosion;
-   maintenance and traffic activities on the pile.

Particulate emissions from waste piles are influenced by the following factors:

-   water conditions (rainfall, wind and temperature);

- duration of storage;
- compaction of pile;
- amount and size of aggregate fines;
- cover materials and other control measures.

A method of estimating particulate emissions from waste piles is available in EPA Publication AP-42. Cuscino et al[121] presents emission factor equations empirically developed by Midwest Research Institute (MRI). Both methods describe emissions of particles smaller than 30 μm in diameter based on a particle density of 2.5 g/cm^3. Rosburg[128] discussed emission equations also.

15 AIR EMISSION MEASUREMENT TECHNIQUES

Sampling approaches for measuring air emissions from hazardous waste sites include the concentration-profile technique, emission isolation flux chamber, transect technique and vent sampling and tracer technique. These are discussed below.

15.1 Concentration-Profile Technique

The concentration-profile technique (CP) was developed at the University of Arkansas by Thibodeaux et al.[103] The specific purpose of the CP technique is to measure volatile emissions from large surface impoundments. The approach is an indirect sampling technique based upon measurements of wind velocity, volatile concentration, temperature profiles and other environmental conditions in the boundary layer above the surface interface.

Sampling equipment consists of the following: a 4 m mast with a wind direction indicator, wind speed sensors, temperature sensors, and air collection probes placed at 6 logarithmic intervals above the surface; a continuous real-time data collection system; a thermo-couple for measuring water temperature; and water sampling equipment. It is recommended that duplicated air samples be collected. Air sample probes are filled with sorbent material selected for specific contaminants to be monitored. Specific meteorological conditions required during the sampling period are: mean wind speed greater than 8 kmh (5 mph); maximum wind speed should not cause disturbances on the surface of the impoundment, and during the sampling period the wind direction should not vary more than 45°. Samples are taken for a 20-30 minute period.

Once the information is collected, the following profiles are computed in order to determine the slopes; temperature as a function of height above the impoundment surface; wind speed as a function of height; and concentration as a function of the log of height. The

emission rates of the selected compounds are calculated from the
profiles.

15.2 Emission Isolation Flux Chamber

The emission isolation flux chamber is a device used to make a
direct emission measurement. This method has been used to measure
emission fluxes of sulphur and volatile organics.[122,112,114] The
technique uses an enclosed chamber that samples gaseous emissions
from a defined surface area. Clean, dry sweep air is added to the
chamber at a fixed, controlled rate. The volumetric flow rate of
sweep air through the chamber is recorded; and the concentration of
interest is measured at the exit of the chamber. Air emission rates
can then be calculated via a simplified equation. All parameters
used in the equation are measured. The flux chamber is applicable to
water and soil samples.

15.3 Transect Technique

The transect technique is an indirect emission measurement
approach used to measure fugitive particulate and gaseous emissions
from line and area sources.[125] The horizontal and vertical arrays
of samplers are used to measure concentrations of volatile emissions
within the effective cross-section of the emission plume. The
emission rate is then obtained from calculations using the measured
concentrations over the plume area. The sampling equipment consists
of a central mast having a number of equally spaced air sampling
probes, wind direction, wind speed, and temperature measurement
devices. The meteorological conditions for the CP method also apply
here. This method is useful for downwind sampling from contaminated
sites.

15.4 Vent Sampling

This method is for measuring emissions from ducted sources.
It requires that the volumetric flow rate of the gas be determined,
typically air measurements of velocity and duct cross-sectional area,
and that the gas concentration be measured. The emission rate can
then be calculated.

15.5 Tracer Technique

This method is a direct method of measuring rates from fugitive
sources. A tracer gas is released at a fixed rate near the source.
The ratio of the measured concentrations of the pollutant to the

tracer concentration is proportional to the pollutant rate to the tracer rate.

16   CONCLUSIONS AND RECOMMENDATIONS

Toxic and flammable gases have been detected at remedial action sites. Their presence in the air may cause health problems for workers at the site and for those off site. The sources of these emissions are disposal operations, storage facilities, treatment units and various types of improper waste management. Concentrations have been reported from the PPB to the PPM level.

As part of a remedial clean-up or an operational hazardous waste facility, a plan for the control and correction of problems associated with toxic and flammable gases must be developed. The plan should address general safety requirements and the following specific areas:

-    sources of gases
-    composition of the gases
-    properties of the gases
-    sampling and analytical aspects
-    protective and safety equipment
-    an evaluation of the health aspects
-    use of operating buildings
-    construction and equipment requirements
-    corrective action measures.

Gas problems are site specific based upon the nature of the waste and the hydrogeologic setting. Complete documentation of the problem is necessary in order to develop control systems. Even though a variety of control systems are available for migration protection, buildings should not be constructed on former landfill or other contaminated sites until the process of gas generation has stopped or the source of gases has been removed or corrected.

Emission models are currently under development. These models are being refined using field data and laboratory data collected under actual/controlled conditions. The use of models and sampling equipment will aid investigators in determining the extent of the problem. When collecting samples, the specific purpose of the collection method must be reported. Personal samples as compared to either ambient air samples or to drum headspace samples will yield different concentrations. Thus the type of sampling employed must be stated so that data can be properly compared and evaluated. Also sampling methods (eg as direct reading instruments as compared to laboratory analysis) must be reported.

Future research should be directed to improve and refine air emission models and sampling and analytical techniques. Reliable data should be generated for an accurate assessment of the problem. Control techniques should be monitored for their effectiveness.

17   REFERENCES

1   T T Shen and T J Tofflemire. Air pollution aspects of land disposal of toxic wastes'. J Environ. Eng Div (Am Soc Civ Eng) 106 (EEI): 211-26 (1980).

2   C L Stratton, K L Tuttle and J M Allan. 'Environmental Assessment of Polychlorinated Biphenyls (PCBs) near New Bedford, MA, Municipal Landfill'. Final Report, EPA 560/6-78-006, US Environmental Protection Agency (1978).

3   T T Shen. Air quality assessment for land disposal of industrial wastes.' Environ. Management (NY) 6 (4): 297-305 (1982).

4   W J Farmer, M Yang, J Letey and W F Spencer. 'Problems Associated with the Land Disposal of an Inorganic Industrial Hazardous Waste Containing HCB.' EPA 600/9-76-015, US Environmental Protection Agency (1976).

5   R Costello and J R Kominsky. Occupational exposure monitoring workers at hazardous waste sites. In: Hazardous Waste Disposal, Assessing the Problem. J H Highland (Editor). Ann Arbor Science (1980).

6   L J Thibodeaux, C Springer, T Heddon and P Lunney. Chemical volatisation mechanisms from surface impoundments in the absence of wind. in: Proc Eighth Annual Research Symposium 161-173, EPA 600/9-82-002, US Environmental Protection Agency (1982).

7   L J Thibodeaux, C Springer, P Lunney, S C James and T T Shen. Air emission monitoring of hazardous waste sites. In: Proc Conf Manage Uncontrolled Hazardous Waste Sites, Washington DC, 70-75. HMCRI, Silver Springs, Maryland (1982).

8   D Brown, R Craig, M Edwards, N Henderson and T J Thomas. 'Techniques for Handling Landborne Spills of Volatile Hazardous Substances'. EPA 600/2-81-207, US Environmental Protection Agency (1981).

9   L J Thibodeaux. Estimating air emissions of chemicals from hazardous waste landfills. J Hazardous Materials. 4 (3): 235-44 (1981).

10   A T McCord and R D Zeigler. 'Study of the Rate of Emission of Gases Leaving Industrial Solid Waste Landfills.' Calspan Corp, Buffalo, NY (1980).

11   J F Walsh and K H Jones. The air quality impact risk assessment aspects of remedial action planning. In: Proc Conf Manage. Uncontrolled Hazardous Waste Sites, Washington DC, 63-66, HMCRI, Silver Springs, Maryland (1982).

12   T T Shen and G H Sewell. Air Pollution Problems of Uncontrolled Hazardous Waste Sites. Ibidem pp 76-80.

13   T T Shen. Control techniques for gas emissions from hazardous
      waste landfills'. Journ Air Pollut Control Association,
      31:132-135 (1981)

14   J J Gushue, J E Ayres, and A J Snyder. Hazardous waste site
      investigation, Sylvester Site, Nashua, New Hampshire. Proc Conf
      Manage. Uncontrolled Hazardous Waste Sites, Washington DC, 359-
      70, HMCRI, Silver Springs, Maryland (1981).

15   G L McGarry and B L LaMarre. Proposed clean-up of the Gilson
      Road hazardous waste disposal site, Nashua, New Hampshire.
      Proc Conf Manage. Uncontrolled Hazardous Waste Sites, Washington
      DC, 291-294, HMCRI, Silver Springs, Maryland (1982).

16   T F Dalton. Case histories in handling unknown hazardous
      materials at dump site locations. Proc Conf Manage.
      Uncontrolled Hazardous Waste Sites, Washington DC, 371-373,
      HMCRI, Silver Springs, Maryland (1981).

17   E D Pellizzari. 'Quantification of Chlorinated Hydrocarbons
      in Previously Collected Air Samples'. EPA 450/3-78-112, US
      Environment Protection Agency (1978).

18   'Ambient Air Monitoring and Health Risk Assessment for Suspect
      Human Carcinogens around BKK Landfill in West Covina'. (Report)
      California Air Resources Board (Haagen-Smit Lab Div), South
      Coast Air Quality Management District and California Department
      of Health Services, Toxic Substances Control Division (1983).

19   D Gominger and D A Lyon. Comprehensive evaluation of the
      abandoned LiPari landfill. In: ASTM Spec Tech Publ 760,
      (Hazardous Solid Waste Test) 321-8 (1981).

20   S R Cochran, M Kaplan, P Rogoshewski, C Furman and S C James.
      Survey and case study investigation of remedial actions at
      uncontrolled hazardous waste sites. Proc Conf Manage.
      Uncontrolled Hazardous Waste Sites, Washington DC, 131-135,
      HMCRI, Silver Springs, Maryland (1982).

21   J D Tewhey, J E Sevee and R L Fortin. Silresim: A hazardous
      waste case study. Ibidem. 280-283.

22   C L McEnery. Uses and limitations of environmental
      monitoring equipment for assessing worker safety in the
      field investigations of abandoned and uncontrolled hazardous
      waste sites'. Ibidem. 306-310.

23   D Wilson, E Smith and K Pearce. Uncontrolled hazardous waste
      sites - A perspective of the problem in the UK'. Chem Ind
      (London) 18-23 (1981).

24   S Wittmann. Methane production, migration, and abatement, A
      case study, City of Burlington, Wisconsin. 4th Annual Madison
      Conference of Applied Research and Practice of Municipal and
      Industrial Waste, 316-329 (1981).

25   F B Flower. 'Case History of Landfill Gas Movement Through
      Soils'. EPA 600/9-76-004, US Environment Protection Agency,
      Office of Research and Development (1975).

26   S C Kim, R Narang, A Richards, K Aldous, P O'Keefe, D Smith,
      B Bush, J Slack and D Owens. Love Canal: Chemical
      contamination and migration'. Proc Conf, Manage. Uncontrolled

Hazardous Waste Sites, Washington DC, 212–219, HMCRI, Silver Springs, Maryland. (1980).

27   M Singal, J Kominsky, P Schulte and P Landrigan. 'Composite Report-NIOSH, Hyde Park Landfill, Niagara County, NY'. (Dec 1980).

28   A H Brunswasser and P L Spence. Ohio River Park, Allegheny County, Pennsylvania; A case history of local government involvement in the assessment of an abandoned hazardous waste disposal site. Proc Conf Hazardous Waste Manage, 121–59 (1981)

29   I R Stegmann. Landfill gas problems – Summary of West German experience. In: Proc Landfill Gas Symposium, Harwell Laboratory, UK (1981).

30   F B Stroud, B G Burrus, and J M Gilbert. A co-ordinated cleanup of the Old Hardin County Brickyard, West Point, Ky. Proc Conf Manage. Uncontrolled Hazardous Waste Sites, Washington DC. 274–279, HMCRI, Silver Springs, Maryland (1982).

31   R M McOmber, C A Moore, J W Massman and J J Walsh. Verification of gas migration at Lees Lane landfill. In: Proc Eighth Annual Research Symposium, 150–159, EPA 600/9–82–002, US Environmental Protection Agency (1982).

32   B W Muller, A R Brodd and J Leo. Picillo Farm, Coventry, Rhode Island: A superfund and State cleanup case history. Proc Conf Manage. Uncontrolled Hazardous Waste Sites, Washington DC. 268–273, HMCRI, Silver Springs, Maryland (1982).

33   D A Buecker and M L Bradford. Safety and air monitoring consideration at the cleanup of hazardous waste site. Ibidem. 299–305.

34   G A Vanderlann. Addressing citizen health concerns during uncontrolled hazardous waste site cleanup'. Ibidem. 321–325.

35   G F Regan and Y Flynn. 'Ambient Air Monitoring for Volatile Organics Near Midco I, Gary, Indiana'. US Environmental Protection Agency, Region V, Chicago, Ill (1982).

36   RNK Environmental Inc, Personal Observations (1983).

37   J Freed, A Twedell, M Christopher, S Cochran, L Adkins and C Ahnell. Environmental and health hazard investigation of the Shenango disposal site. Proc Conf Manage. Uncontrolled Hazardous Waste Sites, Washington DC, 233–238, HMCRI, Silver Springs, Maryland (1980).

38   W C Blackman, B E Benson, K E Fischer and U Z Hardy. Enforcement and safety procedures for evaluation of hazardous waste disposal sites. Ibidem. 91–106.

39   H K Hatayama. Special sampling techniques used for investigating uncontrolled hazardous waste sites in California. Proc Conf Manage. Uncontrolled Hazardous Waste Sites, Washington 149–153, HMCRI Silver Springs, Maryland, 1981.

40   R W Townsend. Air monitoring of hazardous waste sites. Proc Conf Manage. Uncontrolled Hazardous Waste Sites. Washington DC. 67–69, HMCRI, Silver Springs, Maryland (1982).

41 R J Costello, B Fronberg and J Belius. 'Health Hazard
    Evaluation Report HETA-80-232-1055, Allied Chemical, Baton
    Rouge, Louisiana', and 'HETA 81-037-1055, Rollins Environmental
    Services, Baton Rouge, Louisiana'. National Institute for
    Occupational Safety and Health, Cincinnati, Ohio.
    (1982).

42 G A Allcott, R Vandervort and J V Messick. Practical
    considerations for the protection of personnel during the
    gathering, transportation, storage and analysis of samples from
    hazardous waste sites'. Proc Conf Manage. Uncontrolled
    Hazardous Waste Sites, Washington DC, 263-268, HMCRI, Silver
    Springs, Maryland (1981).

43 M L Sproul. Safety considerations at spills and dump
    sites. Proc Control of Hazardous Material Spills Conference,
    Louisville, Ky 1980, 255-258.

44 M S Mathamel. Hazardous substance site ambient air
    characterisation to evaluate entry team safety'. Proc Conf
    Manage. Uncontrolled Hazardous Waste Sites, Washington DC, 280-
    284, HMCRI, Silver Springs, Maryland (1981).

45 R E Montgomery, D P Remeta and M Gruenfeld. 'NATO/CCMS Pilot
    Study on Contaminated Land. Project F: Rapid On-site Methods of
    Chemical Analysis' (USA Report). US Environmental Protection
    Agency, May 1984 (Chapter 10 of this Report).

46 R W Melvold, S C Gibson and M D Royer. Safety procedures for
    hazardous materials cleanup. Proc Conf Manage. Uncontrolled
    Hazardous Waste Sites, Washington DC, 269-276, HMCRI, Silver
    Springs, Maryland (1981).

47 Threshold Limit Values - 1980. Guidance Note EH 15/80 from
    Health and Safety Executive. Adopted by American Conf of
    Governmental Industrial Hygienists (ACGIH).

48 'Handbook for Remedial Actions at Uncontrolled Hazardous
    Waste Sites'. EPA 625/6-82-006, US Environmental Protection
    Agency, Municipal Environment Research Laboratory, Cincinnati,
    OH (1982).

49 K Vershueren. 'Handbook of Environmental Data on Organic
    Chemicals', Van Nostrand, Reinhold Co, New York (1977).

50 J A Riddick and W B Bunger. 'Organic Solvents - Physical
    Properties and Methods of Purification'. 3rd Edition,
    Techniques of Chemistry, Volume II. Wiley Interscience (1970).

51 OSHA Concentration Limits for Gases. Federal Register. Vol 40,
    23072, May 28, 1975.

52 A D Baskin (Edit) 'Handling Guide for Potentially Hazardous
    Materials'. Richard B Cross Co (1975).

53 J Lippitt, J Walsh, A DiPuccio and M Scott. Worker safety and
    degree of hazard considerations on remedial action costs'. Proc
    Conf Manage. Uncontrolled Hazardous Waste Sites, 311-318, HMCRI,
    Silver Springs, Maryland, (1982).

54 G A Gallagher. TAT/FIT Health and safety programs for
    hazardous waste site investigations'. Proc Conf Manage.

Uncontrolled Hazardous Waste Sites, Washington DC, 85-90, HMCRI, Silver Springs, Maryland (1980).

55   T A Kletz.   Some myths on hazardous materials.   J of Hazardous Materials, (2): 1-10, (1977-78).

56   Engineering Science, Inc.   'Landfill Gas Migration Control Systems'.   Arcadia, California (unpublished report 1977).

57   'Procedures for Landfill Gas Monitoring and Control'.   Proc International Seminar, Report EPS 4-EC-77-4, Environmental Impact Control Directorate, Environmental Protection Service, Environment Canada (1977).

58   V P Patel, R L Hoye and R O Toftner.   'Gas and Leachate: Summary'.   EPA 600/9-79-023a, 168-175, US Environmental Protection Agency, Office of Research and Development (1979).

59   A D Astle, R A Duffee and A R Stankunas.   Estimating vapour and odour emission rates from hazardous waste sites.   Proc Conf Manage. Uncontrolled Hazardous Waste Sites, Washington DC, 326-330, HMCRI, Silver Springs, Maryland (1982).

60   C Hagger and P Clay.   Hydrogeological investigation of an uncontrolled hazardous waste site'.   Proc Conf Manage. Uncontrolled Hazardous Waste Sites, Washington DC, 45-51, HMCRI, Silver Springs, Maryland (1981).

61   G J Farquhar.   Evaluation of a pumped gas venting system at a municipal landfill'.   Proc 5th Annual Madison Conf Applied Research and Practice on Municipal and Industrial Waste, September 1982, 36-58 (1982).

62   D Murray.   Case study of leachate and gas management at a sanitary landfill.   Ibidem.   31-35.

63   'Handbook for Remedial Actions at Waste Disposal Sites'. EPA 625/6-82-006, 235-282, US Environmental Protection Agency, MERL, Cincinnati (1982).

64   'Handbook for Remedial Action at Waste Disposal Sites'.   EPA 625/6-82/006, 283-338, US Environmental Protection Agency, MERL, Cincinnati(1982).

65   W J Farmer, M Yang and J Letey.   'Land Disposal of Hexachlorobenzene Waste'.   EPA 600/2-80-119, US Environmental Protection Agency Cincinnati (1980).

66   S S Gross and R H Hiltz.   'Evaluation of Foams for Mitigating Air Pollution from Hazardous Spills.   EPA 600/s2-82-029, US Environmental Protection Agency (1982).

67   J J Walsh, D P Gillespie, H L Rishel and S M Kennedy.   'Cost of Remedial Actions at Uncontrolled Waste Sites'.   EPA 600/9-82-002, 534-549, US Environmental Protection Agency (1982).

68   R J Lofy.   Air injection to control landfill gas migration'. public works, 80-81 (1983).

69   G R Lytwynyshyn, R E Zimmerman, N W Flynn, R Wingender and V Olivieri.   'Landfill Methane Recovery Part II: Gas Characterisation'.   Final Report by Gas Research Institute (GRI-81/0105), GRI (1982).

70   J Ting, T Bierma and J Reed.   Air pollution aspects of gas generation control and recovery at landfills'.   4th Annual

Madison Conference Applied Research and Practice of Municipal
and Industrial Waste, 308-315 (1981).

71   L J Thibodeaux, 'Air Emissions of Volatile Organic Chemicals
      from Landfills: A Pilot-Scale Study', EPA-600/9-84-007, 172-180,
      US Environmental Protection Agency, Cincinnati (1984).

72   S T Hwang and L J Thibodeaux. Measuring volatile chemical
      emission rates from large waste disposal facilities.
      Environmental Progress, 2 (2): 81-86 (1983).

73   C A Moore. Landfill gas generation, migration, and controls.
      CRC, Criteria Rev Environ Control, 9 (2): 157-83 (1979).

74   T T Shen. Estimating hazardous air emissions from disposal
      sites. Pollut Engr 13 (8): 31-40 (1981).

75   B L Murphy. Air modelling and monitoring for site excavation.
      Proc Conf Manage. Uncontrolled Hazardous Waste Sites, Washington
      DC, 331-333, HMCRI, Silver Springs, Maryland (1982).

76   C A Moore. 'Predicting Landfill Gas Movements in Soil and
      Evaluating Control Systems'. EPA 600/9-79-023a, 386-395, US
      Environmental Protection Agency (1979).

77   T Constable, G Farquhar and B Clement. 'Gas Migration and
      Modelling'. EPA 600/9-79-023a, 396-412, US Environmental
      Protection Agency (1979).

78   C A Moore. 'Theoretical Approach to Gas Movement through
      Soils'. EPA 600/9-76-004, 33-43, US Environmental Protection
      Agency (1978).

79   D S Duffala and P B MacRoberts. Implementation of remedial
      actions at abandoned hazardous waste disposal sites. Proc Conf
      Manage. Uncontrolled Hazardous Waste Sites, Washington DC, 289-
      290, HMCRI, Silver Springs, Maryland (1982).

80   S Paige, C Morgan, H Bryson, G Hunt, P Rogoshewski, P Spooner,
      D Twedell and R Wetzel. Preliminary design and cost estimates
      for remedial actions at hazardous waste disposal sites. Proc
      Conf Manage. Uncontrolled Hazardous Waste Sites, Washington DC,
      202-207, HMCRI, Silver Springs, Maryland (1980).

81   H K Hatayama et al. 'A Method for Determining the Compatibility
      of Hazardous Waste'. US Environment Protection Agency (1980).
      (No longer available. An updated version entitled: 'A Proposed
      Guide for Estimating the Incompatability of Selected Hazardous
      Wastes Based on Binary Chemical Reactions' is to be published
      in 1984 by the American Society for Testing and Materials (ASTM)
      D34 Committee on Waste Disposal).

82   R S Ottinger et al. 'Recommended Methods of Reduction,
      Neutralisation, Recovery, or Disposal of Hazardous Waste'.
      Volume I, Summary Report, EPA 670/2-73/053a, US Environmental
      Protection Agency (1973).

83   V Franzius et al. 'List of Types of Toxic and Flammable Gases'.
      Prepared by Gerhard Arendt (Battelle Inst), Dr Rainer Stegmann
      (Technical University of Hamburg-Harburg), Dr Volker Franzius
      (Federal Environmental Agency) (1983).

84   R C Weber, P A Parker and M Bowser. 'Vapour Pressure
      Distribution of Selected Organic Chemicals'. EPA 600/S2-81-021
      US Environmental Protection Agency (1981).

85   R S Ottinger et al. 'Recommended Methods of Reduction
      Neutralisation, Recovery, or Disposal of Hazardous Waste. Volume
      II, Toxicological Summary'. EPA 670/2-73-053b, US Environmental
      Protection Agency (1973).

86   G Lawton, J Whiteman, S Phillips and J Mosher. Contribution of
      the industry occupational health program to the treatment of
      personnel exposed during hazardous material spills. Control of
      Hazardous Material Spills Conference, 250-254 (1980).

87   SCS Engineers. 'Monitoring of the Lees Lane landfill - Data
      Summary'. Performed for Jefferson County Dept of Works,
      (February 1979).

88   R A Markle, R B Iden, and F A Sliemers. 'A Preliminary
      Examination of Vinyl Chloride Emissions from Polymerisation
      Sludges during Handling and Land Disposal', 186-194, EPA 600/9-
      76-015, US Environmental Protection Agency (1976).

89   R S Ottinger et al. 'Recommended Methods of Reduction,
      Neutralisation, Recovery, or Disposal of Hazardous Waste'.
      Volume XI, Industrial and Municipal Disposal Candidate Waste
      Stream Constituent Profile Reports-Organic Compounds (Cont). EPA
      670/2-73-053k, US Environmental Protection Agency (1973).

90   L J Thibodeaux. 'Chemodynamics: Environmental Movement of
      Chemicals in Air, Water and Soils'. John Wiley and Sons, New
      York, (1979).

91   J H Perry (Edit). 'Chemical Engineers' Handbook, 5th Edition'.
      McGraw-Hill, New York (1973).

92   W J Farmer, M S Yang and J Letey. Land disposal of hazardous
      wastes: Controlling vapour movement in soils'. Proc 4th Annual
      Research Symposium, 182-190. EPA 600/9-78-016, US Environmental
      Protection Agency, 1978.

93   R J Millington and J P Quirk. Permeability of porous solids.
      Trans Faraday Soc, 57: 1200-1207, (1961).

94   T T Shen. Emission estimation of hazardous organic compounds
      from waste disposal sites. Presented at 73rd Annual Meeting
      Air Pollution Control Association, Montreal, Quebec, June 1980.

95   L J Thibodeaux. Estimating the air emissions of chemicals from
      hazardous waste landfills'. Journal of Hazardous Materials, 4:
      235-244 (1981a).

96   L J Thibodeaux, C Springer and L M Riley. Models of mechanisms
      for the vapour phase emission of hazardous chemicals from
      landfills'. Presented at Symposium Toxic Substances Management,
      Atlanta, GA, March 1981.

97   O G Sutton. Micrometeorology, McGraw-Hill, New York (1953).

98   T T Shen. Estimating hazardous air emissions from disposal
      sites. Pollution Engineering, 13 (8) 31-40 (1981).

99   J H Arnold. Unsteady-state vapourisation and absorption.
      Transaction American Institute Chemical Engineers, 40: 361-379
      (1944).

100 L J Thibodeaux and S T Hwang. Landfarming of petroleum wastes
     – modelling the air emission problem. Environmental Progress,
     Vol I, (1) (1982).

101 D Mackay and A W Wolkoff. Rate of evaporation of low-
     solubility contaminants from water bodies to atmosphere.
     Environmental Science and Technology, 7 (7) (1973).

102 D Mackay and P J Leinonen. Rate of evaporation of low-
     solubility contaminants from water bodies to atmosphere.
     Environmental Science and Technology, 9, (13) (1975).

103 L J Thibodeaux, D G Parker and H H Heck. 'Measurement of
     Volatile Chemical Emissions from Wastewater Basins'. EPA-
     600/S2-82-095, US Environmental Protection Agency, Cincinnati,
     Ohio (1982).

104 T T Shen. Estimation of organic compound emissions from waste
     lagoons. Journal Air Pollution Control Association, 32 (1)
     (1982).

105 J H Smith, D C Bomberger and D L Haynes. Prediction of the
     volatilisation rates of high-volatility chemicals from natural
     water bodies. Environmental Science and Technology, 14, (11)
     (1980).

106 J H Smith, D C Bomberger and D L Haynes. Volatilisation
     rates of intermediate and low volatility chemicals from Water.
     Chemosphere, 10, (3) (1981).

107 A T McCord. A study of the emission rate of volatile
     compounds from lagoons'. Proc Conf Manage. Uncontrolled
     Hazardous Waste Sites, Washington DC, 129–135, HMCRI, Silver
     Springs, Maryland (1981).

108 W L Dilling. Interphase transfer processes II. Evaporation
     rates of chloromethanes, ethanes, ethylenes, propones, and
     propylenes from dilute aqueous solutions. Comparisons with
     Theoretical Predictions. Environmental Science and Technology,
     11, (4) (1977).

109 P S Liss and P G Slater. Flux of gases across the air-sea
     interface. Nature, 247, (1974).

110 Y Cohen, W Cocchio and D Mackay. Laboratory study of liquid
     phase controlled volatilisation rates in presence of wind
     waves. Environmental Science and Technology, 12, (5) (1978).

111 D Mackay and R S Matsugu. Evaporation rates of liquid
     hydrocarbon spills on land and water. Canadian Journal
     Chemical Engineering, 51 (1973).

112 M Owens, R W Edwards and J W Gibbs. Some reaeration studies in
     streams. International Journal Air and Water Pollution, 8
     (1964).

113 J A Danielson. 'Air Pollution Engineering Manual, AP-40'. US
     Environment Protection Agency, Research Triangle Park, NC
     (1973).

114 American Petroleum Institute, Evaporation Loss Committee.
     'Evaporation Loss from External Floating Roof Tanks'. Bulletin
     2517 (revised), API, Washington DC (1980).

115 American Petroleum Institute, 'Evaporation Loss from Fixed Roof
     Tanks. Bulletin 2518, API, Washington DC (1962).
116 'Emission Test Report-Breathing Loss Emissions from Fixed-Roof
     Petrochemical Storage Tanks'. EMB 78-OCM-5, US Environmental
     Protection Agency, Research Triangle Park, NC, (1979).
117 C C Masser. 'Compilation of Air Pollution Emission Factors,
     Supplement 12'. US Environmental Protection Agency, Research
     Triangle Park, NC, (1981).
118 S T Hwang. Treatability and pathways of priority pollutants in
     the biological wastewater treatment'. American Institute
     Chemical Engineers Symposium Series, Water-1980, 77 (209): 316-
     326 (1981).
119 R A Freeman. Stripping of hazardous chemicals from surface
     aerated waste treatment basins. Proc APCA/WPCF Speciality
     Conference on Control of Specific (Toxic) Pollutants,
     Gainesville, Florida, 1979.
120 R A Freeman, J M Schroy, J R Klieve and S R Archer.
     Experimental studies on the rate of air stripping of hazardous
     chemicals from waste treatment systems. 73rd APCA Meeting,
     Montreal, Canada, June 1980.
121 T A Cuscino, C Cowherd and R Bonn. 'Fugitive Emission Control
     of Open Dust Sources'. Proceedings Symposium on Iron and Steel
     Pollution Abatement Technology 1980. EPA 600/9-81-071 US
     Environmental Protection Agency (1981).
122 F B Hill, V P Aneja and R M Felder. A technique for
     measurements of biogenic sulphur emission fluxes. J Environ
     Sci Health, AIB, (3) 199-225 (1978).
123 D F Adams, M R Pack, W L Bamesberger and A E Sherrard.
     Measurements of biogenic sulphur containing gas emissions from
     soils and vegetation. Proc of 71st Annual APCA Meeting,
     Houston, TX, 76-78, 1978.
124 C E Schmidt, W D Balfour and R D Cox. Sampling techniques for
     emissions measurement at hazardous waste sites'. Proc Conf
     Manage. Uncontrolled Hazardous Waste Sites, Washington DC,
     334-339, HMCRI, Silver Springs, Maryland (1982).
125 C Cowherd, K Axetell, C M Guenther and G A Jutze. 'Development
     of Emission Factors for Fugitive Dust Sources'. EPA -450/3-74-
     037, US Environmental Protection Agency (1974).
126 K Winter. 'Safety Criteria to be Used in the Construction of
     Structures on Landfills in View of the Dangers of Gases, German
     (FRG) Federal Ministry of the Interior, Waste Management Branch,
     Research Report, 103-02-102, (1979).
127 W Farino. 'Evalutation and Selection of Models for Estimating
     Air Emissions from Hazardous Waste Treatment, Storage and
     Disposal Facilities,' GCA Corporation, Bedford, Mass (1983).
128 K Rosburg. 'Handbook of Dust Control at Hazardous Waste Sites',
     US Environmental Protection Agency, Solid and Hazardous Waste
     Research Division, Cincinnati, Ohio (1984).

129 J Lippitt, J Walsh, M Scott and A DiPuccio. 'Costs of Remedial
     Actions at Uncontrolled Hazardous Waste Sites: Worker Health
     and Safety Considerations', US Environmental Protection Agency,
     Solid and Hazardous Waste Research Division, Cincinnati, Ohio,
     (1984).

RAPID ON-SITE METHODS OF CHEMICAL ANALYSIS

R E Montgomery*, D P Remeta** and M Gruenfeld**

*   IT Corporation
** US Environmental Protection Agency

1   INTRODUCTION

The analysis of potentially hazardous air, water and soil
samples collected and shipped to service laboratories off-site is
time consuming and expensive. This Chapter addresses the practical
alternative of performing the requisite analytical services on-site.
The most significant application of such methodology is the initial
appraisal of chemical waste sites and identification of highly
contaminated areas. Hence, the majority of the methods reviewed are
selective for classes of compounds, providing semi-quantitative
and/or semi-qualitative information. These methods are useful as
screening techniques for monitoring the general levels of
contamination in a specific area.

The major advantages of field-oriented chemical anlysis methods
are real time sample analyses and reduced costs. Since field methods
are less intricate than conventional methodology, sample throughput
is increased substantially, resulting in cost savings. Costs are
further reduced through the use of portable instruments that are
relatively inexpensive to operate and maintain.

The use of rapid on-site methodology does not preclude
further characterisation of samples for the presence of specific
toxicants. In this regard, field methods may assist in identifying
certain key analytical parameters which are then examined at a
laboratory facility equipped with the appropriate instrumentation.

As a final consideration, there is a delicate balance between
rapid on-site measurements and the need for legal/technical
defensibility. In most instances, rapid on-site techniques cannot

furnish data of sufficient accuracy for litigation purposes and will
have to be augmented by laboratory-based analyses. The use of mobile
analytical laboratories (as described in Section 9) may be a
practical alternative in such situations.

2   SCOPE

This Chapter addresses the problem of monitoring contaminated
land for the presence of hazardous and/or toxic materials. It is
primarily concerned with methods of chemical analysis that can be
readily applied in the field, can provide real time data and be cost
effective. The study has been restricted to a user survey rather
than an extensive literature search. A user survey supplies
information on available techniques and provides recommendations
regarding their proper application.

The chemical analysis methods reviewed in this Chapter are
presently used for emergency response to hazardous material spills
and uncontrolled hazardous waste sites in the United States. Most of
the techniques are interim in nature and are employed in the field by
several components of the US Environmental Protection Agency (US
EPA)* and their private contractors. These rapid chemical analysis
methods vary in complexity, ranging from non-specific semi-
quantitative screening techniques to compound specific, quantitative,
multi-parametric analytical protocols. The techniques are described
briefly and evaluated in the context of adaptation to field
situations. Owing to the limited scope of the survey a large number
of potentially useable techniques that were identified were not
evaluated but are listed in Tables 11.8 to 11.10. This Chapter
therefore provides a compendium of field methods and analytical
techniques, and provides a critique of their practicality and utility
for the rapid on-site assessment of contaminated land.

3   SAMPLING

Evaluation of the type(s) of contaminant(s) and extent of
contamination at a particular site is dependent on both the
availability of appropriate analytical techniques and the sampling
design employed at the site. The sampling design will ultimately
determine the number of samples to be collected and analysed, and the
matrix and/or matrices to be sampled. Sampling designs will
necessarily vary for each site. The extent and degree of contamination may be more obvious at certain sites, allowing a truncated
sampling program. At other sites, contamination may be more diffuse
requiring more extensive sampling and analysis.

---

* Mention of trade names or commercial products does not
constitute endorsement or recommendation for use.

3.1 Sampling Design

A comprehensive survey of a contaminated area involves the
thorough characterisation of the magnitude and distribution of
chemical contamination. However, investigation of sites can rarely
examine the whole site in detail. Therefore, an adequate evaluation
is largely dependent on the sampling pattern selected. Two principal
types of pattern are routinely applied to the investigation of
contaminated sites:

 (i) non-systematic techniques (eg, simple random sampling);
(ii) systematic methods (eg, regular sampling grids).

Grid sampling (as opposed to simple random sampling) generally
produces a more even coverage of the site and thereby provides a more
reliable estimate of the spatial distribution of contaminants over
the area.[1] In both cases, the fixing of sample locations should
enable the position of trial pits, boreholes, sampling sites, etc, to
be easily identified during site investigations and during and after
site cleanup. Sampling should be carefully designed to suit the
particular needs of each site.

The sophistication of the sampling design and collection
method(s) will depend on the size of the impacted area, the extent of
contamination, the matrix being sampled, and the nature of the
suspect chemical contaminant(s). For most sites, the number of
sampling points should be sufficient to define the contaminated area.
The lateral and vertical distributions of contaminants at many sites
are related to the former use(s) of different parts of the site. The
intensity of contamination may vary between different sites of
similar history, but in general their patterns of contamination will
exhibit similarities. Such relationships are likely at industrial
sites with basic designs conforming to established practice. Other
types of sites may be of an intermediate character (eg, scrap yards,
industrial premises). Where such relationships exist and the site
history is known, it may be possible to make use of this information
in the sampling design.

Determination of what constitutes adequate sample numbers is
more difficult. Preliminary estimates of the numbers of sampling
points required can be obtained by the following methods:

 (i) Establishing a systematic sampling plan and adjusting the
      sampling frequency to suit the area to be investigated. This
      method provides a minimum number based on the dimensions of
      the site. On very large sites, this minimum number may mean
      that large areas of the site are not represented by a sampling
      point.

(ii)   Adjusting the sampling density based upon consideration of
       information on the former uses of individual parts of the
       site. Allowance is made for the expected pattern of the
       contamination, which may be non-uniform. In the remaining
       areas of the site the density of sampling may not be
       adequate.

(iii) Specifying the desired probability of finding isolated areas
       of contamination of given size within the site under
       investigation.

A brief discussion of sampling design patterns and determination of adequate sample numbers based on work[1] in the UK is
presented in Appendix H. The US, EPA has published a document giving
sampling procedures, methods, materials, etc as part of a manual on
the specific requirements of waste site and hazardous spill
investigations.[2]

3.2   Sample Matrices

Determination of the distribution of contamination involves both
sample collection and sample analysis. The sophistication of the
sample collection method depends on the matrix or media being sampled
and the substances of interest. The choice of practical analytical
techniques is based on the nature of the class or classes of
compounds being determined rather than on the sample matrix. Rapid
on-site analytical methodology must successfully deal with the
analysis of hazardous substances extracted from a variety of
matrices including air, soil/sediment, and water, as well as
concentrated wastes from chemical storage containers (eg drums,
tanks, lagoons).

Methods for the analysis of aqueous samples are more well
developed than those currently available for analysing air, soil and
chemical wastes.[3] This is partially due to technological advances
in water quality programs among the NATO/ CCMS member nations.
Consequently, small sample volumes are usually sufficient for
analysis. Sampling of water becomes more complicated when insoluble
or immiscible substances are present and form a separate phase or
layer in the water column. In such instances, solvent extraction of
sample composites followed by chromatographic analysis of the
extracts yields rapid qualitative and quantitative information.
Caution should be exercised when sampling well waters to assess
groundwater contamination. Lee and Jones[4] reviewed and recommended
guidelines for sampling groundwater, noting that well water
composition was frequently a function of the extent to which the well
was pumped prior to sampling, the length of time since water was last
withdrawn from the well, and the sampling methods.

The analysis of soil samples is more difficult than analysis of aqueous samples. The heterogeneous nature of sediment and soils makes it difficult to obtain a homogeneous sample. Analysis of soil and sediment samples involves an extraction procedure to bring the contaminant(s) into solution. These extracts often require extensive cleanup steps to remove substances interfering with the analysis. Several rapid chromatographic and spectro-scopic methods are available for the analysis of soil samples for priority pollutants (see Section 9.1).

Analysis of chemical storage container contents on-site is especially important to facilitate segregation of potentially reactive and incompatible hazardous wastes. Compatibility testing of uncharacterised wastes is an integral component of uncontrolled hazardous waste site cleanup protocols. Section 4 describes several approaches for characterising concentrated wastes by using a chemical compatibility classification scheme. These field tests provide a rapid qualitative and semi-quantitative profile of the chemical waste(s) present. A more quantitative method for analysis of chemically compatible wastes involves sample compositing, and characterisation via multi-parametric analytical instrumentation provided by mobile laboratories operating on-site. The nature and applicability of such methodology is reviewed in Section 9.

Air monitoring differs substantially from the sampling and analysis of soil, water, and/or concentrated waste. A primary concern at spills and contaminated sites is the collection of hazardous and potentially toxic vapours. Owing to the variability in the physico-chemical properties (ie vapour pressure, fugacity, diffusion, chemisorption, etc) of hazardous substances and the open atmosphere, air monitoring requires an array of sampling and analysis techniques. Highly volatile materials are usually detected through the use of organic vapour analysers and colorimetric detector tubes. Low-volatility substances often require composite sampling followed by desorption from the stationary phase into a gas analyser. Section 5 describes air monitoring techniques that are commercially available and routinely applied at contaminated sites (reference should also be made to Chapter 10).

3.3 Quality assurance

Quality assurance (QA) defines the acceptance limits of measurement and monitoring data in terms of sensitivity, reproducibility, detection limits, and accuracy. Quality assurance/quality control (QA/QC) protocols are employed to demonstrate the reliability of sample analysis data and to assure that such measurements are performed in strict compliance with all applicable QA requirements.[6]

Many of the rapid on-site analytical methods reviewed in this Chapter provide either qualitative or semi-quantitative information. Therefore, normal stringent quality assurance procedures may not always be applicable. Prior to use in the field, kits and field instruments should be calibrated to determine their reproducibility and stability under defined conditions using prepared single or mixed standards. Calibration standards may also be prepared for field use, and when warranted, calibration samples can be further evaluated from simulated sample matrices. Sections 5-8 of this Chapter describe field monitoring techniques and provide QA specifications for these instruments.

A routine QA program should monitor the age of the reagents used.[36] In some rapid analytical techniques, reagents are labile or have a limited shelf life (eg hygroscopic or biological reagents). The purity of reagents (including distilled water) should always be monitored to prevent introduction of contaminants and trace impurities during sample preparation. Many analytical procedures are subject to interference from substances which may be present in chemical waste samples. Whenever interference is encountered or suspected, the analyst must eliminate the interference through use of an appropriate sample cleanup technique. A qualitative estimate of the presence or absence of interfering substances in a particular determination may be made by means of a recovery procedure that involves applying the analytical method to:

(i)   a reagent blank;

(ii)   a series of known standards covering the expected range of
        concentration of the sample;

(iii)   the sample itself (analysed in duplicate); and

(iv)   the recovery samples (prepared by adding known quantities of
        the substance analysed to split portions of the sample).

Representative elements of a QA program for on-site monitoring of contaminated land are presented below. This list may be added to or modified in accordance with the specific analytical requirements of each particular site.

(i)   Sampling - Adequate records of sampling and sample numbers
        should be maintained. Sampling techniques should be
        compatible with the analytical technique(s). On occasion, it
        may be advisable to freeze duplicate 'archive' samples.

(ii)   Sample preparation - Is the sample homogeneous or
        heterogeneous? Is the sample representative? Will the sample
        preparation technique alter the chemical species to be
        analysed?

(iii)   Instruments and techniques - When appropriate, calibrate the
        procedure (for procedures in constant use, calibration should
        be performed routinely at defined intervals). When
        appropriate, determine the reproducibility of the measure-ment
        (ie use standards or split subsamples).

        The Oil and Hazardous Materials Spills (OHMS) Branch of the US
EPA has developed an on-site QA/QC program for use aboard mobile
laboratories which is designed to permit rapid dissemination of
analytical results. Several protocols of the QA/QC program are
designed to generate validation data concurrently with sample
analysis activities. These protocols provide sample analysis data of
known quality for establishing specific QC guidelines. The
simultaneous reporting of sample analysis and QA validation data
furnishes analytical results in 'real time'. It is then possible to
determine the need and practicality of repeated or supplementary
analyses. This type of QA management system permits the immediate
release of sample analysis results to those primarily responsible for
implementing appropriate remedial techniques during environmental
cleanup operations.[7]

4   COMPATIBILITY TESTING

        The myriad of chemicals in complex matrices commonly found at
hazardous waste or other contaminated sites exclude most conventional
methods used to characterise samples. Comprehensive identification
and quantification of these materials is not feasible owing to the
cost and time required for such analyses. However, owing to the
nature of the hazards involved in sampling and analysing chemical
wastes, it is important to ensure the safety of personnel involved in
the monitoring and cleanup of contaminated sites. Therefore, rapid
on-site compatibility tests are used to enable the safe, effective
and economic use of available manpower and equipment.

4.1 Description

        Compatibility tests are used throughout a contaminated site
management program to facilitate:

  (i)   the determination of safe proximate storage of wastes on
        site;

  (ii) compositing of samples for reduced comprehensive
        analyses;

 (iii) compositing of wastes for efficient disposal and removal
        from the site;

(iv)   the determination of safe transportation of hazardous
       materials; and

(v)   the determination (and often definition) of the ultimate
       disposal method.

Perhaps most importantly, compatibility tests decrease the possibility of creating a greater environmental hazard by indiscriminate
mixing of incompatible wastes.

     Determining the compatibility of waste materials is not limited
to chemical analyses. In a survey[8] of emergency response workers,
43% stated that no instrumentation was used to identify the hazardous
material. Indirect methods, such as reading placards or shipping
papers or calling CHEMTREC*, were used. Once the materials are
sufficiently identified, the compatibility of the specific compounds
needs to be determined. Hatayama et al[9] described a systematic
method for determining the compatibility of wastes. Compounds are
assigned to specific groups through extensive reference tables, and
the compatibility of binary combinations is predicted using a
hazardous wastes compatibility chart. In contrast, Turpin et al[10]
and Mayhew et al[11] offer compatibility schemes for mixtures of
wastes rather than specific compounds within a mixture (see Figures
11.1-11.4 and Tables 11.1 and 11.2). Although the Turpin[10] and
Mayhew[11] methods are less precise, they are rapid and better suited
for on-site use. These compatibility tests determine which wastes can
be stored, combined and/or shipped together for more efficient
removal, and the types of ultimate disposal options available.

     Compatibility testing is multifaceted and is used in many phases
of contaminated land remediation. This section focuses on specific
on-site analytical methods that comprise the backbone of
compatibility testing.

4.2 Selected Methods

     The US EPA has numerous federal, regional and contractor support
groups to assist in the characterisation of hazardous waste sites.
One of these, the Environmental Response Team (ERT), was established

---

*   CHEMTREC is an acronym for the Chemical Transportation
Emergency Center, a 24-hour toll-free hotline that provides
technical information regarding emergency evaluation and response
to spills of chemicals and hazardous substances. CHEMTREC
maintains a directory of industry experts and co-operatives that
can be contacted to provide technical advice and/or on-scene
assistance.[12]

Table 11.1   Turpin[10] Compatibility Tests

Test	Category
1 pH*	Caustic (NF)
	Caustic (F)
	Acid (NF)
	Acid (F)
2 Water Reactive	
3 Oxidisation/Reduction	Oxidiser (F)
	Oxidiser (NF)
4 Radioactives	
5 Volatile vapour/gases	
6 Flammability	

* pH is the level at which the release of cyanide, sulphide and sulphide gases pose a threat
(F) Flammable
(NF) Non-flammable

Table 11.2   Selection Criteria for the Compatibility Field Testing Methods[10]

Category	High pH9	Low pH9*	HNU or OVA on vapour space	Flammability	Redox	R/A	B/T
A–Caustic (NF)	+	−	−	−	−	−	−
B–Caustic (F)	+	−	+	+	−	−	+
C–Acid (NF)	−	+	−	−	−	−	−
D–Acid (F)	−	+	+	+	−	−	+
E–Water Reactive	−	−	−	−	−	−	+
F–Oxidisers (F)	−	−	+	+	+	−	−
G–Oxidisers (NF)	−	−	−	−	+	−	−
H–Radioactives	−	−	−	−	−	+	−

* At pH9 the release of cyanide, sulphide and sulphide gases pose a threat
(F)   Flammable   R/A Radioactivity
(NF)   Non-flammable   B/T Beaker test for water reactivity

in 1978 to provide technical assistance to multimedia environmental problems. From 1979 to 1982, this group responded to more
than 170 environmental emergencies and hazardous waste sites in the
USA.[13] Turpin et al[10] have described six rapid on-site
compatibility tests (see Table 11.1) that are routinely employed to
assign waste drums into one of the eight categories. Most of the
methods used by Turpin et al[10] are described in the following
sections. These methods are rapid, inexpensive and based on user
experience. Table 11.3 lists these and additional compatibility
methods often used on site. These methods vary greatly in expense
and complexity, ranging from gas chromato-graphic analysis of PCBs to
measuring temperature changes with a thermometer. The specific
character of the hazardous waste site will determine which are the
most appropriate in any particular case. Some of the most rapid and
essential compatibility tests are reviewed below.

4.2.1 Radiation. A test for radiation is one of the first
tests performed to ensure worker safety and to assist in the
determination of ultimate disposal options. In the compatibility
schemes reviewed (eg Figures 11.1-11.3), waste is initially scanned
for radioactivity and isolated from the mainstream if necessary.
Normal background gamma radiation is approximately 0.01 to 0.02
milliroentgen per hour (mR/hr) on a gamma survey instrument, and
employeee exposure should not exceed 10 mR/hr without the advice of a
health physicist.[10]

4.2.2 Flammability. Although a 'Seta^{TM}' flash closed cup
flame tester allows an accurate determination of the degree of
flammability, it is an impractical technique for monitoring large
numbers of samples. Turpin et al[10] have described a rapid method
of determining flammability in which a beaker with 2-5 millilitres of
sample is placed in a sandbox and a propane torch is slowly passed
over the sample. The liquid is considered flammable if a flame is
observed, and non-flammable if none is observed after several passes
of the torch.

4.2.3 pH. The pH of a waste solution affects corrosivity and
can initiate the release of hydrogen sulphide, hydrogen cyanide gas,
ammonia gas or amine vapours. Colorimetric and electrochemical
methods have been used to determine pH, but these   methods are
time consuming and inaccurate with dirty samples. The method selected
by Turpin et al[10] uses a multiband pH paper strip which contains a
reaction zone for colour changes and a zone of reference colours.
Any excess material on the strip is removed by wiping on a clean
surface to reduce interferences from viscous and coloured samples.

4.2.4 Water reactivity. This rapid qualitative test provides a
great deal of information. Small volumes of liquid are added
successively to distilled water in a heavy walled glass test tube.
One can then monitor whether heat is generated, if a gas or a

Table 11.3 Compatibility Tests

	Method	Type of Test	Deliverables
1	Water reactivity and solubility	qualitative	T,G,P miscibility, density (1, 1)
2	Karl Fischer analysis for water	semi-qualitative	% water
3	pH	semi-quantitative	pH
4	Cyanide	qualitative	+/- cyanide
5	Sulphide	qualitative	+/- sulphide
6	Organic halogen	semi-qualitative	% halides
7	PCBs by gas chromatography	semi-quantitative	50 mg/l, 50 mg/l, or ND
8	Compatibility of liquid wastes	qualitative	T,G,P miscibility, order of addition
9	Oxidising agents	semi-quantitative	active oxygen (mg/l)
10	Beilstein's test for halogens	qualitative	+/- halogens
11	Flammability	qualitative	+/- flammability
12	Radiation	quantitative	radioactivity
13	Volatiles	semi-quantitative	see HNU or OVA sections
14	Explosion potential	semi-quantitative	degree of explosion potential

Key: + = present   % = w/v percent
     - = absent   P = precipitate
     ND = not detected G = gas formation
                        T = temperature

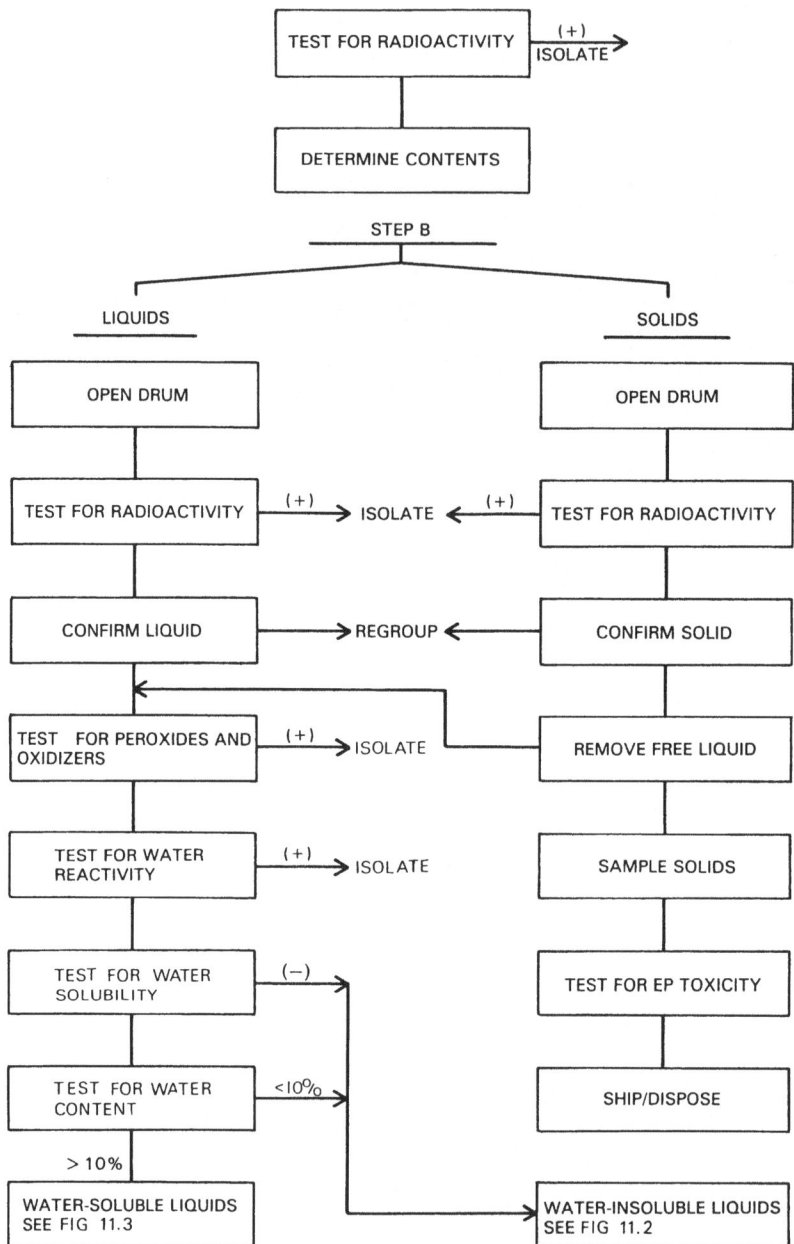

Figure 11.1   Compatibility Testing – Liquid/Solid
Decision Tree (after Mayhew et al[11])

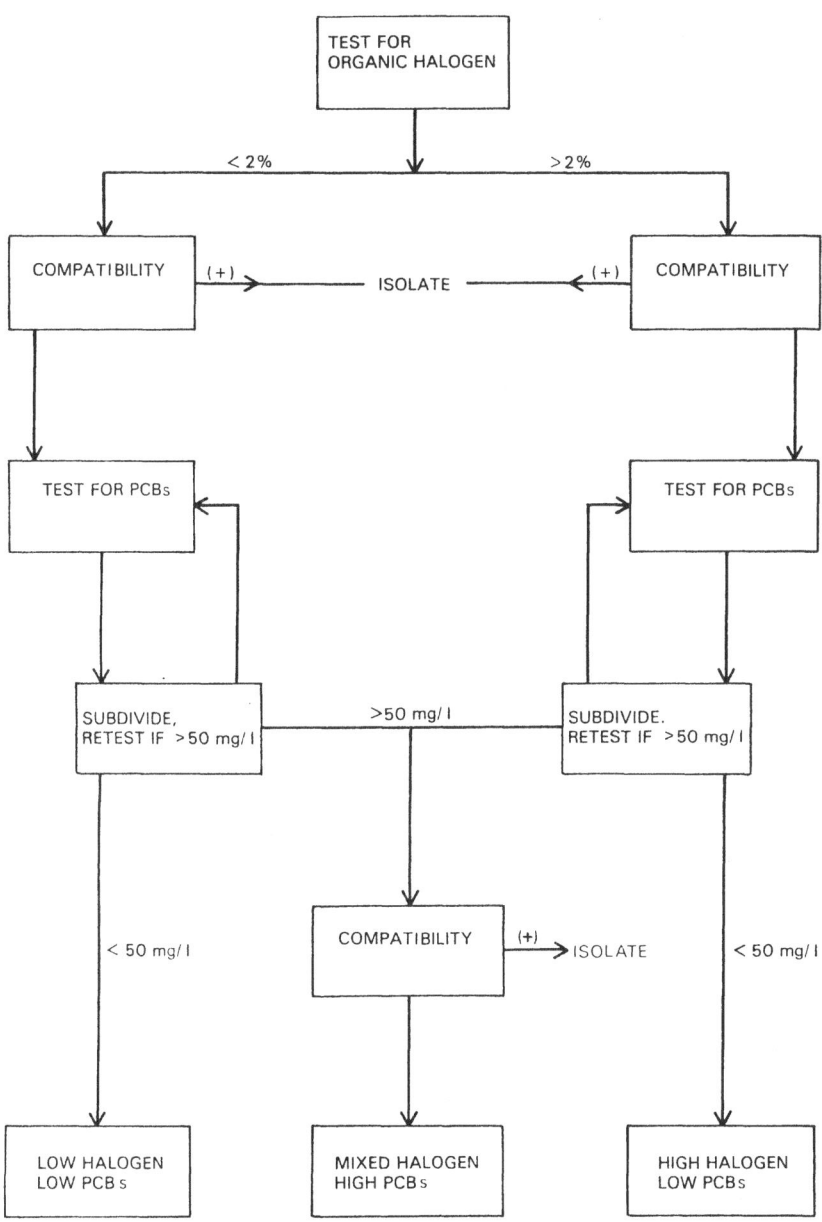

Figure 11.2   Compatibility Testing – Water Insoluble Liquids
Decision Tree (after Mayhew et al[11])

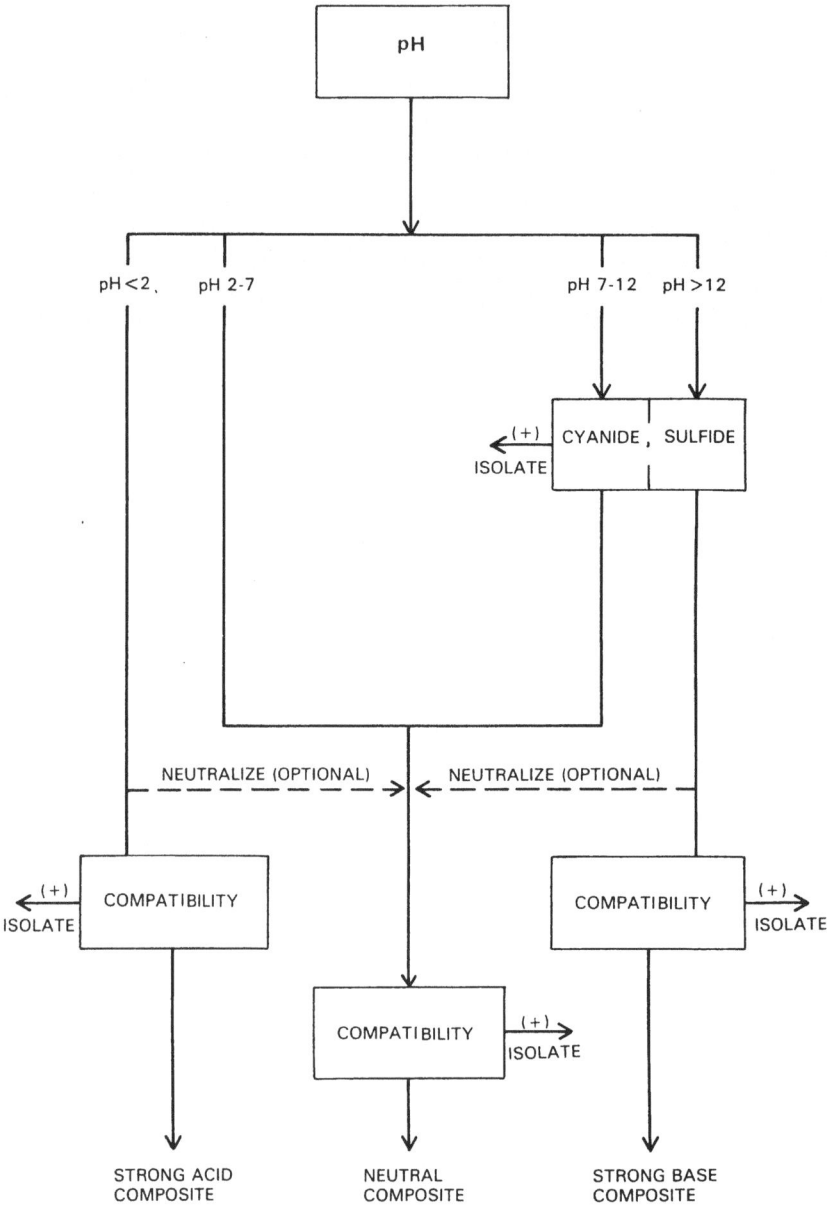

Figure 11.3   Compatibility Testing – Soluble Liquids
Decision Tree (after Mayhew et al[11])

precipitate forms, if the waste is miscible with water, and the
density of the waste (greater than or less than 1 gm/ml).

4.2.5 Beilstein's test. The combination of sensitivity and
simplicity makes this a suitable scanning method for halogens in
waste. A loop of copper wire is heated in a Bunsen burner until the
flame is no longer coloured. The wire is cooled, dipped into the
waste, and then heated in the Bunsen burner (a green flame indicates
the presence of halogens). One drawback to this test is that
volatile halogenated compounds may evaporate before the wire is
sufficiently heated, causing an erroneous result. The sensitivity of
the method is another limitation in that it is susceptible to
interferences when 'dirty' samples are analysed.

4.2.6 Small batch sampling. This is probably one of the
most practical and informative compatibility tests performed before
drums are combined for disposal. Five to ten drum samples of 10 to
15 millilitres each are combined and the order of addition recorded.
After each sample is added, the change in temperature, any precipitation or gas formation, the viscosity, and the miscibility are
noted. This batch sample may then be analysed by gas chromatography
for specific compounds, such as polychlorinated biphenyls.

4.2.7 Oxidation – reduction potentials. The hazards associated with the indiscriminate mixing of oxidisers and reducers make
this an important initial test in characterising chemical wastes. In
a joint venture with a private laboratory the US EPA recently
developed a portable REDOX Field Test Kit for measuring oxidation-
reduction potentials of chemical wastes.[10,36] The kit consists of
a battery operated pH meter capable of monitoring EMF changes, a
platinum sensing combination electrode with a silver–silver chloride
reference electrode, a test solution of 0.001 N ferrous ammonium
sulphate, and a test solution of 0.001 N potassium chromate. A
sample of the chemical waste is added to each of the test solutions.
A change of 50 mV in the positive direction from the standard reading
of 380 mV for the ferrous ammonium sulphate solution, indicates the
presence of oxidisers. A change of 50 mV in the negative direction
from the standard reading of 630 mV for the potassium chromate
solution indicates that reducing species are present.

4.3 Application

Compatibility tests have only recently become widely used. The
need for comprehensive compatibility tests was demonstrated at the
Chemical Control Corporation waste disposal site in Elizabeth, New
Jersey (USA). In the midst of the cleanup operation 40 000 drums of
hazardous waste ignited and exploded during a massive chemical fire.

Table 11.4   Waste Group Colour Codes and Corresponding
Waste Properties (from Rittaler[14])

Colour Code	Waste Properties
White	pH 7, non-flammable
Red	pH 7, torch test properties
Red/orange	pH 7, Seta flash positive
Blue	pH 7, non-flammable
Yellow	pH 7, torch test positive
Yellow/green	pH 7, Seta flash positive
Orange	Water reactive
Green	Oxidiser, torch test positive
Blue/green	Oxidiser, Seta flash positive
Blue/white	Oxidiser, non-flammable
Red/white	Radioactive

Across the United States, environmental service organisations
have been implementing a variety of compatibility tests. Since each
site has its unique problems of sample collection, weather conditions, ultimate disposal, and financing, a variety of compatibility
schemes have been developed.   Three compatibility schemes are
described below.

4.3.1   Callahan Uncontrolled Hazardous Waste Site, St Louis,
Missouri.   In 1980, the Missouri Department of Natural Resources, US
EPA Region VII, and Ecology and Environment Inc undertook a
preliminary investigation of the Callahan abandoned hazardous waste
site.   Monitoring revealed incompatibility and safety tests were
performed on-site as drums were excavated.   The tests made included
measurements for radiation, hydrogen cyanide and organic vapours (HNU
Hazardous Waste Sector – see 5.2).   Samples were analysed for water
reactivity, pH, Redox potential and flammability by the open torch
and Seta[TM] flash closed cup methods.   The results of these analyses
determined the category of the waste drums (see Table 11.4).

4.3.2   Picillo Farm, Coventry, Rhode Island.   In 1978, the Rhode
Island Department of Environmental Management initiated cleanup
operations at this chemical dump site where solid, sludge and liquid
wastes were discovered.[15]   Samples from the drums were transferred
to a laboratory where any fuming or odour was noted and the liquid
and sludge samples were tested for pH.   Those with a pH less than 3
were considered acids, whereas those with a pH greater than 12 were
considered basic.   Following these tests, the flammability
characteristics of the wastes were determined using a portable flash
point tester.   If the flash point was less than 60°C, the waste was

Table 11.5   Analytical Requirements for Disposal Applied at
             Picollo Farm (from Muller et al[15])

---

1   Flammability
2   pH
3   Specific gravity
4   PCB analysis
5   Thermal content (BTU/16)
6   Physical state at 70°F
7   Phases (layering in liquids)
8   Solids (%)
9   Hydrocarbon composition
10 Pesticide analysis
11 Sulphur content
12 Phenols
13 Oil and grease (%)
14 Water (%)
15 Viscosity
16 Organochlorine-percentage
17 Metals analysis
    (a) Liquids were analysed for soluble metals
    (b) Solids were extracted according to the EPA
         Toxicant Extraction Procedure (14 hr) which
         shows leachate metals
    (c) Both liquid and solids were checked for
         concentration of the following metals:

         Arsenic   Mercury
         Barium   Nickel
         Cadmium   Selenium
         Chromium   Silver
         Copper   Zinc
         Lead
18 Both free and total cyanide content were checked
19 Solids were checked for solubility in water,
    sulphuric acid and dimethyl sulphoxide

---

considered flammable. Thereafter, a small amount of sample was added
to distilled water to measure the water reactivity properties of the
waste.

The samples were further examined for specific gravity and
assigned to one of three groups with a specific gravity of: 0.9, 0.9
to 1.1, 1.1. Following the specific gravity test, all samples with
a pH greater than 9 were checked for cyanide, and acids were tested
for oxidation/reduction potentials to determine compatible groups

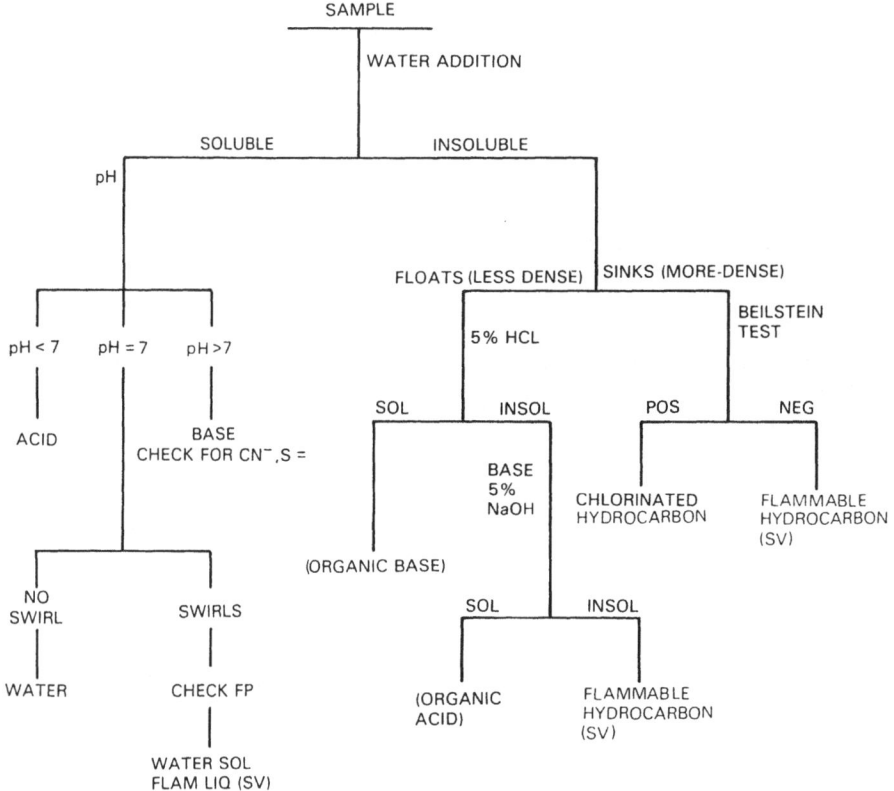

CODES USED FOR MARKING DRUMS:

1. ACID (pH =   )   7.   FLAMMABLE LIQUID (SV)
2. BASE CN FREE (pH =   )   8.   ORGANIC ACID
3. BASE DANGER CN   9.   CHLORINATED HYDROCARBON
4. WATER (pH =   )   10. SOLID (MARK ON PART OF DRUM)
5. WATER SOL SV   11. OILS
6. ORGANIC BASE

NOTE: OILS SEGREGATED AND CHECKED FOR PCB'S INDEPENDENTLY. EACH SOLVENT
COMPOSITE (100 DRUM LOTS) CHECKED FOR PCB'S

Figure 11.4   Analytical Flow Chart Applied to Old Hardin County
             Brickyard, Kentucky (after Stroud et al[38])

for temporary storage of drummed wastes. Prior to disposal, the
contents of 60 to 100 barrels determined to be compatible were
composited, and batch samples from every five barrels were analysed
for polychlorinated biphenyls. This was the most time consuming and
costly analytical task in the compatability testing scheme employed.
Analyses performed for this waste site are listed in Table 11.5.

   4.3.3   Old Hardin County Brickyard Hazardous Waste Site, West
Point, Kentucky.   Investigations started in 1971 by the Kentucky
Department for Environmental Protection revealed approximately 3000
55-gallon (US) drums of liquid, sludge and solid wastes.   In March
1982, cleanup operations commenced with on-scene personnel employing
the analytical flow chart illustrated in Figure 11.4 to segregate the
drums.   The drums were assigned to one of eleven categories, and
disposed of accordingly (eg waste-water treatment, solvent
reclamation, incineration, landfilling, approved hazardous waste
disposal facilities).   When drums were labelled radioactive or
explosive, experts were called upon to assist in hazardous
assessment.

5   AIR MONITORING EQUIPMENT

   Monitoring techniques and devices have been developed that
survey ambient air, generating both qualitative and quantitative
data.

   The primary objectives of an air monitoring program are:

  (i)   to determine the type and approximate quantity of airborne
         contaminants both on-site and in the vicinity of the site;

  (ii)   to establish what level of personal protection should be used
         by on-site workers;

  (iii) to establish whether additional safety measures should be
         initiated for protection of the public, such as an evacuation
         plan; and,

  (iv)   to provide a record of conditions that prevail during the
         course of any cleanup activities.

   Air monitoring methods or techniques that are routinely and
widely used comprise three basic categories:

  (i)   methods that provide immediate results based on direct read-
         out from an analytical instrument or other indicator source;

  (ii)   methods that allow for the collection of an air sample which
         must be analysed; and,

  (iii) passive detection devices that employ a chemical substance
         that reacts with the pollutant and changes its properties in
         some fashion.[16]

   The following section reviews those techniques that have been
evaluated and are currently in use by the US EPA.

5.1 Organic Vapour Analyser

5.1.1 Description. The Model 128 Century Organic Vapour
Analyser (OVA) is a lightweight (5.5 kg) shoulder-borne gas
chromatograph (GC) manufactured by the Foxboro Corporation of
Burlington, Massachusetts. The OVA-GC is capable of monitoring
volatile organic compounds in the air. It is equipped with a flame
ionisation detector (FID) and uses an externally mounted GC column
that enables separation of organic components under ambient
temperature conditions. A portable temperature regulating accessory
pack can be used for operating the unit isothermally. The GC column
can be interchanged, permitting the use of different column lengths
(225 mm, 300 mm, 600 mm) and packing materials.

The instrument can be used in either of two modes of
operation, the total organic mode or the GC mode. In the total
organic mode, a 'sniffer' device is used consisting of a hand-held
output meter attached to a sample probe. The 'sniffer' device is
interfaced to the main OVA module by an electrical cable. In the
total organic mode, a 250 ul sample is semi-automatically introduced
into the unit via the sample probe. The concentration (in ppm) of
non-methane compounds in the air sample is then registered on the
output meter. The total organic concentration response represents
the sum of the percentage relative response for each organic
constituent in the sample. A portable strip chart recorder is used to
obtain a corresponding print-out of the total organic concentration
response.

The GC mode of operation is used primarily for calibration of
the instrument. This must be done prior to using the OVA in the
total organic mode. In the GC mode, a sample is introduced into the
unit via a gas-tight syringe injection through a septum. The portable
strip chart recorder provides a printout of the response, as peaks
similar to those obtained when using a laboratory scale GC. The
output meter, which is used when the OVA is being operated in the
total organic mode, can also be used in the GC mode along with the
strip chart recorder. As a peak is printed out on the recorder, a
total concentration response will register on the output meter. A
maximum read-out (in ppm) on the output meter corresponds to the apex
of the peak on the recorder.

Hydrogen gas serves as a fuel for detector combustion, and
also as the carrier gas. A hydrogen supply tank is contained within
the OVA-GC, and can be refilled from a master tank. The instrument
can be operated using a battery, or a 120 V AC power supply. A
battery charger is also part of the OVA-GC unit.

5.1.2 Evaluation. As part of an evaluation of the Model 128
OVA-GC conducted by the ERT of the US EPA the reproducibility of the
following three instrument parameters was compared:[17] peak height;

Table 11.6   Reproducibility of Three Instrument Parameters
             for 128 OVA-GC

Parameter	Coefficient of Variation (Cv) Total Organic Mode	GC mode
Peak height	5%	5%
Peak retention time	1-4%	1-3%
Relative retention time	5-10%	5-10%

peak retention time; and, relative retention time. The results of
this evaluation are presented in Table 11.6. The detection limit of
the unit is 0.2 ppm.

The OVA-GC does have limitations. It cannot detect inorganic
compounds and air temperature fluctuations affect the operational
performance[19] in the field. Furthermore when monitoring ambient
air in the total organic vapour mode, accurate readings are not
obtained unless the 'sniffer' device is held or placed within inches
of the contamination area(s).

5.1.3 Application. The OVA-GC has been used extensively
by the ERT for the monitoring of air, soil and water for the extent
of chemical contamination, and for the surveying of chemically
contaminated lagoons and waste chemical drum storage sites for the
identification of potential sampling points. It is used in
preliminary field testing operations to determine safety measures to
be taken and levels of personal protection to be used during on-site
activities. The unit also serves as a method of evaluating the
effectiveness of various cleanup operations.[19]

Abandoned Hazardous Waste Site, West Point, Kentucky. The
ERT provided air monitoring and evaluation assistance at this site
during January 1982. The area of concern was an abandoned hazardous
waste site containing 1000 to 2000 drums of unidentified chemical
waste. The initial stages of the air monitoring program utilised the
Model 128 OVA operating in its total organic mode. A survey of the
area was conducted using the OVA to fix locations from which air
samples were subsequently collected. Air samples were taken from
locations giving OVA total concentration readings ranging from 2 to
10 ppm using monitoring pumps equipped with Tenax thermal desorption
tubes (see Section 5.4.2). Laboratory analysis of these samples
revealed the presence of relatively low concentrations of toluene and
xylenes.

Hauppauge, Long Island, New York. The OVA was used as part of
an air monitoring program instituted by the ERT during November 1982
at an incident involving suspected contamination in a school.
Samples were collected both on-site and at off-site stations
established in the vicinity of the building using portable air
monitoring pumps equipped with sorbent tubes packed with activated
carbon and silica gel. The OVA was used in the total organic mode as
a screening device to detect any 'hot spots' on the site, and to
monitor areas other than the fixed sampling stations. Results of
both the OVA screening and air sampling efforts revealed only low
level contamination.

5.2 Hazardous Waste Detector

5.2.1 Description. The Model PI 101Hazardous Waste Detector is
a portable gas analyser manufactured by HNU Systems, Inc, Newton,
Massachusetts. The instrument was designed to detect the presence of
potentially harmful chemical vapours.

The Model PI 101 operates on the principle of photoionisation.
Gases are drawn into a sample probe, and then into a lamp/ion
chamber. Molecules are then ionised as a result of the absorption of
ultraviolet light. This type of detection system is sensitive to
vapours of both organic and inorganic composition.

The instrument is composed of two modules, the read-out unit and
the sample probe, connected by an electrical cable. The read-out
unit houses a DC rechargeable battery, an AC recharger and power
supply, and an ion chamber. A direct concentration (ppm) linear
scale read-out is located on the face of this module. The sample
probe houses the lamp-ion chamber where photoionisation occurs, and
a pump which transfers the gas sample from the probe to the ion
chamber of the read-out unit. The sample probe also has a 300 mm tip
extension.

The unit is sensitive to a wide range of compounds using three
different sample probes. Each probe is equipped with a lamp having a
different energy level (9.5 eV, 10.2 eV, 11.7 eV). Thus, suspect
compounds are positively identified based on correlation of the known
ionisation potential of the suspect compound through use of the
appropriate probe.

The Model PI 101 is portable, having a total weight of less than
4.1 kg, and requiring no additional pumps or gas cylinders for
operation. The unit can also be powered from 110 V AC, for use in
the stationary mode. An optional portable recorder is available.
The instrument is calibrated on benzene or vinyl chloride by the
manufacturer, but can be calibrated on any compound as specified by
the buyer.

5.2.2 Evaluation. Accuracy of the PI 101 is 96% at the 5 ppm concentration level, while instrument results are reproducible to ± 1%. The instrument is sensitive within the range of 0.1 to 2000 ppm.

The model PI 101 has several limitations in that it cannot be used to monitor for specific gases and vapours and it does not detect the presence of hydrogen, cyanide, or methane gases. The latter may be an advantage, however, when investigating municipal/industrial landfill sites where the methane produced can totally mask the presence of other more toxic vapours when a flame ionisation detector is used. An additional disadvantage is that the operational performance deteriorates in an uncontrolled environment since weather conditions (particularly increases in humidity) may limit the accuracy of results.

5.2.3 Application. Practical applications of the Model PI 101 include a broad range of situations because of the instrument's wide dynamic operating range. They include:

(i)   the detection of chemical leaks;

(ii)   the location of site 'hot spots' and below-ground level waste tanks and barrels;

(iii) the determination of plume spread at hazardous material spill sites;

(iv)   the classification of work site hazard levels as an indication of personal protection requirements for workers at the site;

(v)   screening samples of soil, water, and other materials prior to laboratory analysis.[19]

Niagara Falls, New York. The Model PI 101 Hazardous Waste Analyser was used as an air monitoring device at a landfill site in Niagara Falls, New York. The instrument was used in the portable mode for the detection of 'hot spots' during an aquifer survey of the site perimeter in order to determine necessary safety precautions for on-site workers and to fix specific locations for air sampling.

5.3 Portable Photoionisation Gas Chromatograph

5.3.1 Description. The Photovac Inc 10A10 Portable Photoionisation Gas Chromatograph is designed for on-site monitoring low concentrations of volatile compounds in air.[37] The unit uses a photoionisation detector and air as the carrier gas. The gas chromatographic columns within the instrument operate at ambient temperature. Column lengths can range from 150 mm to 6 m, and

flexible capillary columns can be installed with adapters. Samples
(1 ul to 1 ml) are injected directly without any preconcentration.
Two options are also available for the rapid screening of samples: a
dual column arrangement and a carrier gas backflush valve.

The instrument weighs approximately 11 kg and is self-contained
except for the recorder. The unit operates on rechargeable batteries
or on 115 V AC. The air is supplied by a lecture bottle mounted on
the top of the instrument which can deliver carrier gas for 10 days
in typical operations.

5.3.2 Evaluation. The greater sensitivity of the Photovac
10A10 when compared with the HNU Model PI 101 makes it more
appropriate in situations where volatile compounds in the low ug/l
concentrations are anticipated.

The photoionisation detector responds to a broad range of
organics and inorganics such as phosphine, arsine, ammonia and nitric
oxide. The two available options, dual column operation and
backflush valves, make the 10A10 versatile. The dual column option
allows for more precise identification of compounds when two columns
of different packing materials are used, or for rapid screening and
selective quantification when a long and a short column are used.
The backflush valve routes the sample directly to the detector which
records the total volatile content of the sample.

The instrument's linearity and sensitivity for specific
compounds demonstrates its utility for on-site use. Hydrogen
sulphide is detected at 0.1 ppb, with excellent linearity between 0.1
and 1 ppb and between 1 ppb and 5 ppb. In ambient air samples, where
benzene was detected at approximately 4.5 ppb, the signal to noise
ratio was 45. The coefficient of variation for benzene between 1 ppb
and 1 ppm was less than 4%.

The limitations of the instrument are the power source, the
carrier gas, and the ionisation potential of the detector. The
batteries provide for 4 hours of operation and can be recharged while
operating the instrument on AC. The lecture bottle of carrier gas
lasts for approximately 24 hours of continuous operation. The
photoionisation source uses photons of slightly less than 11 eV and
should not detect molecules with a greater ionisation potential. In
practice however, some molecules with a theoretical ionisation
potential above this value are detected.[20]

The instrument's primary limitation is operation of the gas
chromatographic column at ambient temperature. Most volatiles can be
detected at ambient temperatures. Higher boiling compounds (eg
xylene, trichlorobenzene) can also be detected, but with reduced
sensitivity. Ambient air columns are more prone to baseline drift
and require additional stabilisation when the temperature changes.

This could occur when moving the instrument from a vehicle to the
sampling area, or during daily temperature cycles. If high
sensitivity (ppb) is required, the instrument requires a location
free from temperature fluctuations with approximately 30 minutes to
stabilise. At lower sensitivities (10 ppb to 1 ppm), acceptable
performance can generally be achieved within 5 minutes.

5.3.3 Application. The Photovac 10A10 is most useful for
screening air samples containing compounds in the low ug/l
concentration. Typical situations can include the screening of
volatiles emanating from drinking water downwind from the
contamination source, tracking the front of a contamination plume,
and ambient air monitoring. Since preconcentration is unnecessary,
the instrument is quite portable.

Municipal Landfill. Monitoring of a municipal landfill in a
large metropolitan area (since litigation is pending, the site shall
remain unidentified) has been carried out. As parts of the landfill
were filled, vent pipes were installed to provide venting of gas and
vapours. Preliminary field testing determined the presence of
methane and other volatile substances, including vinyl chloride. The
Photovac 10A10 was used on-site to further determine the extent of
contamination.

Using a gas-tight syringe, grab samples were taken from the
emission stream of several vents. Samples of ambient air in and
around the site were also taken. Several vinyl chloride standards
were used to establish the relative retention times and to determine
concentration. Concentrations of vinyl chloride in the vent
emissions ranged from 10 to 50 ppm. Concentrations in the ambient
air ranged from a few ppb to 100 ppb.

Laboratory analysis of the vent emissions corroborated field
results. Vinyl chloride was not detected by laboratory analysis of
the ambient air samples, although the detection limit for the method
used (adsorption by activated charcoal followed by solvent
desorption) is usually 1 ppm. The Photovac 10A10 therefore provided
a more sensitive means of detecting vinyl chloride in air.

5.4 Air Monitoring Tubes - General

Air monitoring and sample collection can be accomplished using a
system consisting of a portable air sampling pump equipped with one
or a series of air sampling tubes. The pumps are calibrated by
delivering a known volume of air through the sample tube per unit
time. Determinations of which tube types to use are based on the
types and suspected concentrations of chemical substances
anticipated. A series of different sampling tubes is often used to
maximise collection and monitoring efficiency. Two types of tubes

are most commonly used in environmental testing, namely detector
tubes and sorbent tubes.

5.5 Air Monitoring Using Detector Tubes

5.5.1 Description. Detector tubes are glass tubes filled with
a packing material treated with a chemical indicator. The detector
tube is specific for the identification and semi-quantification of a
chemical gas or vapour, depending on the chemical indicator used.
When the sampled air contains that specific substance, a pre-
determined colorimetric change occurs within the tube.[21]

The specific design of a detector tube determines the air
sampling procedure. Some detector tubes are manufactured for use
with a portable air-monitoring pump. The tube is attached to a
calibrated pump, and a specific volume of air is pulled through the
tube per unit time. Other detector tubes are manufactured for use
with a calibrated plunger. Upon pulling the plunger back, a specific
volume of air passes through the detector tube. This type of tube is
manufactured in different sizes to allow for sampling of different
fixed air volumes. Other types of tubes are available as personal
monitoring devices to monitor worker exposure to hazardous
vapours.[16]

Detector tubes are manufactured in the USA by National Draeger,
Inc. of Pittsburgh, Pennsylvania ('Draeger tubes'), Matheson Gas
Products of East Rutherford, New Jersey and Bendix/GASTEC of
Lewisburgh, West Virginia. Detector tubes are not available for all
gas types.

5.5.2 Evaluation. The utility of detector tubes is somewhat
limited by:

 (i)   their low accuracy (approximately ± 25%);
 (ii)   tendency for leaks to occur during pumping of the air sample
        through the tube; and,
(iii)   the need for subjective reading of colour charts for deter-
        mination of semi-quantitative results.

Despite these inherent limitations, detector tubes are easy to
use, inexpensive, and can provide qualitative and semi-quantitative
information. In this respect, they are useful as a supplement to
more sensitive and accurate instrumental analysis techniques.

5.5.3 Application. Detector tubes may be used effectively as
qualitative confirmation of the presence or absence of certain
extremely toxic inorganic gases (ie H_2S, HCN, PH_3, SO_2). They
may also be used on-site to screen the headspace of containers.

Detector tubes are usually more useful however in a confined industrial workplace than at hazardous material sites. Inter- ferences from other gases often present at hazardous incident sites tend to affect detector tube readings,[22] usually resulting in higher concentrations being measured than are actually present.

5.6 Air Monitoring Using Sorbent Tubes

5.6.1 Description. Sorbent tubes are widely used for monitoring ambient air providing both qualitative and quantitative information. To collect a sample, a known volume of air is pumped through the tube. The contaminants in the sample are adsorbed and concentrated on the sorbent material in the sampling tube. The sorbent is then subjected to either thermal desorption or solvent extraction before conclusive analyses are performed. Two companies that manufacture sorbent tubes in the USA are Mine Safety Appliances Company (MSA) of Pittsburgh, Pennsylvania, and Matheson Gas Products.

5.6.2 Evaluation. Specificity in the use of sorbent tubes is dependent upon the sorbent medium with which the tube is filled, of which there are three principal categories:

(i)   porous polymers

(ii) inorganic adsorbents

(iii) activated charcoal and carbon.

The porous polymers most frequently used are Tenax GC, XAD resins, Porapak and Chromosorb series adsorbents. These polymers adsorb a wide-range of compounds, including organophosphorous compounds, some hydrocarbons, organic nitrogen and sulphur compounds, inorganic acids, and vapours of certain pesticides. Once a sample has been collected, thermal desorption is commonly applied to remove these materials from the sorbent. Subsequently, qualitative and quantitative analyses are performed. When a porous polymer is required, Tenax is usually preferred as its high thermal limit (approximately 350°C) permits thermal desorption of the adsorbed compounds.[23] An additional advantage is that Tenax adsorbs organic materials of low molecular weight more efficiently than carbon at equal temperatures.

Problems associated with the use of solid sorbents include the irreversible adsorption of some compounds, limited retention capacity, sampling flow limitations, and atmospheric moisture content. Porapak and Chromosorb series adsorbents are particularly limited by their tendency to absorb water vapour during field use.

Inorganic adsorbents, such as alumina and silica gel, can be used to collect a wide range of contaminants based on their polarity. Silica gel is often used to collect haloacid gases, aromatic amines, and polar substances. However, field use of the inorganic adsorbents is limited by the tendency to retain water vapour, which reduces the number of active sites on the adsorbent medium.

Sorbent tubes containing activated charcoal or carbon will efficiently collect most high molecular weight organic compounds from an air sample, including benzene, carbon tetrachloride, chloroform, and xylenes. Once the sample has been obtained, the sorbent material is extracted with carbon disulphide and the extract is then analysed by an appropriate laboratory method. Activated charcoal and carbon tubes are very practical for use during field activities since their water absorption properties are significantly less than those of other sorbent tubes.

Sorbent tubes can be arranged in series to provide for maximum efficiency in the collection of chemical contaminants from an air sample. A simple arrangement of sorbent tubes might consist of a porous polymer tube containing Tenax followed by an activated carbon tube.

5.6.3 Application. The monitoring of ambient air using air monitoring tubes at contaminated sites frequently indicates the type and extent of chemical contamination present. Air monitoring serves as a preliminary investigation of a site prior to any large-scale investigation or sample collection, and it can be used to test the effectiveness of cleanup measures and remedial measures.

Abandoned Hazardous Waste Dumpsite, Pennsylvania. Air monitoring was performed by the US EPA at an abandoned hazardous waste dumpsite located in Pennsylvania in February 1983. As part of the air monitoring program, samples were collected from eleven on-site sampling stations, as well as from a nearby residential area. Sampling was accomplished using personal air sampling pumps equipped with air sampling tubes. Three different types of sorbent tubes were used; activated carbon tubes to collect organic vapours, silica gel to collect inorganic acids and aromatic amines, and Florisil to collect PCBs.

Chemical analysis of the sorbent tubes yielded qualitative and quantitative information regarding the levels and types of personnel protection required for workers operating at the site. The results also indicated the need for further sample collection at the site.

Abandoned Chemical Company, Texas. The ERT performed air monitoring operations at the site of an abandoned chemical company in Texas during August 1982. The ambient air sampling program was developed specifically to monitor for total organics, organic

solvents, and aromatic amines. Three types of sorbent tubes were
used; activated carbon for organics; silica gel for aromatic amines
and Tenax for the collection of total gas/vapours. Subsequent sample
analyses revealed low level contamination from which recommendations
were made regarding the appropriate protective levels for workers at
the site

5.7   Other Air Monitoring Techniques

Air monitoring equipment in frequent use by the US EPA during
hazardous material incidents and at waste sites was dis-cussed in the
preceding sections. Table 11.7 lists other instrumental monitoring
techniques that may be used for air analyses. The compounds
detected vary for each instrument, as does the instrument's sophistication and cost. Some of these instruments or kits may provide
distinctive or supplemental information for on-site air monitoring.

6   WATER MONITORING EQUIPMENT

This section addresses the analysis of aqueous samples using a
variety of portable field instruments and kits that permit rapid
qualitative and semi-quantitative measurement of hazardous and toxic
materials.

6.1   Toxicity Monitor

6.1.1 Description. The Beckman Microtox System[24] can be
used to evaluate the acute toxicity of aqueous samples. The system
consists of an analyser and chart recorder to measure the
bioluminescence. There are four bottles of Microtox reagents:

(i)   bacteria (lyophilised photobacterium phosphorum);

(ii) reconstitution solution to rehydrate and suspend the bacteria;

(iii) diluent for preparing serial dilutions of the sample(s); and,

(iv) osmotic adjustment solution.

To conduct the test, the bacteria are reconstituted and allowed
to stabilise for fifteen minutes. Ten cuvettes are needed for each
sample: nine for serial dilutions and one for a normalisation blank.
The light emitted from the bacteria is measured before the addition
of the sample, and at defined time intervals thereafter. From this
data, the normalised percentage light decrease is determined and
plotted against the concen-tration on semi-log paper to establish the
concentration at which there is a 50% reduction in the light emitted

Table 11.7 Air Monitoring Techniques*

Instrument/Manufacturer	Deliverables	Other	Reference
Hydrogen Sulphide Analyser kit SKC, Inc, Eighty Four, PA	Hydrogen Sulphide	Tutwiler Principle, starch solution indicator	2
Model 715 Oxygen Monitor Beckman Instruments, Inc	Measures oxygen in gaseous samples or DO in aqueous or non-aqueous solutions	Battery powered system accuracy $\pm$ 1% at given sample temperature $\pm$ 6% for temperature variations	1
Instantaneous Vapour Detector Model 38D Sunshine Instruments, Co. Philadelphia, PA	Mercury vapours, acetone, aniline, benzene, benzyl alcohol, diethyl acetal, illuminating gas, naptha, pyridine, toluene, xylene	Includes batter-powered inverter unit, accuracy of $\pm$ 5% of full scale 3.6 kg	3
Halide Detector GasTech, Mountain View, CA	Compounds containing halogens	Includes 3 calibration charts and a maintenance kit. 5.9 kg requires 120V 50/60 Hz. Accuracy: $\pm$ 5% Precision: $\pm$ 3% 15.9 kg requires 110V, 60 cycle AC Accuracy: $\pm$ 10%	1
Halide Meter Scott Aviation-Davis Instruments, Charlottesville, VA	Halogenated hydrocarbons in air particularly perchloroethylene, trichloroethylene, CCl_4 &	15.9 kg requires 110V, 60 cycle AC Accuracy: $\pm$ 10%	1
Calorimetric Analyser CEA Instruments, Inc	Acrylonitrite, ammonia, bromine, Cl, F, formaldehyde, hydrazine, hydrogenchloride, hydrogen cyanide, hydrogen fluoride, hydrogen sulphide methylhydrazine, nitrogen dioxide, oxides of nitrogen, phosqene, sulphur dioxide		3
Niran-IA Foxboro Analytical, Inc Burlington, MA	Portable dispersive Ir analyser, intended for quantitative analysis of multi-component mixtures	15.9 kg Precision and Accuracy: 2%	1,3
Model 511 Flame Ionisation Gas Chromatograph, Analytical Instrument Development, Inc Avondale, PA	Intended for air pollution and environmental health studies	Separate sample into individual components, concentration determination possible	3,1

Unico PGC-10 Portable Gas Chromatograph National Environmental Instruments, Inc. Warwick, RI	Developed for on-the-spot field analysis of trace gases in the ppm range	22 kg requires 115V, AC-60 Hz power source provided with a recorder, the Unicorder 20, TCD	1,3
Custom made Multi-Parameter Test Kit for Air, LaMotte Chemical, Chestertown, MD	Ammonia, Bromine CO, Cu, cyanide, hydrazine Lead, SO_2, HSO, MO_2	Kits are custom made to buyer's specifications	4
Ultra Gas-U3S SO_2 Analyser Calibrated Instruments, Inc. New York, NY	Sampling and analysis device for measuring SO_2 in air. Needs same type of protection from adverse weather conditions	27 kg requires 100 Watts Detection limit: 0.005 ppm interference by HCl, H_2S, N, H_3, not affected by CO_2	1
Series 9000 Portable Gas Analyser System, Devco Engineering, Inc	Toxic gases or vapours in atmosphere or trace concentrations of contaminants in process streams	115V VAC. 60 Hz Accuracy: 2% at calibration point 5% over range	1
Series 2000 Ecolyzer Portable Co Monitor, Energetic Sciences, Inc, Elmsford, NY	Measurement of CO in ranges from 0 to 3000 ppm	Self-contained rechargeable batteries, 4 kg. Precision and Accuracy: ± 1% full scale	1
Model 03T Ozone Recorder Ozone Research and Equipment Corp. Phoenix, Arizona	Designed for ozone measurement	32 kg 110V, 400 Watts Accuracy: 3% of scale	1
Portable Flame Ionisation Meter Model 11-654 Scott Aviation/ Davis Instruments, Charlottesville, VA	Detects trace hydrocarbons in air, monitoring for toxic concentrations of solvents or process chemicals	13.6 kg range from ppm to volume %	1
Triton Beta-Gas Monitors, 1055B Johnston Laboratories, Inc Cockeysville, MD	Monitors radioactive gases, ie H^3, C^{14}, Kr^{85}, A^{41}, Rn^{222}	10.9 kg Accuracy: ± 10% of full scale	1

References:

1   American Conference of Governmental Industrial Hygenists, 1978.   Air sampling instruments for evaluation of atmospheric contaminants.
2   SKC Catlogue and Guide to Air Sampling Standards, 1981.
3   Cross, Harris, Lachs, and Dillman, 1980.   Instrumentation for detecting hazardous materials.
4   LaMotte Chemical Products Co. Catalogue, 1983.

* This tabular compilation of Air Monitoring Techniques is necessarily restricted to field portable instruments manufactured in the USA although similar kits and devices may be commercially available in other nations.

by the bacteria. This is known as the EC50(t,T), which refers to the effective concentration in terms of time (t) and temperature (T) two important parameters that must be noted when reporting results.

6.1.2 Evaluation. The Microtox System may offer a cost and time efficient alternative to bioassays such as the 96 hour fish toxicity test. The results have a statistical advantage over the bioassays of higher organisms owing to greater standardisation within the population and a sample population size exceeding one million. The relative toxicity data achieved with the Microtox System compares well with rat toxicity (correlation coefficient of 0.90) and fish toxicity (correlation coeffficient of 1.0) studies for certain compounds.

According to Beckman Instruments, Inc,[24] the major limitation of this technique is that it 'cannot fully define the toxicity of a particular substance in relation to other life forms and therefore results should be considered as an indicator of potential hazard'. In addition, coloration in samples requires an additional calculation during analysis. High turbidity can be accounted for by a correction factor if particulates are removed prior to analysis. There is also a 2% variability from geometric differences in sample cuvettes.

6.1.3 Application. The Microtox System was developed for monitoring waste treatment and industrial influent and effluent streams as well as toxic substances in the environment. The technique can also be used to study the synergistic and antagonistic effects of toxic mixtures and to determine the potential hazards of new chemicals. Microtox is not a chemical analysis method, and as such cannot identify or quantify toxicants. Hence, its major application is to determine the relative toxicity of contaminants present in an aqueous sample. In this regard, the Technical Assistance Team of the US EPA has used the Microtox for a variety of sample analyses and has found it useful for screening the relative toxicity of hazardous materials.

Greenville, Mississippi. Ethylene dichloride (EDC) leaked from a storage tank into an oxbow lake of the Mississippi River, settling at the bottom. Pumps were used to remove most of the EDC but the toxicity of the remaining residue required further determination. Samples were collected at various locations and the Microtox responses of the samples were compared to a dose-response calibration curve for EDC. The data indicated there were no signs of toxicity from any residual EDC present.

Stroudsberg, Pennsylvania. A tank containing penta-chlorophenol (PCP) contaminated oil leaked into the surrounding area. Sampling wells were drilled to recover the oil, but there was the possibility that the oil would migrate a nearby waterway. A retaining wall was built to stop further migration of the oil and samples were taken at

Table 11.8   Water Monitoring Techniques*

Intrument/Manufacturer	Deliverables	Other	Reference
Spectronix Mini-20 Bausch and Lombe Rochester, NY	Spectrophotometer w/ wavelength range of 400-800 nm	Battery operated, compact, easy to operate	1
Mini Conductivity Meter Hach Chemical Co. Loveland, CO	3 ranges; 0-100, 0-1000, 0-10000	0.7 kg, includes batteries and carry case	5
Model 16300 Conductivity Meter Hach Chemical Co. Loveland, CO	+ 1% accuracy 5 ranges 0-2, 0-20, 0-200 0-2000, 0-20000	3.9 kg includes battery pack, ac	5
Model DR/1 Portable Colorimeter Hach Chemical Co. Loveland, CO	pH, 50 different water and wastewater tests	Batteries, AC	5
Model DR-EL/4 Portable Laboratory Hach Chemical Co. Loveland, CO	Spectrophotometer, titrator, cartridges		5
Model CYN-2 Cyanide Test Kit Hach Chemical Co. Loveland, CO	0-2.6 ppm free cyanide	Colorimetric test	6
Hazardous Materials Detector Kit Hach Chemical Co. Loveland, CO	pH, heavy metals, benzene, phenol, cyanide, conductivity, turbidity, nitrate, nitrogen, colour, sulphate, phosphate, NH_3, fluoride	qualitative tests of non-specific nature to analyse for classes of	3 7
Mark XV Quality/Microprocessor Martex Instruments, Irvine, CA	temperature, salinity, conductivity, depth, pH, specific ions		2
Water Analysis Laboratory LaMotte Chemical, Chestertown, MD	alkalinity, NH_3, Cl, Br, I, Chromium Copper, iron, F, nitrate, nitrite, phosphate (low), silica, sulphate, sulphide, pH, turbidity, TDS, conductivity, CO_2, chlorine/salinity, DO		4
Luminescence Photometer Analytical Luminescence Lab, Inc San Diego, CA	bacterial biomass in sludge, toxic levels of waste, salt water constituents	1.1 kg maximum 100 samples/hr	2
8500 DO/BOD Meter, L G Nester, Co. Millville, NJ	range of 0-20 ppm, LCD readout, for use in wastewater, BOD, Environmental monitoring applications		2
Custom Made Multi-Parameter Test Kit for Water, LaMotte Chemical, Chestertown, MD	amine, CO_2, cyanide, formaldehyde, heavy metals, iodine, Fe, Pb, phenol	Kits are custom made to buyers' specifications	4
Cam-4, Midwest Research Institute, Kansas City, MO	monitor the levels of organophosphate and carbonate pesticides in water	12V DC maximum of 8 continuous hours 110V AC 13.6 kg	8

References:

1   Snyder, Tonkin and McKissick, 1980. Development of hazardous/toxic wastes analytical screening procedures.
2   Pollution Equipment News, April, 1983.
3   Gross, Harris, Lachs, and Dillman, 1980. Instrumentation for detecting hazardous materials.
4   LaMotte Chemical Products Co. Catalog, 1983.
5   Hach Chemical Co. Products for Analysis, 1980.
6   Hach Chemical Co. Water Test Kits, 1971.
7   US EPA, 1982. Environmental emergency response unit capability.
8   US EPA, 1980. Cam-4, A portable warning device for organophosphate hazardous material spills.

* This tabular compilation of water monitoring techniques is necessarily restricted to field portable instruments manufactured in the USA although similar kits and devices may be commercially available in other nations.

wells known to contain the oil and at various other areas of
suspected leaching. Microtox analysis confirmed the presence of PCP
in known wells and demonstrated that the retaining wall had halted
migration of the oil.

Laskin's Waste Oil Service, Jefferson, Ohio. At a waste
oil service site in Jefferson, Ohio, a number of ponds and lagoons
contained an oil layer laden with polychlorinated biphenyls and
other hazardous materials. A mobile carbon treatment unit was used
to pump the contents of the lagoon through activated charcoal to
remove contaminants, and the Microtox System was used to monitor the
toxicity of the effluent as the cleanup progressed.

6.2 Water Monitoring Techniques

Field test instruments and kits that could be applied to the
analysis of water (and possibly soil) samples are listed in Table
11.8. Although these kits have not been used or evaluated by the US
EPA to date, they may serve to provide additional field monitoring
capabilities. The majority of these kits are designed to measure
specific contaminants in aqueous solution and thereby facilitate
rapid qualitative and semi-quantitative measurements of on-site and
surrounding area water samples. Analysis of surrounding area waters
can indicate any natural background conditions which may possibly
interfere with on-site analyses. It may also be feasible to use
these kits to survey contaminated land by simply analysing aqueous
extracts of oil/sediment. However, these procedures are currently
unverified, and the methods will require further validation.

7   SOIL MONITORING TECHNIQUES

The complex composition of soil with its inherent inorganic and
organic background, and differential ability to sorb and desorb
contaminants, makes soil analyses difficult. There are relatively
few simplified procedures, techniques or kits available that can
analyse contaminated soil directly. Almost all current procedures
incorporate an extraction step (eg Soxhlet or acid digestion) and
sample cleanup. However, by performing a rapid aqueous extraction of
the soil it may be possible in many instances to use kits and
techniques that normally measure contaminants in water. In some
instances, an aqueous extract of soil might be a better indicator of
the relative 'bioavail-ability' of a particular contaminant. If a
contaminant is bound so strongly to the soil that it will not desorb
into water, then it may not be readily available to exert acute toxic
effects. The applicability of a kit or technique to analyse an
aqueous soil extract should be evaluated before actual field use to
verify that background soil constituents do not interfere with the
analysis.

Table 11.9 Soil Monitoring Techniques*

Instrument/Manufacturer	Deliverables	Other	Reference
Custom Made Multi-Parameter Test Kit for Soil, LaMotte Chemical, Chestertown, MD	nitrate, phosphorous, K, Ca-Mg, organic matter, boron, cobalt, Cu, zinc, pH, dissolved salts dissolved solids	Kits are custom made to buyers' specifications	1
Soil Test Laboratory Model 1700 Hach Chemical Co. Loveland, CO	soil nutrient, pH, ammonia, nitrate, K, lime, phosphorous	includes a battery operated photo-electric colorimeter and built in pH meter	2
Soil Extraction Kit Hach Chemical Co. Loveland, CO	used to prepare soil sample extracts for testing with water test kits		2

References:

1   LaMotte Chemical Co. Catalog, 1983.
2   Hach Chemical Co, Products for Analysis, 1982.

* This tabular compilation of Soil Monitoring Techniques is necessarily restricted to field portable instruments manufactured in the USA although similar kits and devices may be commercially available in other nations.

A relatively new instrument which was developed for field use is the portable X-ray fluoresence unit. X-ray fluorescence is a commonly used laboratory technique for determining the composition of samples. An X-ray source irradiates the sample, causing the sample to emit X-rays at various energies characteri-stic of the chemical elements in the sample. A spectrometer measures the energy levels of the X-rays emitted to analyse the sample composition. Several versions of these units are available from Martin-Marietta, Inc and Texas Nuclear Division, Ramsey Engineering Company. The manufact-urers anticipate that these portable instruments should detect elements in concentra-tions as low as 10 ppb. This technique might also be used to detect fixed organochlorine compounds (ie PCBs) and organo-metallics (ie tetraethyl lead, arsenates) in soil samples.

Simple soil tests that yield additional information about soil conditions are also commercially available in the USA. Several soil analysis kits that are relatively inexpensive and simple to use are listed in Table 11.9. These kits yield background information on critical soil parameters which ultimately determine the habitability of the soil in conjunction with the presence or absence of toxic contaminants. In addition, Sections 4, 6, and 9 describe techniques and equipment which may be applicable for soil analyses with appropriate modifications. The most applicable techniques are those that measure contaminants directly in water (ie aqueous soil extracts). Selection of the appropriate technique may also be influenced by the types and quantities of contaminants present and the ability of the technique to adequately resolve the analysis.

8   SPECIALISED SCREENING TECHNIQUES

Sampling techniques, compatibility tests, and air, water and soil monitoring equipment have been discussed above. Other techniques, employed in special situations requiring the screening for particular contaminants, are the subject of this section, which addresses two problems of frequent concern namely, determination of the presence and abundance of (i) phenolic materials and (ii) polychlorinated biphenyls (PCBs).

8.1   Total Phenol Analysis

8.1.1 Description. Aqueous phenolics can be quantified spectro-photometrically using 3-methyl-2-benzothiazolinone hydrazone (MBTH) as a coupling agent. The method described here adapts EPA Method 420.3 (Methods for Chemical Analysis of Water and Wastes) to the requirements of on-site analysis. The major modification to Method 420.3 is exclusion of the distillation process. The distillation is included in Method 420.3 to remove interferences, but with sufficiently clean samples, the distillation step can be omitted.

8.1.2 Evaluation. Although the MBTH spectrophotometric method uses a large number of prepared solutions, it can be a valuable on-site method of analysis. A portable spectro-photometer (such as the Bausch and Lomb spectronic mini 20) and a portable pH meter (or pH paper) can be used. When many samples require testing, an assembly line approach can yield an analysis rate of approximately 10 samples per hour.

An inherent inaccuracy in the method results from quantifying the samples on a phenol calibration curve. Many phenolic wastes contain a variety of phenols which do not have the same colour response to MBTH coupling. When this mixture of phenols is quantified on the phenol calibration curve, the results will represent the minimum concentration of phenolic compounds present. Although it is not possible to differentiate between the various types of phenols, the method is capable of measuring phenolic materials from 50 to 100 ug/l in water. There is a concentration/extraction step with chloroform which can increase the sensitivity of the technique to 2 ug/l.

8.1.3 Application. Most phenols exhibit some toxicity to humans and eleven phenolic compounds are considered priority pollutants by the US EPA. Phenols are often an indication that other pollutants are present. This method can be used to analyse for phenolic compounds in drinking, surface and saline waters, and in domestic and industrial wastes. However, the method is limited to quantifying the minimum concentration of phenolic compounds in a mixture.

Laskin's Waste Oil Service Site, Jefferson, Ohio. The ERT used this method at the Laskin's Waste Oil Service site in Jefferson, Ohio. A number of ponds and lagoons on the site contained an oil layer (laden with polychlorinated biphenyls and other hazardous materials) and a water layer. After the initial characterisation of the water indicated the presence of phenolic compounds, the ERT used the MBTH spectrophotometric method (without distillation or concentration) to monitor the influent and effluent streams of a mobile carbon treatment unit. The analyst was able to quantify ten samples per hour and experienced no significant problems with the on-site methodology.

8.2 Polychlorinated Biphenyl Field Test Kit

8.2.1 Description. The Centec Analytical Services TR 201 is a lightweight, battery operated, fully portable field testing kit designed for semi-quantitative analysis of PCBs in transformer fluid.[25,26] To analyse PCBs, chlorine atoms are chemically stripped from the biphenyl rings and analysed electronically with an ion selective probe. A digital display in millivolts corresponds to the concentration of PCBs present. This analysis is a three step

procedure and averages five minutes per sample. The kit is
calibrated at the start of the analysis and hourly thereafter. It
must also be calibrated each time it is moved to a new location.

8.2.2 Evaluation. Major advantages of the PCB field test kit
are: (i) that it does not require the skills of a chemist to perform
the analysis or to interpret the data, and (ii) it is a cost-
effective technique (ie approximately 90% less expensive than GC-ECD
analysis). In laboratory studies using known Aroclor standards, the
results of the TR 201 and GC-ECD analysis were comparable (ie correlation coefficient of 0.96 at levels above 7 ppm). The company
claims a precision of $\pm$ 5% and accuracy of $\pm$ 10% in the analysis of
PCB-contaminated oil. However, with unknown Aroclor samples, the
accuracy of the method drops to $\pm$ 17% as a results of the use of an
average percentage Aroclor standard for instrument calibration.
Consequently, samples approaching the disposal category limits of 50
and 500 ppm (see below) require quantitative analysis by gas
chromatography.[25]

The kit's major limitation is that it cannot identify or quantify PCBs in soil or water matrices, and the presence of certain
solvents (eg alcohols) negates the chlorine stripping reaction. In
addition, the ion selective probe cannot differentiate between
chlorine from PCBs or from other chlorinated compounds such as
pesticides. The major use for the kit is to analyse oil samples
where the presence of PCBs is expected, and when other chlorinated
compounds are not present to cause interferences. However, the kit
may be used as a screening tool for qualitatively determining the
presence and approximate concentration of chlorinated compounds
contained in drums at hazardous waste sites.

8.2.3 Application. The US EPA has taken regulatory action
to prohibit the resale, distribution, and manufacture of PCBs after
established deadline dates, and has restricted their disposal and
handling. The following criteria were set to categorise the
dielectric fluid in transformers, the primary source of PCB usage:

Greater than 500 ppm PCBs: PCB transformers
50-500 ppm PCBs: PCB contaminated transformers
Less than 50 ppm PCBs: Non-PCB transformers
PCBs not detected: No restrictions

The Centec Model TR 201 PCB Field Kit is useful for the mandated
screening of fluids in transformers requiring repairs and in the
inspection of transformers that are 'critically located' (US Federal
law requires the removal of any transformer categorised as a PCB
transformer or PCB contaminated transformer located near water
supplies, crops, or livestock.) Quantification of PCBs in
transformer fluids is essential in determining the appropriate
disposal methods to comply with federal standards and restrictions,

since any untested transformer is considered to be a PCB transformer
unless proven otherwise. The kit is also useful in the on-site
analysis of transformer content spills.[26] The primary users of the
TR 201 are utility companies and transformer repair companies.

9   MOBILE ANALYTICAL LABORATORIES

     The increasing number of hazardous material spills and the
continuing discovery of abandoned chemical waste disposal and other
contaminated sites requires both rapid and effective methods for on-
site chemical analysis. Immediate analytical data is often needed
to:

  (i)   assess the extent of environmental contamination;

  (ii) make decisions regarding methods of cleanup and disposal;
        and,

  (iii) efficiently monitor ensuing cleanup efforts.

     The use of mobile analytical laboratories effectively
accomplishes this by eliminating the delay associated with the transport of samples to distant laboratories for comprehensive analysis.

     As noted in the preceding sections of this Chapter, the utility
of field test kits and portable instruments is somewhat limited in
that these techniques do not afford a complete qualitative and
quantitative characterisation of the contaminants present in a
particular matrix. Mobile laboratories on the other hand provide
for: (i) versatile capabilities based on the array of instruments
available; (ii) rapid sample turn-around time and cost-effective
analyses; and (iii) accurate qualitative and quantitative
multicomponent analysis.[7]

9.1 EPA Mobile Laboratory

     9.1.1 Description. The US EPA's Oil and Hazardous Materials
Spills (OHMS) Branch and Environmental Response Team (ERT) operate a
mobile laboratory which provides rapid on-site analytical services in
support of multimedia hazardous waste cleanup activities. Housed in
a 35 foot (10.7 m) semitrailer van, the mobile laboratory is equipped
with numerous analytical instruments, including:

  (i)   two computerised gas chromatographs equipped with flame
        ionisation, electron capture, and nitrogen-phosphorus
        detectors;

  (ii) a computerised gas chromatograph/mass spectrometer;

(iii)   an emission spectrometer;

 (iv)   infrared and fluorescence spectrophotometers; and

  (v)   a total organic carbon analyser.

This broad range of analytical capabilities permits accurate chemical
analysis of most organic and inorganic substances encountered at
hazardous waste sites.[27]   For the safety of laboratory personnel,
the mobile laboratory is furnished with glove boxes, a fume hood,
vented solvent locker, explosion-proof refrigerator, fire blanket,
eye wash, and safety shower. The heating and air conditioning system
is designed to provide totally non-recirculated air.

    The mobile laboratory is readily dispatched to any part of the
United States within a few hours of an official request for emergency
analytical services. The laboratory is staffed with two to four
chemists, depending on the type and number of analyses requested.
Contact is generally maintained with technical co-ordinators in the
central laboratory in New Jersey, through the use of telephone lines
and telefacsimile units.[28]

    9.1.2 Evaluation. Specialised rapid and direct methods of
analysis are used in the mobile laboratory that avoid time consuming
and tedious extraction, evaporation, concentration, and chromatographic cleanup steps and enable rapid processing of numerous samples
during field use.

    Sample preparation procedures. Several rapid, cost effective
sample preparation and pretreatment procedures were developed for
use in the mobile laboratory that require a minimum amount of space.
Rapid extraction methods have been developed for a variety of
hazardous materials and petroleum oils in aqueous media. A single
step extraction is used, in which a small amount of solvent is added
directly to the water sample and the mixture is agitated with a
magnetic stirrer for 30 minutes. The layers are permitted to
separate, and a portion of the organic phase is removed for analysis.
The use of a single step extraction procedure also reduces potential
contamination from multi-step extractions.

    Contaminated soil and sediment samples are extracted by using
a single step mechanical shaking technique. The extraction solvent
is added to an ertenmayer flask containing a weighed sediment sample
which is then placed on a gyratory device for approximately 20
minutes. This rapid extraction procedure significantly reduces
sample analysis time when compared with conventional methods
incorporating exhaustive soxhlet and alcoholic-KOH extraction
procedures.

Table 11.10   Selection of Chromatographic Cartridges

Compounds	Recommended Chromatographic Cartridge	Precautions & Limitations
Pesticides	Florisil	
PCBs	Florisil	
		Extracted phenols are derivatized with penta-flourobenzylbromide (PFB) before the extract is
Phenols	Silica gel	cleaned with silica gel.
Phthalate	Florisil or	
Esters	Silica gel	
		N-Nitrosodiphenylamine decomposes in the GC injection port and is detected as diphenylamine. Therefore all dipheny-lamine in the sample must be removed by
N–Nitrosodium		Florisil to prevent
phenylamine		interference.
Organochlorine	Florisil	The extract may be
Pesticides		cleaned with Florisil
Haloethers	Florisil	to prevent pthalate
Chlorinated	Florisil	interference.
Hydrocarbons		
2,3,7,8 – Tetra	Silica Gel	Because of the extreme
chlorodibenzo-	cartridge	toxicity of this
p-dioxin	followed by an	compound the analysis
(TCDD)	Alumina cartridge	method should be
Herbacides	Florisil	reviewed thoroughly
Aflatoxins	Florisil	before it is used.
Mycotoxins	Florisil	
Lipids	Florisil	
Alkaloids	Florisil	
Steroids	Florisil	
	Florisil or	
Benzidine	Silica gel	
Diphenylhydrazine	C18	
Toluene	C18	Pre-rinse cartridge
Nitrosamines	Florisil	with distilled water and elute with methanol.

A sample consolidation procedure is employed when a large number of drummed chemical wastes are analysed. Since complete characterisation of the contents of each drum is neither practical nor cost efficient, compatible waste samples are composited, in order to decrease the number of analyses.[29]

Environmental sample cleanup is facilitated by the use of SEP-PAK[TM] chromatographic cartridges which simplify sample preparation procedures by eliminating the need for column and separatory funnel extractions. Sample extracts are injected onto a cartridge, followed by a known amount of solvent, to selectively isolate (elute or retain) components of interest from interferences. SEP-PAKs[TM] are available containing alumina, C_{18}, silical gel, or Florisil packing material. Table 11.10 provides guidelines for selecting the appropriate sorbent material for a particular class of compounds.

Instrumental analysis procedures. The analytical methodology used aboard the mobile laboratory specifically addresses the analysis of US EPA designated priority pollutants. These compounds comprise a broad range of hazardous materials including pesticides, polychlorinated biphenyls, base neutrals, phenolics, purgeables, and heavy metals. The priority pollutants represent a template of substances potentially encountered at contaminated sites. The methodology employed permits accurate quantification of these materials down to the low part per billion level. Table 11.11 provides a listing of the analytical methodology developed by the OHMS Branch.

The methods used most frequently employ gas chromatographic techniques. The gas chromatograph provides several capabilities for the quantitative determination of organic priority pollutants in air, water, sediments, and the contents of chemical storage tanks and drums. Quantification of sample extracts containing pesticides, polychlorinated biphenyls, and base neutrals is accomplished through the use of a mixed phase column consisting of 1.5% SP-2250 and 1.95% SP-2401 coated on 100/120 mesh Supelcoport using electron capture detection (ECD). Sample extracts containing phenolic compounds are analysed using a column containing 1% SP-1240 DA on 100/120 mesh Supelcoport and a flame ionisation detector. Purgeable priority pollutants are analysed using a purge and trap sampler, a column packed with 1% SP-1000 on 60/80 mesh Carbopack B, and a flame ionisation detector. Automatic samplers on the gas chromatographs permit continuous overnight operation thereby facilitating rapid sample throughout.

The gas chromatographic techniques employed by the mobile laboratory are augmented through the use of computerised gas chromatography/mass spectrometry. This enables separation of complex mixtures into their individual constituents for compound identification. Interpretation of mass spectra is accomplished by internal library searching and matching for priority pollutant

Table 11.11   OHMSB Chemical Analysis Methods and Special
              Techniques (from Gruenfeld et al[30])

A   GAS CHROMATOGRAPHY

1   Rapid quantification of pesticide mixtures in water
2   Rapid quantification of polychlorinated biphenyls in water*
3   Rapid quantification of basic and neutral materials in water
4   Rapid quantification of polychlorinated biphenyls in
    sediments
5   Rapid extraction of oils from sediments for GC profile
    analysis

B   FLUORESCENCE SPECTROSCOPY

1   Rapid quantification of hazardous materials directly in
    water
2   Rapid quantification of oil directly in water
3   Rapid quantification of fluorescing organics in water
4   Rapid quantification of hazardous materials in sediments
5   Rapid quantification of oils in sediments

C   INFRARED SPECTROSCOPY

1   Rapid quantification of water dispersed petroleum oils
2   Rapid quantification of petroleum oils in sediments
3   Silica gel treatment for the separation of hetero-organics
    from hydrocarbons

D   SAMPLE PREPARATION PROCEDURES

1   Weathering small amounts of oil
2   Removal of sulphur from sediment extracts

E   SAMPLE ANALYSIS PROTOCOL

F   QUALITY ASSURANCE/QUALITY CONTROL

1   Chemistry staff quality assurance program
2   Protocol for preparing synethetic samples of hazardous
    materials in water
3   Protocol for preparing synthetic samples of hazardous
    materials and petroleum oils in sediment

---

identification. Files of spectra are maintained on disks to
facilitate retrieval and storage.

    Cost effective analysis of inorganic compounds is performed by
plasma emission spectrometry. Emission spectrometry enables analysis
of over 70 elements in complex or concentrated environ-mental
samples, with reduced sample preparation. A multi-channel optical
cassette containing 20 pre-selected exit slits (including the
priority pollutant metals), provides the capability to analyse for 20
elements simultaneously. After calibration of the instrument, one
sample can be analysed every 2 minutes with the concentration of 20
elements displayed on a printer.

    Synchronous excitation (SE) fluorescence spectroscopy is also
used in the mobile laboratory for the analysis of organic priority
pollutants and petroleum oils. The development[29] of several
analytical methods based upon the SE fluorescence technique has
considerably reduced sampling preparation and analysis time.[30,31,32]
Direct in-situ quantification of hazardous materials in aquous media
enables detection of fluorescing organics at the part-per-billion
level.[33]

    Infrared spectroscopy is almost exclusively employed for the
identification and quantification of petroleum oils extracted from
various matrices. Several methods have been developed to rapidly
extract and quantify petroleum oils in water and sediments.[34,35]
An unusually time efficient method for analysing petroleum oils in
sediments entails extracting the samples with carbon tetrachloride
via gyratory agitation. The extract are then diluted for infrared
spectroscopic measurement which provides accurate quantification of
petroleum oil in the sediment.

    Total organic carbon (TOC) analysis is a useful technique for
measuring the total organic content of an aqueous sample. The US EPA
has employed TOC analysis in conjunction with a mobile carbon
treatment unit designed to remove organic contaminants from aqueous
media. TOC measurements of influent and effluent samples facilitate
on-site monitoring of the cleanup and removal efficiency of organics
from large bodies of water.

    9.1.3 Application. Since its development in 1976, the EPA
mobile laboratory has responded to 19 emergencies (see Table 11.12).
Examples of three emergency responses are presented below.

    Niagara, New York. Perhaps the most noted incident in which the
rapid extraction methods were applied was the emergency response
activity conducted near Niagara, New York, in October 1978.
Over 200 families living along the abandoned Love Canal waste
disposal site had to permanently evacuate their homes when
toxic chemicals seeped through the ground into their basements

Table 11.12 EPA Mobile Laboratory Emergency Response Actions

Location of incident	Date	Compound(s) analysed	Technique
Haverford, Pennsylvania	Nov 1976	Pentachlorophenol; Oil	GC; IR
Dittmer, Missouri	Apr 1977	Oil, PCBs	GC
Kernsville, N Carolina	Jun 1977	Oil, mixed chemicals	FS
Oswego, New York	Oct 1977	PCBs	GS
Niagra, New York	Oct 1978	Chlorinated benzenes, PCBs, hexachlorocyclo-hexanes (BHC)	GC
Pittson, Pennsylvania	Aug 1979	c-dichlorobenzen	GC
Sharptown, Maryland	Feb 1980	Oil	IR
Crosby, Texas	Feb 1980	PCBs; oil	GC; IR
Manassas, Virginia	Mar 1980	Total fluorescing organics	FS
Elizabeth, New Jersey	Apr 1980	PCBs, o-dichlorobenzene	GC
Kingston, New Hampshire	Jan 1981	Volatile organics, PCBs Chlorinated pesticides; Oxidation-reduction potentials	GC; Redox kit
La Marque, Texas	Sep 1981	1,2-dichloroethane, 1,1,2-trichloroethane	GC
Fort Lawn, S Carolina	Oct 1981	PCBs, chlorinated pesticides	GC
Kingston, New Hampshire	Nov 1981	PCBs	GC
Old Forge, Pennsylvania	Apr 1982	PCBs	GC
Epping, New Hampshire	Apr 1982	Volatile organics; Total organic carbon	GC; TOC
Rock Creek, Jefferson, Ohio	Sep 1982	Volatile organics	GC
Epping, New Hampshire	Dec 1982	Volatile organics	GC
Kent, Washington	Apr 1983	Volatile organics, PCBs, heavy metals	GC; ES

Key: ES = emission spectroscopy
     FS = fluorescence spectroscopy
     GC = gas chromatography
     IR = infrared spectrophotometry

and gardens. Included with the more than 20 000 metric tonnes of industrial wastes in the Canal was an estimated 300 different toxic chemicals, nearly 10% of which were classified as mutagens, teratogens, or carcinogens. The mobile laboratory was assigned to analyse leachate and sediment samples for priority

pollutants. The samples were extracted according to the rapid
extraction techniques developed for the analysis of base
neutrals, phenolics, and PCBs. Gas chromatographic analysis of
the extracts revealed the presence of chlorinated benzenes,
hexachlorocyclohexanes (BHC), and PCBs.

Manassas, Virginia. In March 1980, a massive quantity of
kerosene spilled into a small river leading to the Occoquan
Reservoir, threatening the drinking water supply of Manassas,
Virginia. Water samples were analysed aboard the mobile
laboratory to monitor the concentration of total fluorescing
organics. The samples were cooled to $4^{\circ}C$ to prevent
volatility losses and then rapidly extracted with cyclohexane
using a spinning technique. The extraction step was
incorporated into the method to avoid interferences encountered
during direct in-situ measurements of the river water. The
extracts were then quantified by synchronous excitation
fluorescence spectroscopy.

Epping, New Hampshire. An emergency cleanup activity utilising
a carbon treatment unit was conducted in March 1982. A
contaminated solvent lagoon located in Epping, New Hampshire was
in danger of overflowing. The water from the lagoon was pumped
through a carbon treatment unit and discharged into holding
tanks. The mobile laboratory monitored the cleanup efficiency
of the unit by analysing influent and effluent samples for
purgeable priority pollutants and total organic content.

9.2 Mobile MS/GC System

9.2.1 Description. The TAGA^{TM} (Trace Atmospheric Gas Analyser)
6000 Mobile Mass Spectrometer System, manufactured by Sciex, Inc, is
a self-contained computerised mobile laboratory providing real time
air monitoring for the analysis of trace levels of organic and
inorganic pollutants. The laboratory is equipped with a tandem mass
spectrometer capable of providing both identification and quantification of contaminants in the atmosphere with minimal sample
preparation.

The system is designed to monitor low levels of contaminants in
air under rapidly changing conditions. As the laboratory travels
through a contaminated area it continuously monitors the ambient
atmosphere by drawing air through a sample probe mounted on the roof
of the vehicle. The high sampling flow rate minimises memory effects,
resulting in a rapid response to concentration changes at the sample
inlet. The computer provides concentration contour maps of hazardous
material plumes. Figure 11.5 illustrates the plume of aniline
concentrations in the atmosphere near a chemical plant as seven
transits of a downwind area were made during a twenty minute period.

effects, resulting in a rapid response to concentration changes at the sample inlet. The computer provides concentration contour maps of hazardous material plumes. Figure 11.5 illustrates the plume of aniline concentrations in the atmosphere near a chemical plant as seven transits of a downwind area were made during a twenty minute period.

9.2.2 Evaluation. Although the TAGA^{TM} 6000 system is expensive, the manufacturer claims that the system's flexibility, reliability, sensitivity, and rapid analysis capability provide exceptionally cost effective analyses. Hazardous waste site

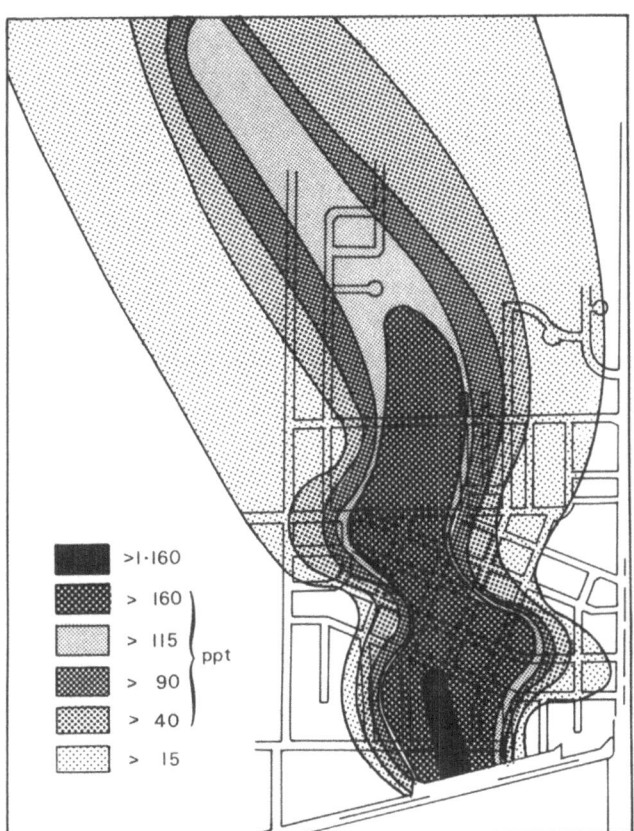

Figure 11.5   Plume Contour Plot of Aniline Concentrations in Atmosphere Near Chemical Plant as Measured by the TAGA 6000 Mobile Mass Spectrometer System (after Karasek[39])

investigations are normally conducted by on-site sampling followed by off-site laboratory analysis, creating a delay of one to two weeks for results. This delay is eliminated by the real time monitoring capability provided by the TAGA^{TM} system.

9.2.3 Application. The scope of applications for the TAGA^{TM} 6000 in environmental areas is broad. It has been used in emergency monitoring of toxic spills and in investigating dump sites. Special probes allow the screening of soil samples to determine the extent of contamination at a particular site. Air monitoring can be performed continuously during the cleanup of a site, providing additional safety measures for on-site workers. Because direct sampling of smokestack emissions is possible, the unit can also be used for monitoring combustion by-products. Several examples citing the use of the TAGA^{TM} 6000 are presented below.

Niagara, New York. In October of 1978, the TAGA^{TM} 6000 was employed by the Canadian Environmental Protection Service and New York State officials to develop information on the location of buried drums of hazardous waste that were leaching into the ground at Love Canal. The system was used as a rapid means of identification of contaminants present and as an alarm system to alert workers to dangerous levels of chemical releases into the immediate atmosphere surrounding the excavated areas.

Saskatchewan, Canada. In July 1979, Sciex responded to an accidental underground pipe leak in Saskatchewan, where PCBs were spilled. The TAGA^{TM} 6000 was able to analyse the surrounding soil to determine the extent of the spill and the concentration od PCBs present.

Mississauga, Ontario. When a train carrying chlorine gas and propane gas derailed in Ontario, the TAGA^{TM} 6000 traversed streets in a wide area downwind of the wreckage. Chlorine concentration profiles were developed to help officials decide which areas should be evacuated. The burning of propane resulting from the crash, in combination with the chlorine gas, had the potential to react to form deadly halogenated hydrocarbons. The TAGA^{TM} 6000 analysed for these compounds during cleanup efforts, and checked for remaining chlorine pockets before allowing re-entry into evacuated areas (see Figure 11.6).

10   CONCLUSIONS

The Pilot Study Groups' definition of contaminated land and classification of the hazards and problems associated with contaminated land were presented in Chapter 1. Contaminated land was defined as land where toxic substances are present that in sufficient

quantities or concentration pose a threat to the health and safety of
users or occupiers of land or of workers engaged in its redevelopment
or reclamation. Numerous analytical methods have been developed over
the past decade in response to the increased need for rapid on-site
assessment of contaminated land. Rapid on-site methods of chemical
analysis facilitate monitoring of toxic and hazardous substances and
thereby minimise exposure of potential targets (ie, on-scene
personnel, site users, etc) to these hazards. In addition, by
monitoring the general level of contamination at a particular site,
these analytical techniques assist on-scene co-ordinators in
identifying unacceptable and unhealthy conditions that may require
immediate remedial response.

A large number of analytical techniques that are currently in
use by the US Environmental Protection Agency and its private
contractors for monitoring contaminated sites in support of cleanup

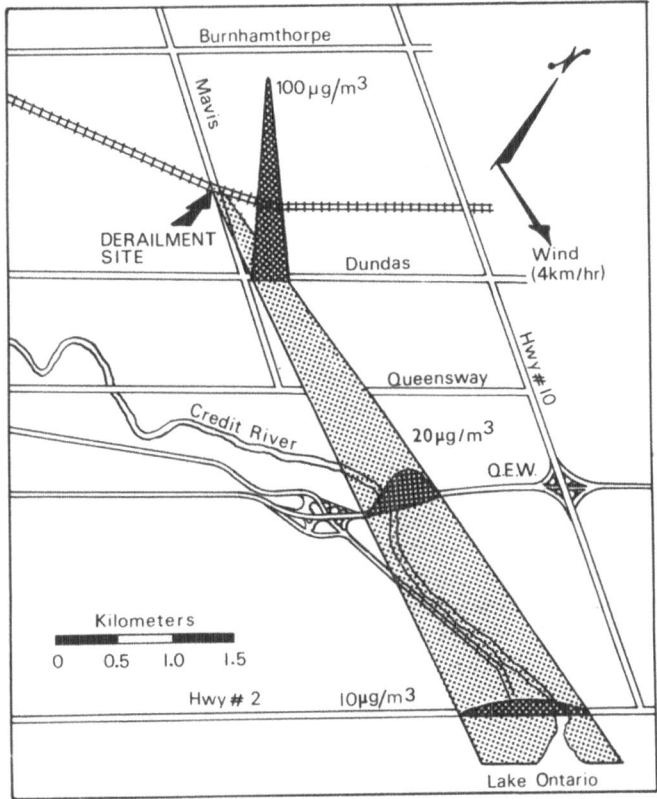

Figure 11.6   Three Plume Cross Sections taken by the TAGA[TM]
              6000 Mobile Mass Spectrometer System Following a
              Chlorine Spill (after Lane and Thomson[40])

and reclamation efforts have been reviewed in this Chapter. The
reader is advised that the techniques described herein represent a
user survey and are not meant to be all-encompassing, since other
techniques are either commercially available or being developed as
research programmes progress. The methods surveyed vary in
complexity, ranging from non-specific, semi-quantitative screening
techniques to compound specific, quantitative, multi-parametric
protocols. These methods are extremely useful when applied to the
initial appraisal of chemical waste sites and for monitoring the
general levels of contamination at a particular site. Major
advantages realised through the use of such field-oriented
methodology are real time sample analyses and reduced costs of site
investigations. Disadvantages include the limitations inherent in
key analytical parameters (ie, accuracy, selectivity, sensitivity,
etc) when portable instruments and field test kits and devices are
employed.

The ultimate choice of on-site techniques for evaluating
contaminants will be determined by many factors. Typical decision
criteria include:

(i)   the type and size of the site and associated environmental
       factors;

(ii)   the types and abundances of contaminants anticipated at the
       site; and

(iii) the level of effort required at the site (ie reconnaissance,
       preliminary and full surveillance, monitoring, and remediation
       and follow up activities). These efforts are further affected
       by the intensity of response required (ie spill and emergency
       response vs reclamation efforts).

There is a continuing need for the development of new techniques
and refinement of others that can provide on-site analyses for the
vast array of hazardous substances and sample matrices (ie, sludges,
concentrated wastes, etc) encountered at contaminated sites.
Application of practical chemical analysis methodology lends valuable
support to management decisions involving the mitigation of immediate
and long-term environmental hazards, particularly recommendations
regarding appropriate remedial measures and ensuing reclamation
efforts.

11   ACKNOWLEDGEMENTS

Special acknowledgement is extended to the members of the
Analytical Support Group, IT Corporation, for performing the user
survey and assisting in the preparation of the manuscript. Members
of this group include Mr Walter Berger, Mr Daniel A Bingham,

Ms Janet J Cullinane, Mr Anthony J Lombardo, Mr Thomas J Mancuso, Ms Jeanne-Marie O'Donnell, and Ms Kathleen M Vasile. The authors also gratefully appreciate the support and co-operation of the Environmental Response Team (US EPA) and the Technical Assistance Team (Roy F Weston, Inc).

12   REFERENCES

1   Environmental Advisory Unit, Liverpool University, UK, (private communication 1983).

2   US EPA. 'Characterisation of Hazardous Waste Sites – A Methods Manual, Volume II: Available Sampling Methods,' EPA 600/X-83-018, Environmental Monitoring and Support Laboratory, Las Vegas, NV (1983).

3   US EPA. 'Technical and Financial Impact of Monitoring Releases of Hazardous Substances,' EPA 600/X-82-016, Environmental Monitoring and Support Laboratory, Las Vegas, NV (1982).

4   G F Lee, and R A Jones. Guidelines for sampling groundwater. Journal of the Water Pollution Control Federation, $\underline{55}$, (1), 92–96 (1983).

5   US EPA. Handbook for Analytical Quality Control in Water and Wastewater Laboratories, EPA 600/4-79-019, Environmental Monitoring and Support Laboratory, Cincinnati, Ohio (1979).

6   US EPA. Methods for Chemical Analysis of Water and Wastes. Environmental Monitoring and Support Laboratory, Las Vegas, NV (1979).

7   M Gruenfeld, U Frank, and D P Remeta. Specialised methodology and quality assurance procedures used aboard mobile laboratories for the analysis of hazardous wastes. in: Proc American Chemical Society Conference, Las Vegas, Nevada, 1982 (AC5).

8   G J Gross, D E Harris, G Lachs and R M Dillman. Instruments for Detecting Hazardous Materials (1980).

9   H K Hatayama, J J Chen, E R de Vera, R D Stephens and D L Storm. A Method for Determining the Compatibility of Hazardous Wastes. California Dept of Health Services, Berkeley, California (1980).

10   R D Turpin, J R Lafornara, H L Allen and U Frank. Compatibility field testing procedures for unidentified hazardous wastes. Proc Conference Management Uncontrolled Hazardous Waste Sites, Washington, DC, 110-113, HMCRI, Silver Spring, Maryland (1981).

11   J J Mayhew, G M Sodaro and D W Carrol. A Hazardous Waste Site Management Plan. Chemical Manufacturers Assoc, Inc, Washington, DC (1982).

12   J C Clow and J C Zercher. The Coast Guard's National Response Center, and CHEMTREC, of the Chemical Manufacturers Association. Proc Hazardous Materials Spills Conference, 1980. Louisville, Kentucky, 358-360.

13   US EPA. Environmental Response Team, Report HW-2, Washington, DC (1982).

14   W E Rittaler. Callahan uncontrolled hazardous waste site
     during extreme cold weather conditions. Proc Conference
     Management Uncontrolled Hazardous Waste Sites, Washington,
     DC, 254-258, HMCRI, Silver Spring, Maryland (1982).
15   B W Muller, A R Brodd and J Leo. Picillo Farm, Coventry,
     Rhode Island: A superfund and state fund cleanup case
     history. Ibidem, pp 268-273. Waste Sites, Washington, DC,
     268-273, HMCRI, Silver Spring, Maryland (1982).
16   S Miller. A monitoring report: Personal monitors.
     Environmental Science and Technology, 17 (8), 343A-346A
     (1983).
17   M Gruenfeld, J Quimby and B DeMaine. Limited evaluation of
     a portable gas chromatograph. EPA Quality Assurance
     Newsletter, 4 (1), 9-10 (1981).
18   J Quimby, R Cibulskis and M Gruenfeld. Evaluation and use
     of a portable gas chromatograph for monitoring hazardous
     waste sites. Proc Conference Management Uncontrolled
     Hazardous Waste Sites, Washington, DC, 36-39, HMCRI, Silver
     Spring, Maryland (1982).
19   T M Spittler and P F Clay. The use of portable instruments
     in hazardous waste site characterisations. Ibidem, pp 40-44.
20   R C Levenson and N J Barker. A portable multi-component air
     impurity analyser having sub-part per billion capability
     without sample preconcentration. Instrument Society of
     America. Analysis Instrument Symposium (1981).
21   C L McEnery. Uses and limitations of environmental
     monitoring equipment for assessing worker safety in the
     field investigations of abandoned and uncontrolled hazardous
     waste sites. Proc Conference Management Uncontrolled
     Hazardous Waste Sites, Washington, DC, 306-310, HMCRI,
     Silver Spring, Maryland (1982).
22   R U Townsend. Air monitoring of hazardous waste sites.
     Ibidem, pp 67-69.
23   S Crisp. Solid sorbent gas samplers. Ann Occup Hyg, 23,
     47-76 (1980).
24   Beckman Instruments, Inc. The Microtox System Instrument
     Manual, Microbics Operations, 6200 El Camino Real, Carlsbad,
     California.
25   Centec Analytical Services, Inc. Performance data for the TR-
     201 (1983).
26   Centec Analytical Services, Inc. Technical Bulletin No 83-1
     (1982).
27   M Urban and R Losche. Development and use of a mobile chemical
     laboratory for hazardous material spill response activities.
     Proc 1978 Hazardous Material Spills Conference, Miami Beach,
     Fla, 311-314 (1978).
28   M Gruenfeld, U Frank, D P Remeta and R Losche. Management of
     analytical laboratory support at uncontrolled hazardous waste
     sites. Proc Conference Management Uncontrolled Hazardous Waste
     Sites, Washington, DC, 96-102, HMCRI, Silver Springs, Maryland
     (1981).

29   U Frank, M Gruenfeld, R Losche and J Lafornara. Mobile
      laboratory safety and analysis protocols used at abandoned
      chemical waste dump sites and oil and hazardous chemical
      spills. Proc Hazardous Material Spills Conference,
      Louisville, Kentucky, 259-263 (1980).

30   M Gruenfeld, U Frank and D P Remeta. Rapid methods of chemical
      analysis used in emergency response mobile laboratory
      activities. Proc Conference Management Uncontrolled Hazardous
      Waste Sites, Washington, DC, 165-173, HMCRI, Silver Springs,
      Maryland (1980).

31   U Frank. Quantification of petroleum oils by fluorescence
      spectroscopy. Proc Conference Prevention and Control of Oil
      Pollution. Washington DC, 1975, pp 87-91, American Petroleum
      Institute (1975).

32   U Frank. A review of fluorescence spectroscopu methods for oil
      spill source identification. Toxicological and Environmental
      Chemistry Reviews, 2 (3), pp 163-185 (1978).

33   U Frank and M Gruenfeld. Use of synchronous excitation
      fluorescence spectroscopy for in-situ quantification of
      hazardous materials in water. Hazardous Material Spills
      Conference, 1978, Miami Beach, Fla, pp 119-123 (1978).

34   M Gruenfeld. Extraction of dispersed oils from water for
      quantitative analysis by infrared spectrophotometry.
      Environmental Science and Technology, 7 (7), pp 636-639 (1973).

35   M Gruenfeld. Quantitative analysis of petroleum oil pollutants
      by infrared spectrophotometry. ASTM Water Quality Parameters,
      ASTM Spec Tech Publication, No 573, pp 290-308 (1975).

36   American Public Health Association, 1976. Standard Methods for
      the Examination of Water and Wastewater, 14th Edition, New York,
      New York.

37   Photovac, Inc, Photovac 10A10 Operating Manual, Thornhill,
      Ontario, Canada.

38   F B Stroud, B G Burrus and J M Gilbert. A co-ordinated cleanup
      of the Old Hardin County Brickyard, West Point Kentucky, Conf
      Management Uncontrolled Hazardous Waste Sites, Washington DC,
      1982 pp 274-278. HMCRI, Silver Spring, Maryland (1982).

39   F W Karasek. A mobile laboratory for ultra-trace environmental
      monitoring, Industrial Research Development, Montana, USA
      (1978).

40   D A Lane and B A Thomson. Monitoring a chlorine spill from a
      train derailment, Journ Air Pollution Control Association, 2
      (2), (1981).

FORMER IRON AND STEELMAKING PLANTS

D L Barry

W S Atkins and Partners

United Kingdom

1   INTRODUCTION

1.1 Historical, Economic and Social Factors

The world steel industry grew at a remarkable rate in the 30 years after World War II; crude steel production increased from about 140 million tonnes (1943) to about 710 million tonnes (1974). Recent years have shown a decline in production owing to a general economic recession and, while worldwide production is predicted to increase in the long term, there is considerable change in the share and size of the market held by North America and Western Europe. Table 12.1 gives some recent values of crude steel production (the figures in brackets reflect the production potential).

While these figures show the considerable gap between actual and potential production in many countries they do not reveal the real changes in the size of the industry. This is because in the late 1970s, restructuring and rationalisation measures produced a significant increase in plant productivity. Thus, even if production figures had remained steady, improved productivity would have resulted in many plant closures and reduced workforces (see Table 12.2) particularly in the USA and in the UK. Many of these plants were major local employers and the effects of closure have generally been very significant. All of these sites require reclamation. Those close to, or within, the existing urban fabric offer considerable opportunities for redevelopment. It is in the process of reclamation or redevelopment, rather than due to concern for public environmental health, that contamination is usually first encountered as a problem.

Table 12.1   Crude Steel Production (million tonnes)

	1967	1974	1977	1981		1982	% change 1967–82
USA	117.4	134.6	115.4	111.8		69.0	– 41
FRG	37.4	54.3	39.8	42.4	(69.0)*	36.6	– 2
Italy	16.2	24.3	23.8	25.3	(40.6)	24.5	+ 51
France	20.1	27.5	23.6	21.7	(30.2)	18.8	– 6
UK	24.8	22.8	20.9	15.8	(26.0)	14.0	– 44
Canada	9.0	13.9	13.8	15.1		12.1	+ 34
Belgium	9.9	16.5	11.5	12.6	(19.5)	10.1	+ 2
Luxembourg	4.6	6.5	4.4	3.9	(6.5)	3.6	– 22
Netherlands	3.5	5.9	5.0	5.6	(8.8)	4.5	+ 29

*   Production potential

1.2   Redevelopment Problems

     The problems with redevelopment take many forms, eg hazards to
investigation and construction workers, chemical attack on building
materials and services, toxicity to plant life, and hazards to future
site users.   In addition there are very considerable engineering
problems.   However, generalisations about iron and steelmaking sites
are difficult because of the wide range of processes that can be
carried out and the technical and operational changes that have taken
place in the industry.   Also site locations can vary from marshy
coastal areas to inland mining areas.   Works range from modern,
complex and integrated facilities to older, modest-sized sites built
early in the century.   What they all have in common, however, is the
potential problems they present for redevelopment in terms of both
contamination and engineering factors.   These problems, which can
vary greatly from site to site and within each site, can only be put
into a proper perspective when the nature of development or planned
reclamation is taken into account.   Expansive slags or coal tars, for
example, will present different problems for different developments.

     The most significant chemical ground contamination problems
are likely in the coking byproducts area, pickling plant area and any
waste disposal areas such as lagoons and slag heaps.   Slags can also
present considerable engineering problems due, on the one hand, to
their potential instability in chemical terms and, on the other hand,
to the very large volumes involved.   Nowadays, however, little slag
is stockpiled in Western European countries where virtually all the
current production is used and where blastfurnace slag is a well
established re-usable material. Any stockpiles that do exist (and

Table 12.2   Employment in the Steel
Industry (thousands)

	1974	1981	1982
USA	512	391	289
FRG	232	187	176
Italy	96	96	91
France	158	97	95
UK	194	88	75
Canada	54	55	50 (E)
Belgium	64	44	42
Luxembourg	23	13	12
Netherlands	25	21	20

Source: Ref 3a   (E = Estimate)

they are frequently very large) tend to be very old and consist of mixtures of slags and other wastes.

Much of the contamination may extend to considerable depths, particularly where works have been built in marshy or open-cast mined areas reclaimed with slag and other wastes (eg when works are being extended). Iron making has traditionally been located on or close to coal mining areas and so underground workings and old mine shafts can add greatly to redevelopment difficulties.

2   IRON AND STEELMAKING PROCESSES

2.1   Introduction

The principal processing steps in making iron and steel products from iron ore are as follows:

-    ore preparation (including agglomeration and sintering)
-    coke making
-    iron making
-    steel making
-    rolling, finishing and surface treatment.

Not all of these steps are carried out at one site; for example iron making may only involve the first three steps. Also coke making is not always carried out at the iron or steelworks (this is particularly true in Germany, where coke is usually produced at the

coal mine site). Thus the relative sizes of works and the nature and
extent of processes can vary greatly.

2.2 Principal Process Descriptions

A general outline of the principal processes and how they interact is given in Figure 12.1.

2.2.1 Ore preparation. Iron ores are crushed and screened and
fed into the blast furnace. Fine ores, coke breeze and, frequently,
limestone are fed to a singering machine where they are made into
agglomerates (pelletising is a recent agglomeration process). This
process allows greater use to be made of fine raw materials that
would otherwise lack mechanical strength.

2.2.2 Coke making. There are basically two methods of
manufacturing metallurgical coke, namely the more efficient byproduct
process and the beehive process which causes extensive air pollution.
The very large majority of current plants use the byproduct process
(also known as the retort or chemical recovery process) while early
in the century the beehive process was used extensively.

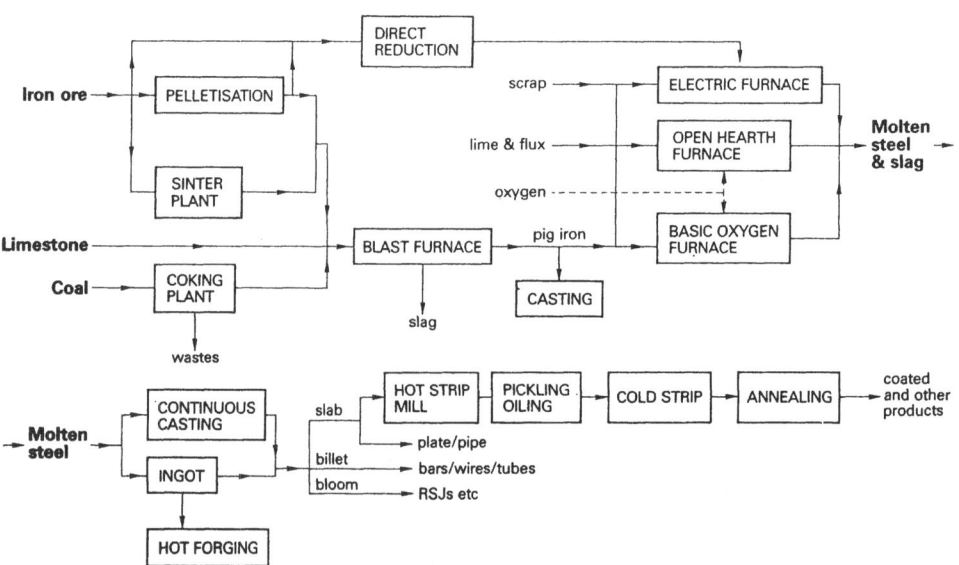

Figure 12.1 Simplified Process Diagram for Modern
Iron and Steel making

In the byproduct process air is excluded from the coking chambers and the necessary heat for distillation is supplied from external combustion. The volatiles are collected and hot gases are cooled by spraying with a flushing liquor, the resultant liquor being decanted giving an ammonia liquor and tar which contains a large proportion of the phenols produced in the ovens.

In the beehive process, air is admitted to the coking chamber in controlled amounts for the purpose of burning the volatile products distilled from the coal and to generate heat for further distillation.

2.2.3 Iron Making. The blast furnace is the most common current method of manufacturing iron (the other principal techniques are based on electrodes). The furnace melts iron ore (or sinter pellets) mixed with fuel/reductant and a flux (eg coke and limestone) to assist in the melting and separation of impurities from the ore (see Figure 12.2). The furnace is, essentially, a vertical steel cylinder lined with fire brick and the materials are fed in at the top. Heated air and fuel (in recent practice, gas, oil, or pulverised coal) to achieve the necessary high temperatures are blown in at the bottom. The blast burns part of the fuel to produce heat for the chemical reactions involved and for melting the iron, while the balance of the fuel and part of the gas from combustion, removes the oxygen combined with the metal. As the coke burns away, the

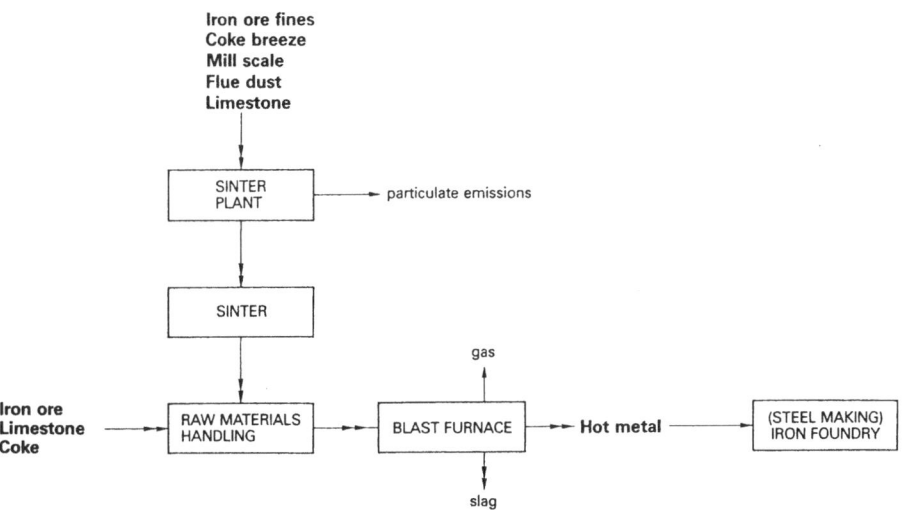

Figure 12.2 Simplified Activity Diagram of Blastfurnace Area

Table 12.3 Materials Balance for 1 tonne of Pig Iron

Input	Tonnes	Output	Tonnes
Iron ore	1.7	Pig iron	1.0
Coke and other fuel	0.5-0.6	Slag	0.2-0.4
Fluxes	0.25-0.3	Blast furnace gas	2.5-3.5
Scrap	0.33-0.34		
Air	1.8-2.0	Flue dust	0.05

materials descend in the furnace against the upward-rushing gas resulting in the extraction of the iron. The iron being the heaviest substance in the furnace, sinks in molten droplets to collect in the hearth from where it is tapped through a hole.

Another hole is provided higher up the hearth for running off the slag. The whole process is continuous with raw materials being constantly fed in at the top. The metal product from the furnace is known as pig-iron and contains over 90% of iron.

The average quantities of raw materials used and the byproducts produced in manufacture are given in Table 12.3. The figures refer to a modern plant using high grade ore. The ratios are different for plants using lower grade fuel and in general for plants operating in times past.

Blast furnace slags result from the fusion of limestone with ash and coke together with the iron ore residues. One of their important functions is the removal of sulphur from the molten iron, the chemical composition being appropriately adjusted. Slags can finish in one of four forms (air-cooled, foamed, pelletised or granulated) depending upon the rate and manner of cooling.

2.2.4 Steel Making and Casting. Steel making processes and techniques have undergone radical changes in the past 100 years. Before 1856 wrought iron was used for plates, etc. Bessemer then discovered how to make steel from molten pig iron by an 'acid' process. A 'basic' process was developed some 20 years later.

The four main processes of steel making are:

-    Bessemer
-    Open hearth
-    Electric furnace
-    Basic oxygen

Over the past 20 years or so the relative use of these processes
has changed very significantly so that currently the vast majority of
steel is produced using the latter two. The Bessemer process is all
but extinct in all major industrialised Western countries.

In the Bessemer process oxidation is achieved by blowing air
through the molten iron, which is poured from a ladle into the
converter. The oxygen 'burns' out the carbon and oxidises the
manganese, silicon and phosphorus which go into the slag. Since the
process eliminates all the carbon, a specified amount must be added
at the end. The acid (rather than basic) process was, at one time,
used to produce most of the world's steel. In air pollution terms
the process was considered the worst polluter in steel making.

The open hearth process is also very old and consists of a
shallow bath in which the metal is subjected to an intense flame from
above. When the materials are completely melted the slag floats on
top.

The two types of electric furnaces are electric arc and electric
induction. The electric arc furnaces are essentially in two forms,
open bath and submerged arc, the latter producing the great majority
of electrically melted steel. The melting process is achieved by use
of electrodes. In electric induction furnaces, currents are induced
in the steel causing it to melt.

The basic oxygen furnace (BOF process), which was developed in
the early 1950s, involves pure oxygen being blown through a lance
onto the surface of a charge of molten iron and steel scrap. All the
heat required for refining is generated by the reactions between the
oxygen and the metalloids in the hot metal. There are also acid type
oxygen processes.

2.2.5 Rolling, finishing and surface treatment. Molten steel is
either cast into ingot moulds or, in the continuous casting process,
is converted directly into billets. A plate mill rolls ingots or
slabs which are progressively reduced in thickness. Hot strip
rolling is usually followed by any of the following: pickling, cold
rolling, annealing, electrolyte tinning and galvanising, hot dipped
galvanising, or the application of non-metallic coatings (eg, paints,
enamels, plastics) by dipping, painting or spraying.

3   PRINCIPAL SOURCES OF POTENTIAL CONTAMINATION

3.1 Introduction

The principal sources of potential contamination are very
extensive and cover a wide range both in volumes, concentrations and
degrees of hazard. The likelihood of particular types of ground

contamination must, in the first instance, be related to the
particular processes and activities that were carried out on the
various parts of the works. In general terms, and especially before
environmental controls were exercised, wastes were recycled only if
they did not need treatment before re-use.

3.2 Ore Composition

Iron ores can consist of the following substances to varying
degrees (these are listed merely to illustrate the origin of some
sources of ultimate contamination: the ores in their natural form
cannot be considered as contaminating):

iron compounds	vanadium
silica	zinc
alumina	lead
lime	copper
magnesium	chromium
manganese oxides	nickel
phosphorus	arsenic
sulphur	tin
titanium	boron

3.3 Coke making

3.3.1 Introduction. Only certain types of bituminous coal are
suitable for the production of metallurgical coke since the coke

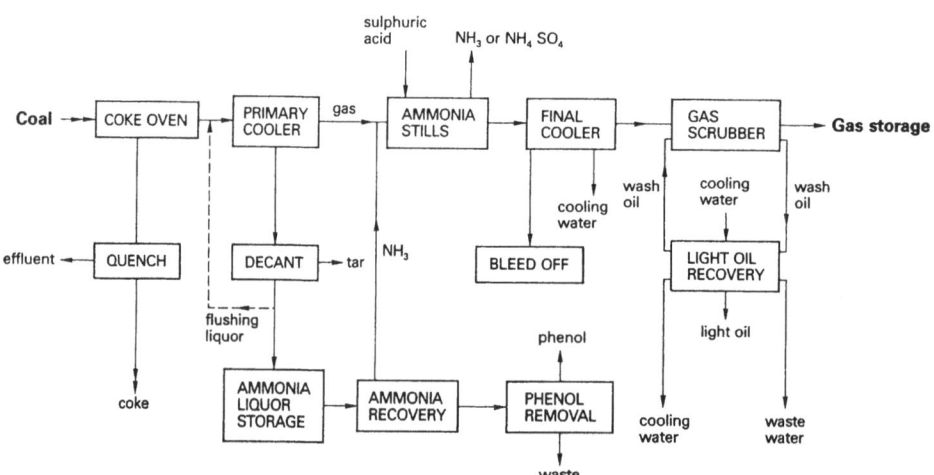

Figure 12.3   Simplified Process Diagram of Byproduct
              Coke Manufacture

produced requires specific characteristics to fulfil its function in
iron making.

A good metallurgical coke is considered to contain

- fixed carbon   85-90%
- ash   5-9%
- sulphur   0.6-1.5%
- phosphorus
- volatile matter   0.5-2%

Sulphur content is kept to the minimum possible as it is the chief
source of sulphur in pig iron.

Coke production is carried out in similar structures to the
vertical retorts used in town gas manufacture. Although the need for
high quality gas in the iron and steel industry is not as critical as
for town gas, the initial removal of byproducts follows a similar
process (see Figure 12.3 for process diagram). These byproducts
generally form the most critical compounds giving rise to ground
contamination.

3.3.2 Byproducts and Wastes. Typical yields of coke and
byproducts from 1 tonne of coking coal are given in Table 12.4.
Since about 0.47 tonnes of coke are used in the production of 1 tonne
of iron, very considerable quantities of byproducts result from a
normal iron making works. In addition to the primary products of
coke and coke gas the principal byproducts are; coke breeze, phenol,
ammonium sulphate, ammonia liquor, sulphuric acid, cyanides,
chlorides, sulphides, tars, light oils. Crude tar contains a vast
range of chemical compounds; the principal ones with an average
occurrence in tar are listed in Table 12.5. Tars and light oils are
frequently refined to a variety of compounds including; toluene,
benzene, naphthalene, pyrene, phenol, pyridine, anthracine.

Table 12.4 Typical Yields of Coke and Byproducts
from 1 tonne of Coking Coal

Coke	544-634 kg
Coke breeze	45-91 kg
Gas	270-325 m^3
Tar	30-45 litres
Ammonium sulphate	9-13 kg
Ammonia liquor	57-132 litres
Light oil	9-15 litres

Table 12.5 Typical Composition of Coal Tar[6]

Pitch	62%
Anthracene-heavy oil	10-12%
Crude napthalene	8-12%
Total tar acids (phenols, creosols, zynols)	2-4%
Total tar bases (eg pyridine)	1-2%
Methylnaphthalene oil	2-3%
Liquor	2-6%
Fluorene oil	2-4%
Acenanaphthene oil	1-3%
Naptha	1-2%
Biphenyl oil	1-2%
Benzol	1%
Toluol	1%
Xylol	1%

The principal constituents of coke plant effluents are:

phenol	cyanides
ammonia	chlorides
sulphates	oil
sulphides	

Other wastes include effluents from:

-   light oil recovery plant (containing, for example, phenols, cyanide, ammonia, oil, chlorides and sulphur compounds)
-   coke quenching (see below)
-   cooling (see below)
-   acid washing.

Chemicals potentially present[5] in waters used in coke quenching and cooling are listed below; some are frequently present in high concentrations (see Table 12.6):

phenol	xylene
chlorides	thiophenes
sulphur acid	mercury
ammonia	lead
ammonium sulphate	selenium
formaldehyde	nickel carbonyl
pyridine	hydrogen cyanide
benzene	ammonium cyanide
toluene	ammonium thiocyanate

Table 12.6   Concentration of Major Constituents in Coke
Quenching Water for some US Coking Plants[4]

Phenol	776 ppm
Sulphates	1066 ppm
Chlorides	1954 ppm
Total ammonia	2517 ppm
Cyanides	98 ppm
Total solids	5214 ppm

3.4   Iron Making

In producing pig iron the principal wastes from the blast
furnace are:

- slag
- flue dust and slurries
- blast furnace refractories (zinc and some lead)
- waste waters.

Of these the most significant in terms of mass is slag (see 3.7).
Chemical composition of flue dusts (dry) are given in Table 12.7).
About 42 kg of dust and sludge can arise in the production of 1 tonne
of metal.

The main contaminants of waste waters are: ammonia,* phenol,*
cyanide,* fluorides, lead, zinc, suspended solids. (* These are
mainly a carry over from coke quenching with still waste.)

They also contain the following;

nitrogen	aluminium
phosphates	oil
nitrates	manganese
sulphates	sodium
sulphides	potassium
chlorides	iron

The ranges of pollutant content of blast furnace waste water
for old, typical and advanced plants in the United States are
given in Table 12.8.

Sinter plants where operated are typically part of blast furnace
departments to agglomerate fine ores, flue dust (from blast furnace)
and mill scale. Solid waste products do not arise during sintering
but the gases have significant amounts of dust and sulphur dioxide
and high chloride levels can be expected due to water spillages, for

Table 12.7   Range of Composition of Flue Dust for Blast Furnaces[6]

COARSE DUST

Fe	25-38%	K_2O	0.5-0.6%
C	24-34%	MgO	1%
SiO_2	6-7%	Mn	1%
CaO	5-7	S	1%
Al_2O_3	2-3	PbO	0.1-0.6%
Na_2O	0.25-2.5%	ZnO	0.06-0.6%

FINE DUST

Fe	10-32%	K_2O	0.8-4.8%
C	10-30%	MgO	1.3-2.5%
SiO_2	3-8%	Pb	0-10%
CaO	3-7%	S	1-6%
Al_2O_3	2-4%	PbO	0.2-1.4%
Na_2O	0.3-1.3%	ZnO	0.6-5.7%

Dusts may also contain; copper, nickel, chromium and arsenic

Sinter plants where operated are typically part of blast furnace departments to agglomerate fine ores, flue dust (from blast furnace) and mill scale. Solid waste products do not arise during sintering but the gases have significant amounts of dust and sulphur dioxide and high chloride levels can be expected due to water spillages, for example. Some plants have desulphurised gases, resulting in a waste collected in the form of gypsum ($CaSO_4 . 2H_2O$) and ammonium sulphate (($NH_4)_2SO_4$). Elemental sulphur is sometimes produced as an end product in some plants. Generally these products would be disposed of to other industries.

Table 12.8   Range of Contaminant Concentrations in Waste
             Water for some US Iron Making Plants[4]

Suspended solids	28-38
Phenols	0.007-0.008
Cyanides	0.009-0.010
Fluorides	0.015-0.016
Ammonia	0.007-0.008

Gas cleaning plant can produce cyanides, heavy metals, phenols, ammonia and, occasionally, thiocyanate.

3.5 Steel Making

3.5.1 Introduction. The main difference between iron and steel is the carbon content. In general terms, pig iron may contain the following principal elements (in addition to iron, that is), and to convert pig iron to steel all of these must be either removed almost entirely or at least reduced drastically:

3.0-4.5%	carbon
0.15-2.5%	manganese
0.2%	sulphur
0.025-2.5%	phosphorus
0.5-4%	silicon

The fundamental raw materials for steel making are:

(i)   hot pig iron or scrap iron. (Reduced iron ore can also be used in electric furnaces.) Scrap iron can be a major source of heavy metal contamination.

(ii) lime and flux

(iii) oxygen (except for electric furnace).

Lime and fluxes comprise; burned lime (CaO), limestone ($CaCO_3$), dolomite (MgO and CaO), fluorspar (CaF_2).

3.5.2 Byproducts and Wastes. The principal byproducts and (non-gaseous) wastes from steel making are:

- slag
- arising scrap
- slurries and dust
- refractory materials
- waste waters.

Quantitatively the most significant byproduct or waste is slag, ranging from 70 to 170 kg per tonne of liquid steel. The chemical composition of these slags is dealt with in 3.7.

Some 12-16 kg of dust or slurry can arise per tonne of steel. The chemical composition of wet dust from a BOF primary fume-cleaning plant (dry basis) is given in Table 12.9. Although high in iron content, the relatively high levels of lead and zinc prevent re-use of these dusts in the blast furnace. Dusts from electrical steel making furnaces are even higher in lead and zinc (20-45%) and

Table 12.9   Chemical Analysis of Wet Dust from a BOF Primary
             Fume Cleaning Plant[2] (dry wt basis)

Fe	43–65%	Mn	1–2%
SiO_2	1–3%	CaO	5–10%
S	1%	Zn	2–7%
Pb	1–4%		

processes have been devised for their extraction.[10] The presence
of Ni and Cr oxides prevents the recycling of dust in alloy and
stainless steel making.

Refractory linings represent a considerable waste volume and
these are high in lead and zinc in particular. Fluoride and zinc are
the major pollutants entering waste waters, which also contain: iron
oxides, lead, nitrates, manganese, phosphates, sulphates, sulphides,
sulphites, iron, oil, aluminium, silicon, lead and copper. Iron
oxides and lime (in the form of suspended solids) form a very
significant part of waste waters. These are very fine and red in
colour.

3.6   Cold Finishing

The cold finishing process is shown in Figure 12.4. The
principal sources of contamination relate to the following:

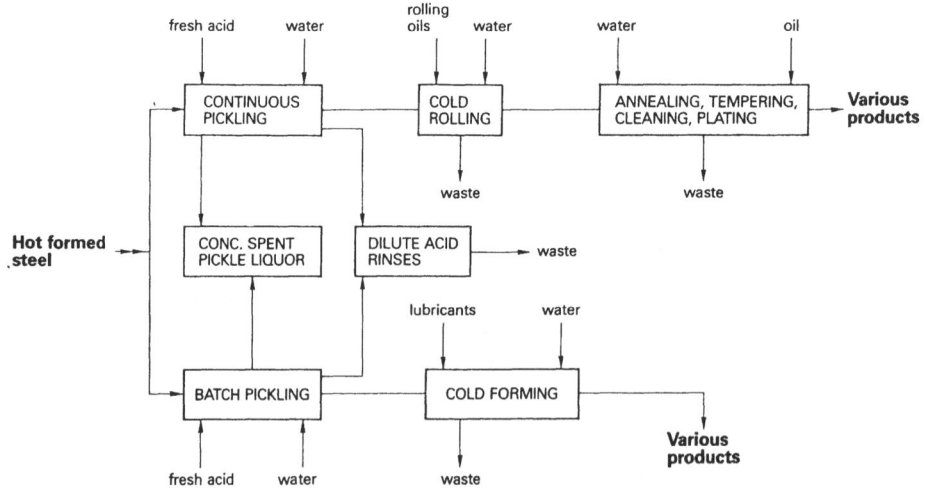

Figure 12.4   Process Diagram for Cold Finishing

- concentrated spent pickle liquor
- dilute acid rinse waters
- spilled acids, oils and lubricants
- acid or alkaline electrolytes
- tin plating wastes
- galvanising wastes
- cyanide wastes
- metallic coating wastes
- cleaning, annealing wastes
- cooling water wastes.

The most general pickling agents are:

- sulphuric acid (10-25% conc)
- hydrochloric acid (5-13% conc).

The regeneration of sulphuric acid can give rise to ferrous sulphate monohydrate (insoluble) or ferrous sulphate hepta hydrate, the former usually being dumped. Mixtures of hydrofluoric acid and nitric acid are used for special steels. In certain circumstances, iron sulphate and iron chloride are reaction products which may be converted to hydroxide sludges on neutralisation of the acid.

Metallic coatings or speciality steels require the use of a variety of metals that can give rise to subsequent contamination of the sites including: tin, molybdenum, zinc, vanadium, nickel, copper, chromium, aluminium, cadmium and lead.

Waste waters typically contain: oil and grease, tin, chlorides, zinc, cyanides, chromium, sulphates, phosphates, sulphides, cadmium, phenols, nickel, ammonia, iron, nitrogen and copper.

Emulsified oils from cold rolling present particular treatment problems because proprietary mineral oils have generally replaced palm oil in cold rolling. Solid wastes such as neutralised pickle liquor sludge or furnace (BOP) dust tend to accumulate principally for economic reasons.

3.7 Blastfurnace and Steelmaking Slags

3.7.1 Introduction. Since iron and steel making processes have changed radically over the last 100 years, the composition of slags produced has also changed. Any existing stockpiles or slag banks will often be heterogeneous and contain slags of different types deposited over a significant period of time and may also contain deposits of other process wastes relating to iron and steel production. Slags of recent origin found in sites are unlikely to be suitable for re-use, the better material having already been taken

Table 12.10 Utilisation of Blastfurnace Slags
             in some NATO Countries

Country	% utilised (approximate)
USA	95
Canada	100
Germany (FR)	100
France	90
Luxembourg	100
UK	100

for commercial uses. Figures for the utilisation of blastfurnace
slag in various countries are given in Table 12.10.

   3.7.2 Blastfurnace slags. Slag can be considered as the molten
silicate complex formed by the combination of agglomerated earthy
matter with the ore, fuel and fluxes. Modern slags are typically
hard, dense, crystalline or glassy materials and are frequently used
for roadstone and aggregate. Also their generally innocuous nature
allows them to be used as filter media in water purification.

   In the latter half of the 19th Century slags were extensively
used for the production of blocks (eg bridges, paving) slag sand,
bricks and mortar and slag wool (eg fireproofing insulant). Large
quantities of cement were manufactured by mixing it with 25% lime.
However, in 1903 it was reported that for every 10 million tonnes of
slags produced in the UK, about 1 million were used as broken stone
for (road) metalling, walling and the like, less than 75,000 tonnes
were used for cement, and a few hundred tonnes for slag wool. This
means that about 89% of slag did not have a use at that time (50
million tonnes of slag were produced in 1907 and 7.2 million
tonnes in 1972)[7]. Since then, however, good quality slag has
become a very remarkable commodity and, as indicated above, it is
almost entirely totally re-used nowadays in many countries.

   The slag produced per tonne of iron is a function of the iron
content of ores. Australian and US ores have, typically, 60% or 50%
iron, while British and European have around 30%. Thus, British ores
would create about 1.25 tonnes of slag as opposed to 0.25 tonnes for
rich imported ores.

   By varying the rate and manner of cooling of the molten slag,
several different types of solid slag are produced, namely air
cooled, foamed, pelletised or granulated slag.

Table 12.11 Comparative Chemical Compositions of Blastfurnace slags[8] (Weight %)

	USA	Germany	Norway	UK	France
SiO_2	34–38	35	34–38	31–36	31–36
Al_2O_3	11–15	12.2	7–10	9–20	11–21
CaO	45–47	41	40–48	33–45	39–45
MgO	1–3	8	6–13	4–15	4–8
TiO_2	–	–	0.7–1.1	0.8	0.4–0.7
FeO	1.3–4.5	0.25	–	0.5	0.1–1
MnO	–	0.5	–	0.8	0.1–1.2
Na_2O	–	1.2	0.1–0.4	0.7	0.2–0.8
K_2O	–	–	–	0.8	0.2–1.5
S	–	0.6	–	0.5–2	0.7–1
P_2O_5	–	–	–	0.6	–

The crystalline materials that form slag are compounds of the oxides of calcium and magnesium with silica and alumina. Sulphur is also present as sulphides and polysulphides, as sulphates and thiosulphates and as elemental sulphur in small quantities. There are also small quantities of compounds of iron and manganese, and traces of manganese, titanium and fluorine.

Chemical compositions of modern blastfurnace slags from a number of countries are given in Table 12.11 and Table 12.13. The slags are all broadly similar in composition. Nowadays many of the industrialised countries, especially in Western Europe, import much of their iron ore and the slag analyses are influenced mainly by the composition of this parent ore.

3.7.3 Steel Making Slags. The composition of steel slags is fundamentally different from that of blastfurnace slags and chemically and mineralogically they are very variable (see Tables 12.12 and 12.13); even material from the same ladle can vary considerably. (This is in contrast to blastfurnace slags where the compositions vary relatively little.) In particular, large and varying amounts of iron (included in the total Fe_2O_3 figure) are retained in the slag. The slags are, however, very low in sulphur components. The most significant change in steel making, in terms of the nature of slag, is that from open hearth furnaces to the basic oxygen process (see Table 12.12).

The mineralogies of steel slags reflect their chemical analyses in being much more variable than blastfurnace slags and usually higher in lime. In general, the compositions of the

Table 12.12   Comparative Chemical Analyses of Slags from
              Different Steel Making Processes[8] (% Weight)

	Basic oxygen slags*	Open hearth (N America)	Electric Arc (UK)	Blastfurnace (NATO countries)
SiO_2	9-22	16-26	11-24	31-38
Al_2O_3	4	2-9	5-18	11-21
CaO	33-60	25-36'	31-50	33-48
CaO free	13.3	2.1	-	-
MgO	1-7	10-11	2-8	1-15
TiO_2	1	0.8	0.3-1	1.1
FeO	-	-	-	4.5
MnO	-	-	-	1.2
Na_2O	0.1	-	0.3	0.8
K_2O	0.1	-	0.4	1.5
S	-	-	-	2.0
P_2O_5	2.5	-	1.8	0.6
Fe_2O_3 (total iron)	10-45	18-26	5-30	-
Mn_2O_3	3-10	13	6-22	-
SO_3	0.4	-	0.9	-
Sulphide	0.15	-	0.4	-
F	0.5	-	0.1-2.6	
CaO/SiO_2	1.9-5.4	1-2	1.9-2.6	1.2

* USA, Canada and UK

basic oxygen slags are the most variable and these slags tend to have
higher basicities and a greater tendency to contain free lime.
Electric arc slags are more homogeneous, contain less free lime and
are considered to make a good roadstone after weathering.

3.8   Some Specific Ground Contamination Levels

Table 12.14 gives data for samples taken from the coking plant
area of a steelworks site. The samples, taken at the surface, at
0.2 m - 0.5 m, at 0.9 m - 1.1 m and 1.5 m - 2.0 m depths show the
considerable variation that occurs spatially and with depth. The
wide ranges merely reflect the heterogeneous nature of these sites
where 'hot spots' can abound. It is also of interest to note the
great variations in pH values, which can have a profound influence on
the degrees of hazards, eg toxicity and gas generation can be greatly
enhanced in acidic conditions.

Table 12.13   Trace Element Analyses of Some Ferrous Slags[9]

Sample	Blastfurnace slag				Steel slag		
	A	B1	B2	C	D	E	F

TOTAL CONCENTRATIONS (mg/kg)

	A	B1	B2	C	D	E	F
Antimony	ND	ND	ND	ND	210	64	220
Arsenic	ND	ND	87	3	ND	ND	ND
Barium	200	100	200	400	700	70	100
Cadmium	10	6	8	9	7	4	9
Chromium	162	65	55	60	420	96	752
Copper	188	90	60	4400	32	17	20
Fluorine	9	1	0.8	0.4	9.5	4.8	0.7
Lead	63	56	48	61	70	38	80
Magnesium	37360	44040	54800	60980	4542	20902	20370
Manganese	18800	2580	2210	3620	44900	12600	43800
Mercury	1.7	ND	ND	ND	ND	ND	ND
Nickel	50	20	30	17	140	50	120
Selenium	24	ND	80	57	ND	ND	16
Thallium	23	95	108	90	53	34	58
Zinc	780	20	25	1160	38	41	54

EXTRACTABLE CONCENTRATIONS (mg/kg)

		A	B1	B2	C	D	E	F
Hot water:	Boron	5.5	3.0	3.0	4.0	3.5	3.0	3.5
0.5M Acetic	Copper	0.3	0.5	0.8	0.6	0.7	–	0.3
Acid	Nickel	0.2	0.4	1.6	1.0	3.0	5.0	0.4
	Zinc	3.6	2.0	3.2	3.4	2.8	8.2	3.2
0.5M EDTA	Copper	0.2	1.3	0.9	1.3	0.1	–	0.7
	Nickel	1.3	0.9	1.4	1.0	1.8	–	2.0
	Zinc	8.9	2.7	4.8	2.9	1.0	–	1.2

Notes: B1 and B2 are 'blind' repeat analyses
       ND not detected
       – not determined

4   HAZARDS FROM CONTAMINATION

4.1   Introduction

When former iron and steel making sites are redeveloped chemical
substances can attack a number of targets:

Table 12.14 Samples From Steelworks Coking Plant Area

Sample	Surface samples						0.02–0.5 m depth			1.5–2.0 depth	
	A	B	C	D	E	F	G	H	J	O	P
pH	9.3	2.6	7.2	13.0	4.3	6.8	8.9	12.6	8.7	7.8	12.8
Sulphate	2300	1.0%	3000	6000	1400	1000	6000	3000	800	4000	300
Sulphide	7	1	40	1	1	4	25	13	6	25	8
Chloride	100	100	50	100	100	150	300	100	100	150	100
Total cyanide	20	6000	52	2	1	8	1	1	4	60	4
Free cyanide	4	120	5	—	—	—	—	—	—	8	—
Phenols	320	28	1400	2	2	2	2	2	2	45	2
Toluene extract	21.6%	3.9%	30.2%	5000	1000	500	500	1000	2000	11.8%	4000
Cyclohexane extract	1.5%	1.1%	5.2%	500	—	—	—	—	—	3%	—
Coal tar	18.0%	2.0%	25.0%	500	—	—	—	—	—	9%	—
Total lead	800	6500	450	180	1500	410	950	60	360	100	470
Total cadmium	4.8	1.5	4.5	3.5	1.2	34.8	3.0	3.0	1.3	1.0	2.0
Total zinc	900	140	720	230	550	640	970	130	510	140	860
Total nickel	85	35	10	60	70	100	35	85	50	20	30
Total copper	200	20	15	120	150	300	85	70	25	80	
Available zinc	160	30	60	5	10	120	240	25	65	20	300
Available nickel	5	5	5	15	5	10	5	10	5	5	10
Available copper	25	15	5	10	15	60	10	10	25	5	35

- site investigation and construction workers
- site users after development
- building materials
- flora and fauna
- off site effects (on above targets).

The nature of hazards or effects can be generally listed as follows:

- skin irritation (from solids and liquids)
- inhalation of toxic dusts, asbestos, gases, etc
- ingestion of toxic materials
- water pollution
- explosions and fire
- degradation of materials
- phytotoxic effects.

4.2 Skin Effects

The principal contaminants that can present a hazard through skin contact include: coal tars* (including cyclohexane and toluene extractable materials), phenols, spent oxides (neat), nickel (metallic), nickel salts, zinc salts, boron compounds (greatly enhanced attack if skin is broken), free cyanide, ferro/ferri cyanide, sulphides, sulphates, acids and ammonium salts (liquids).

(* Coal tars are carcinogenic owing to the presence of certain polynuclear aromatic hydrocarbons (PAH).)

4.3 Inhalation

There are also many potentially lethal and very harmful substances that can be inhaled, particularly by investigation or construction workers in confined spaces such as trenches and pits. Some of the principal substances likely to occur are:

- hydrogen sulphide hydrogen cyanide, carbon dioxide (production enhanced under acidic conditions)
- coal tars (vapours and dusts)
- phenols (vapours)
- asbestos (dust)
- sulphur (vapour)
- boron compounds (dust)
- manganese (dust)
- ammonia (vapour)

4.4   Ingestion

The chemicals considered most harmful if ingested include: free cyanide and potassium cyanide, phenols (particularly if ingested by children), coal tars, ferro/ferricyanides (spent oxides), thiocyanate, sulphates, sulphides, zinc salts, copper salts (some) and chromium (hexavalent).

4.5   Water Pollution

Phenols are particularly important compounds because they contaminate groundwater supplies and can also infiltrate polymeric materials such as those frequently used for transporting drinking water supplies. Coal tars and mineral oils can also cause drinking water problems. By virtue of the wide range of liquid chemicals and the soluble solid chemicals used on iron and steel making sites, there is a legion of potential sources of groundwater contamination, provided, of course, that these contaminants can be carried into the saturated zone as liquids or leached downwards through surface water ingress. By far the most likely areas of significant pollution are the coking byproducts area and waste disposal areas. The most significant pollution sources are likely to include: mineral oils, coal tars, phenols, free cyanide, thiocyanate, sulphides, sulphates, ammonia/ammonium salts, and acids.

4.6   Degradation of Materials

The most common materials used in construction are attacked by the following:

concrete: sulphates (ammonium sulphate is particularly aggressive); chlorides; mineral oils; acids.

metals: sulphates; sulphate reducing bacteria; sulphide-oxiding bacteria; scrap metal; acids, alkalis, chlorides.

'plastics'   phenols, sulphates; (some) acids, alkalis and hydrocarbons.

rubbers   oxidising acids; oils, organic solvents; phenols; salts of copper, iron and manganese.

4.7   Explosion and Fire Risks

The risks of explosion or fire can relate either to the demolition of existing facilities or in the construction of the new

development. The following are examples of some features/locations
where such risks exist:

-     storage of:
                    liquid butane
                    gaseous oxygen
                    gaseous nitrogen
                    liquid nitrogen
                    fuel oils
                    natural gas
                    liquefied petroleum gas (LPG)
-     gas holders
-     mineral oils
-     explosive vapours
      mineral oils
      coal tars
      free cyanide
      sulphur
      sulphide
      ammonia salts
      acidity
-     combustibility of:
                    soil
                    mineral oils
                    coal tars
                    sulphur
                    coal/coke residues

4.8 Phytotoxicity

     Plants may be very susceptible to the presence of toxic
substances in the soil, including:

-     sulphates (including ammonium sulphates)
-     sulphides
-     spent oxides
-     free cyanide
-     complex cyanides
-     phenols
-     coal tars
-     acids
-     cadmium
-     zinc salts
-     nickel
-     copper (available) - greatly increased by low pH
      and low organic matter
-     manganese (usually if pH   5.5)
-     boron compounds

5   ENGINEERING PROBLEMS IN REDEVELOPMENT

5.1   Introduction

The engineering problems in redeveloping iron and steel making
sites can vary greatly from site to site and within the site itself.
On the one hand many works are situated in extensively mined areas or
in reclaimed coastal areas, while on the other hand, sites will
invariably hold many and extensive foundations, the full nature of
which cannot always be readily determined owing to successive
developments, for example. This range of factors can result in
stability problems as well as costly demolition and excavation
problems, in some instances.

5.2   Ground Stability

Many UK steel making sites are in coal mining areas.
Accordingly general subsidence can be a critical consideration,
unless, that is, the mining activities have long since ceased.
However, there may well be many mine shafts within the site boundary
owing to expansion of the original works. The locations of such
shafts can be critical to redevelopment proposals.

Also, many sites are in coastal marshy areas some of which are
often reclaimed using slag from the original works. Depending on the
use to which particular parts of the reclaimed areas were put, the
general stability of such sites is likely to be satisfactory owing to
the generally high loading imposed by such works. Nonetheless, this
does not reduce the require-ment to investigate the ground conditions
as appropriate to the redevelopment under consideration. Lagoons
used for the tipping of liquid wastes are likely to need particular
attention both in respect of removing liquids and stabilising or
removing sludges.

5.3   Old Above-Ground Structures

The nature of above-ground structures is, of course, very varied
and extensive. Their proper dismantling and demolition is not of
interest here except in the context of the potential chemical hazards
they hold. This interest has two principal facets: (a) there can be
numerous hazards to workers involved in site clearance operations (eg
from acids, phenols, gases, dusts and asbestos); and (b) the lack of
care during such operations frequently causes more extensive and
intensive ground contamination than that relating to the normal
operation of the plant. Overall, therefore, the method of removal or
alteration of above-ground structures (whether they be furnaces,
tanks or pipes) can be the most significant determinant in ground
contamination.

5.4 Old Foundations and In-Ground Structures

There is a very wide range of features that have extensive foundations or other structures underground. These include mass concrete, reinforced concrete, concrete/steel piles, cellars, pits, pipelines and cables. The physical dimensions of these features can be very large, depending on both the supported structure and the bearing capacity of the ground. The following are some of these features:

blastfurnace	milling shops
sinter plant	melting shops
crushing plant	mill cellars
coking plant	soaking pits
gas holders	slag pits
steel furnaces	boiler house
fuel storage	motor rooms
lime kilns	pump rooms
cooling towers	pickling plant
crane tracks	oil reclamation
drainage system	conveyors
acid tanks	

Floor slabs in relevant areas can vary from 150 mm to 1500 mm in thickness. Moreover, these might well be underlain by considerable fill, comprising hardcore, ash and slag. If such slag is disturbed, then expansive reactions could occur in some instances.

5.5 New Structures

5.5.1 Introduction. In addition to any inherited problems of ground stability, whether resulting from old mining activities or poor bearing capacity or possible differential settlement exacerbated by old foundations, one further particular engineering problem that could exist relates to the expansive nature of some slags. These problems are due to the presence of potentially expansive phases such as free lime and magnesia, the variability of their physical and chemical properties and the presence of varying, sometimes considerable, amounts of free iron. The presence of phases which hydrate and expand when wet rule out the use of steel slag in concrete. It is, however, possible to use the slag successfully for applications in which it is unbound and where some disruption can be accommodated or where it is bound by a material impervious to water such as bitumen, although suitable precautions to minimise the expansion have to be taken.

5.5.2 Causes of Instability in Slag. There are four commonly recognised chemical reactions giving rise to volumetric instability in slags:

<u>Free lime hydration</u>: calcium oxide commonly found in steel slag as a remnant of limestone added during the steel making process, hydrates to calcium hydroxide with a large (100%) volume increase. The hydroxide subsequently carbonates, but this reaction is not nearly so expansive. There is normally sufficient ground moisture to hydrate any free lime present in fill in a matter of months, but often the lime is embedded in otherwise stable slag, inaccessible to moisture, and remains unreacted for years, or until mechanically disturbed.

<u>Periclase hydration</u>: similarly periclase, or magnesium oxide, hydrates to form brucite (magnesium hydroxide), again associated with volume increase of more than 100%. However, this reaction is slower to begin and develop than the hydration of free lime. Disruption can take place several years after the deposition of periclase-bearing fill, and all that is certain, if periclase is present, is that the expansive hydration reaction will eventually occur. Free MgO can be found in high-magnesia steel slags (usually basic open hearth slags). Both free lime and periclase hydration are almost impossible to inhibit since all that is required is the potentially unstable fill, and moisture.

<u>Sulphoaluminate formation</u>: it is known that blastfurnace slag glass and sulphate can, under alkaline conditions react with moisture to form sulphoaluminates. A volume increase of about 130% relative to the glass is associated with this reaction. The process can be temporarily halted by the local absence of a reactant, eg glassy blastfurnace slag or gypsum, or a reaction condition, eg alkaline environment. However, extensive movement and remixing of fill, such as is inevitable in large-scale site reconstruction, will almost certainly restart such inhibited reactions.

<u>Iron unsoundness</u>: iron unsoundness is properly the term used to describe a specific hydrolysis reaction involving an iron sulphide, liable to occur in some blastfurnace slags. However, it is conveniently used to describe the hydrolysis of any iron sulphide, metal or low oxidation state oxide. Such hydrolysis reactions usually require acid conditions, and are associated with a characteristic 'exfoliation' texture of the slag.

The two expansion mechanisms most likely to be found in basic open hearth slag are free lime and periclase hydration.

6   REDEVELOPMENT: UK EXAMPLES

6.1 Introduction

A large number of former iron and steelworks sites in the UK
have been or are currently being redeveloped for uses that include
industry, housing, sports facilities, amenity areas, agriculture
and, in one instance, a garden festival site. These uses reflect
both normal land use restraints and attractions and also the range of
chemical and engineering difficulties on various parts of the site.
On some sites no reclamation has taken place but a range of the old
buildings have been converted to other industrial uses which have a
much smaller physical scale.

Most derelict works are in regions of economic decline and the
demand for new development, industrial or otherwise, can be low.
Accordingly many sites were identified as requiring specific capital
investment by central government to deal both with site clearance and
rehabilitation as well as providing new infrastructure, principally
roads and sewers. However, demolishing steelworks and creating new
building land with a view to replacing the old industry has many
complications in social, economic and technical terms. For a
successful redevelopment scheme, it is necessary to have a practical
understanding of the site conditions and to consider the redevelopment potential of the site as an integral part of the reclamation
process. Brief accounts of the developments on a number of sites are
given below. These are in part based on information presented at a
conference held in the UK in 1983 and partly on private information
sources. Information on other sites is also available.[11-17]

6.1.1 Workington Ironworks: The area surplus to the steel
industry's needs measured about 100 ha and covered the former coke
works and byproducts plant, a 10 million tonne slag bank and a marsh
site. The redevelopment proposals,[13] currently under way, include
three industrial areas, the reclamation of the slag banks incorporating leisure facilities, a housing area and a landscape improvement
programme for the whole site. Although one of the primary objectives
of the reclamation was to reduce the high levels of unemployment
following the works closure, there was a considerable desire to
redeem the site for recreational purposes. The estimated cost of the
reclamation proposals, which are phased to last through to 1988, is
about £15 million.

6.1.2 Consett Steelworks: This site covers some 290 ha.
Industry, housing, agriculture, public open space and forestry are
the principal proposed land uses.[13] These uses were selected
having given due consideration to such factors as:

-    possible retention and re-use of existing structures and
     features;

-   physical and chemical restraints;
-   landform and drainage;
-   site attractiveness (internally and externally);
-   inter-relationship of land use groups.

6.1.3 Corby Steelworks: Some 140 ha of land, which had
accommodated the blastfurnace and sinter plants, coke oven complexes
and steelmaking plant, have been cleared of all buildings and
virtually all foundations in advance of a detailed redevelopment
proposal. The majority of the site is destined for light industrial
(unspecified) uses with a modest area proposed for commercial and
shopping functions.

6.1.4 Hartlepool: The most significant reclamation on this site
has concerned the slag banks which covered some 40 ha in area and
which had been in use for over 100 years.[14,15] As with many such
tips, site investigations showed that they contained a mixture of
iron and steel slags, flue dust and general industrial rubbish. The
site also included a number of old slurry ponds (or recent and older
origins) containing coke oven, gas cooler and water softener wastes.
Again, as with similar sites, the materials encountered were very
variable in nature with widely differing degrees of compaction and
strength but also chemical instability. One significant reason for
the reclamation was the fact that the tip was a very prominent
feature on a main approach to the local town. The reclamation
included not only the toxic waste disposal but also landscaping and
planting, provision of playing fields and preparing an area for
eventual industrial redevelopment. This is technically most
important aspect of the redevelopment as the future industrial area
has been formed by reworking and properly compacting an area of
steelmaking slags shown in laboratory tests to be chemically
unstable. The physical level of the site is being carefully
monitored. Considerable uplift has occurred and the area is not yet
considered sufficiently stable for building purposes.

7   REFERENCES

1   'Problems Arising from the Redevelopment of Gas Works and
        Similar Sites'. AERE Harwell, UK, 1981.
2   Environmental Control in the Iron and Steel Industry,
        International Iron and Steel Institute, Brussels (1978).
3   World Steel in Figures 1983, International Iron and Steel
        Institute, Brussels (1983).
4   Industrial Process Profiles for Environmental Use: Chapter 24,
        The Iron and Steel Industry, 1972. US Environmental Protection
        Agency, EPA 600/2-77-023X.
5   Potentially Hazardous Emissions from the Extraction and
        Processing of Coal and Oil. US Environmental Protection Agency,
        EPA 650/2-75-038 (1984).

6   M Perch and R E Muder.   'Coal Carbonisation and Recovery of
     Coal Chemicals'.   In: Riegel's Handbook of Industrial Chemistry,
     7th Ed. New York, Van Norstrand Reinhold, pp 193-206 (1974).

7   A R Lee.   'Blastfurnace and Steel Slag: Production, Properties
     and Uses'.   Edward Arnold (Publishers) Ltd, London (1947).

8   W Gutt and P J Nixon.   'Use of Waste Materials in the
     Construction Industry'.   Building Research Establishment,
     Reprint R2/79, BRE Garston, Watford, UK (1979).

9   M A Smith, Building Research Establishment (private
     communication).

10   H Maczek and R Kola.   Metals, 1980 (Jan) 53-58.

11   Proc Conf Reclamation of Former Iron and Steelworks Sites,
     Windermere, UK, 1983, Cumbria County Council/Durham County
     Council, Durham (1983).

12   W M Stephens. 'Workington Ironworks Reclamation Scheme' in Ref
     11, pp G1 to G7.

13   C J Connally and P J Veitch.   'Computer aided landscape
     design': in Ref 11 pp F1 to F14.

14   W R G Eakin.   'Hard Development Potential and Design
     Techniques.   Part II - Some Technical Problems: in Ref 11, pp E7
     to E17.

15   B Clouston and R D McLean.   'Role of the Landscape Architect
     in Landscape Reclamation' Proc Con Reclamation of
     Contaminated Land, Eastbourne 1979, Society of the Chemical
     Industry, London (1980) pp G1/1-8.

16   A Gilchrist.   'New Industrial Land from Steelworks Tip Land'.
     Proc Conf Reclamation of Contaminated Land, Eastbourne 1979,
     Society of the Chemical Industry, London (1980), pp B9/1-5.

17   Ward Ashcroft and Parkman/Joint Unit for Research on the
     Urban Environment.   'The Redevelopment of the Bilston
     Steelworks Site, JURUE, Aston University, Birmingham
     (1980).

13

OVERVIEW, CONCLUSIONS AND RECOMMENDATIONS

M A Smith

Building Research Establishment

United Kingdom

1 INTRODUCTION

This Chapter presents an overview based on the discussions of the Study Group during its various meetings leading into the overall conclusions and recommendations arising from the study. The more detailed recommendations from each of the separate Projects are then summarised. These last two Sections formed the basis of the Summary Report submitted to the NATO Committee on Challenges of Modern Society which is reproduced in Appendix J.

2 OVERVIEW

The Study was primarily concerned with reviewing existing and potential technologies. Nevertheless it is believed that the Study will help in the solution of problems in three principal ways:

(i) by clarifying some of the interactions that must be taken into account,

(ii) by emphasising the need to consider the long-term effectiveness of remedial measures, and in so doing, pointing out the limitations of current knowledge on this aspect,

(iii) by drawing attention to positive developments in finding both permanent treatments and improvement in the likely long-term effectiveness of other treatments.

The subject of contaminated land is developing rapidly and during the course of the Study a number of important manuals have

341

been produced, particularly in the USA and the Netherlands.[1-5]
These, and the present report are largely based on reviews of
potential technologies and the practical experience incorporated in
them is limited. It is important therefore that remedial schemes are
carefully evaluated and that the results of this evaluation, together
with the results of the research that is continuing in a number of
countries, are used in developing guidance in the future.

The Study has undoubtedly made authorities in the participating
countries more aware of alternative approaches adopted in other
countries and useful technology transfer is occurring as a result.
It is hoped that the Report will act as a spur to tackle the
technical problems that must be overcome to remedy contamination from
both deliberate and accidental acts in the past.

'Contaminated land' was deliberately defined to embrace a
spectrum of sites which differed widely in size and in the nature and
extent of contamination. A 'typical' contaminated site might,
however, be characterised as follows: a volume of material deposited
in a hole in the ground or a volume of soil that has become
contaminated; water-soluble and non-soluble contaminants are moving
from the site into groundwater towards an aquifer used for water
extraction; some gaseous pollutants are also entering the atmosphere
from the site. The essential questions that the Study Group
addressed were (i) what can be done to prevent further environmental
impact, (ii) can the site be brought back into beneficial use? and
(iii) how effective in the long term will the remedial action be.

The Study Group started from the premise that excavation and
removal of the contaminated material for deposition elsewhere is not
always an acceptable or practicable option. There can be major
environmental impacts whilst the work is in progress and the problem
may be simply transferred elsewhere to re-emerge later. Nor is
excavation a simple process or one that can be carried out without
recourse to other remedial operations such as creation of barrier
walls and treatment of groundwater. In retrospect the Study Group
could usefully have looked at the practical problems encountered
where excavation had been employed. Excavation is the first step in
application of on-site treatment processes, such as those discussed
in Chapter 3, and thus further attention to this might have yielded a
better idea of the overall value of such processes.

On-site treatment processes, with the exception of those
involving stabilisation of the contaminated soil, are designed to
provide a permanent or ultimate solution at the site. Overall
the technical prospects look promising. Some processes involving
thermal treatment and separation have already been applied successfully and others are at an advanced stage of development. In general
on site treatment processes will rely on the application of

established technology from chemical engineering and mineral processing and it is to be expected that any technical problems would be capable of solution. In the medium-term, microbiological treatment systems look promising. There is considerable commercial interest in the development of such systems for the treatment of waste streams and as a part of mineral processing and chemical production systems.

There have, as yet, been few successful applications of in-situ treatment of the type dealt with in Chapter 4 and there are significant technical difficulties to be overcome. Nevertheless some interesting and promising developments are taking place. As in-situ treatments concept can in many cases offer the possibility of permanent solutions it would be worthwhile perservering with research and development in this area. Microbiological treatment systems, in particular, have considerable potential.

At present, in most countries, attempts to solve a 'contaminant' problem are likely to involve cover, barriers, and hydraulic measures with provision for treatment of groundwater and leachate. Such measures may, in any case, be required to supplement on-site and in-situ treatment processes. Long-term effectiveness is of paramount importance for such remedial measures. The designer, in assessing the likely overall effectiveness of any scheme, has to consider the likely installed effectiveness of each individual component, its interaction with other components, and any changes that there may be with time. The designer must therefore carry out a systematic analysis to identify the risks (to the system) and to quantify these where possible (eg one in one hundred years rainfall events are commonly taken into account in the design of drainage/sewerage schemes). This procedure is analogous to what should be done for any major engineering work. It may be aided by the use of modelling of groundwater and contaminant movement including the effects of barrier and hydraulic systems (see Chapter 9). Efforts are currently being made by the US Environmental Protection Agency to apply modelling to the whole remedial process.

Whilst Chapter 2 draws attention to the current lack of information on the effectiveness of containment systems when used to treat contaminated land there are, nevertheless, promising aspects. The construction of vertical barriers is a well established engineering technique and the potential limitations with regard to interaction with contaminants have been recognised and work is in progress to produce improved systems. Similarly most covering systems are expected to serve several different functions. This is now recognised and they are being designed progressively on a more rational basis. Thus whilst containment, with associated hydraulic measures, may not provide a once and for all solution, it can provide in many cases a solution that is likely to remain effective for a considerable period of time.

The discussion of volatile organic compounds (VOC's), in Chapter 9 illustrates just how difficult it can be to assess the potential health and environmental impacts of contamination. It shows clearly that volatile substances can enter the environ-ment from many different sources by a variety of routes, that both sources and routes may differ at different stages of investigation and remedial treatment and that the impact may occur at significant distances (several kilometres) from a site. The last means that it may be difficult and expensive to conduct a survey that will identify all impacts and conversely it may be difficult to link an overt health or environmental impact to a remote and unsuspected source.

In the case of VOC's and other gaseous contaminants, field measurements are an essential part of the investigation process. Chapter 10, in describing methods of on-site chemical analysis including mobile laboratories, demonstrates the benefits already obtained from the use of such techniques. The Chapter deals only with United States experience. Information was sought from other countries but little was forthcoming. This reflects perhaps the different scale and nature of the problems in the different countries. The occurrence of highly damaging sites in Europe is not so high as to have generated such systems to date although the members of the Study Group recognised a need for them. The experience of the United States Environmental Protection Agency in this respect will be of help to agencies in other countries faced with developing their own approach to rapid site assessment.

Chapter 11 is important in drawing attention to the dereliction and the social consequences that result from the large-scale run-down of an industry such as the iron and steel industry. These are such as to make it imperative that the land be brought back into beneficial use. The other Chapters are all concerned with that task and take as their starting point the idea that the elimination of an immediate environmental impact is not enough; the land has to be restored to use in a way which requires only minimal long-term attention.

3   GENERAL CONCLUSIONS

The success of the Pilot Study has been in part due to the fact that the participating countries have each tended to approach the problem from slightly different angles. As a result they have con-centrated their technological efforts on different forms of remedial action, and there has been mutual advantage in sharing knowledge and experience with other participants. Most countries are engaged in the production of guidance for national use which has been an additional pressure on the individuals concerned but it has meant that the Study has already been of benefit to participants and will rapidly influence thinking throughout the participant countries.

The development of on-site and in-situ processes resulting in the removal or destruction of contaminants is to be encouraged as providing once and for all solutions.

Very few of the technologies described have been sufficiently proved in applications specific to the treatment of contaminated land although they may have been well tried for other purposes. Consequently, it is essential that long-term evaluation studies are established.

Retrospective studies of already reclaimed sites should be encouraged in order to obtain information on the performance of the treatment strategies adopted. It is recognised, however, that the lack of baseline data and the unwillingness of responsible authorities to have any doubts cast upon the 'success' of a completed reclamation scheme may create difficulties.

The long-term nature of research into remedial measures designed for containment or stabilisation must be recognised. Some projects may need several years for completion and it may take longer to accumulate sufficient confirmatory evidence. Research funding should reflect this time-scale. New methods of treatment cannot be proved satisfactorily under laboratory conditions alone. There is a need for properly designed field trials and demonstration projects. The natural reluctance of local communities to be guinea-pigs for novel solutions can be eased by the underwriting of such schemes, and any subsequent remedy required, by responsible authorities.

Selected monitoring of completed or on-going remedial actions should be viewed as an essential component part of the remedial scheme. Such monitoring is an essential part of many civil engineering projects and should not be viewed as casting doubt on the effectiveness of the selected solution. This point will have to be presented with care in some cases where public concern is high.

It should be recognised that maintenance must be considered and is likely to be required whenever a non-permanent treatment option is adopted.

Funding of remedial actions should take account of the need for monitoring, maintenance and the need sometimes for a phased approach to remedial action.

Records of contaminated sites and of the treatments carried out should be kept unless the contamination is removed or destroyed. Otherwise the site remains contaminated, even though any environmental risk is either eliminated or reduced to an acceptable level for the forseeable future. Such records are important since the use of the site might change, the remedial measures may deteriorate, or our knowledge about the effects of the contaminants may change thereby changing our perception of the risks. The future development of such sites should be subject to adequate control by the appropriate authorities.

The development of contaminated land for residential use, recreational or other uses likely to bring people into close contact with it, requires special care and the responsible authorities will wish to satisfy themselves that controls are adequate.

Policies for the disposal of wastes to land should take after use of the land into account.

Industries should be encouraged to practice good house-keeping and to avoid contaminating their sites. The potential for contamination that might inhibit after-use should be a material consideration when development takes place.

Remedial actions on contaminated sites require a multi-disciplinary approach. It is important to recognise that scientific and professional judgement will need to be exercised and that it may, in consequence, be difficult to convince those with political responsibility and the community about the effectiveness of proposed actions. The last two will tend to ask for assurances of absolute safety and to seek 'fail safe' remedies. The responsibility to provide advice should not be shirked by the professional specialist even in circumstances where that advice may not be entirely welcomed and may be eventually ignored.

4   CONCLUSIONS AND RECOMMENDATIONS OF THE INDIVIDUAL
      CHAPTERS IN THE REPORT

4.1   Long-term Effectiveness of Remedial Measures

The principal conclusions of the discussion in Chapter 2 on the long-term effectiveness of remedial measures have provided the basis of the conclusions of the study cited above: the need to view monitoring as an essential part of remedial action, the need to evaluate the performance of remedial measures carried out in order to advance knowledge, and the need for long-term research.

Chapter 2 also drew attention to the time dependent and dynamic nature of contamination, remedial systems and the environment in which the remedial system has to operate.

Remedial actions are seen as falling into three main groups:

(i)   those that remove contaminants or render them harmless,
(ii) those that prevent the release of contaminants, and

(iii)   those that decrease the release of contaminants.

If the first group of actions are carried out properly, they provide an ultimate solution and the question of long-term effectiveness does not arise. On-site processing and certain forms of in-situ treatment may provide such solutions and the further development of the technologies should be encouraged. All other technologies offer a solution of only limited or uncertain duration unless other mechanisms, such as microbial attack, are such as to reduce the level of contamination.

In general treatment systems based on isolation of the contamination (eg covering systems) are vulnerable to loss of effectiveness with time. Like many other engineered projects they will have a finite life. They need maintenance and renewal as long as the contaminants inside are present and their release would be considered harmful. Despite a potentially limited lifetime such measures may prevent or reduce the release of contaminants for as long as is considered necessary or may provide time in which longer-term solutions can be developed. The development of materials more resistant to chemical and microbial attack should make macro-encapsulation more economic as well as improving its long-term effectiveness.

The evaluation of remedial actions requires the establishment of an adequate set of baseline data before and immediately after treatment and the monitoring of parameters that will

   (i)   describe the behaviour of the remedial system,
  (ii)   describe conditions within and outside the contaminated area,

4.2   On-site Treatment

In principle there are several different procedures which might be used for on-site treatment. These include extraction, thermal treatment, chemical treatment, mechanical and physical separation, steam-stripping, micro-biological treatment, stabilisation and flotation. In practice, only thermal treatment, sludge farming, certain stabilisation methods and a number of methods involving treatment with a liquid phase (extraction, flotation, classification) are developed to the stage where application would be possible in some cases at present. The thermal treatment of soil with the object of evaporating the contaminants and treating the exhaust gases in an afterburner has been operational in the Netherlands since 1982.

Other methods such as extraction/classification, flotation and removal of hydrocarbons by jetting them loose from soil have been operational for only a few months or are being tested on a pilot scale at present (Spring 1984). One special mode of soil treatment,

namely sludge farming, has been well known for many years and is
particularly suitable for contaminants that are easily biodegradable.
Although little information is available concerning the other modes
of soil treatment discussed in Chapter 3 they do have valuable
potential applications and merit more detailed investigation and
further evaluation. The most interesting on-site processes should be
the subject of an intensive study in the near future.

4.3   In-situ Treatment

     In-situ treatment of contaminated land offers a number of
attractive options for (i) removal or destruction of contaminants,
(ii) stabilisation of the contamination (iii) solidification to
achieve some engineering objective such as improved ground stability.
Two main treatment methods are possible:

  (i)   surface application of the treatment agent,
  (ii) injection of the treatment agent.

     The latter is analagous to the well established engineering
practice of grouting, and indeed grouting is one available treatment
option.

     The major technical difficulties are: how to ensure intimate
contact between treatment agent and contamination particularly with
the inherent chemical and physical heterogeneity of many contaminated
sites; possible unwanted interactions between treatment agents and
contaminants; difficulties in ensuring that the treatment has been
fully effective; difficulties in applying injection techniques at
depths of less than about 2 m; and production, in many cases, of a
liquid waste stream requiring treatment.

     Even if such measures can be properly developed they are
unlikely to be sufficient on their own and will tend to form part of
a remedial system including supplementary measures (eg in-ground
barriers).

     The thermal treatment of soil by direct application of
electrical energy to cause melting is an option of a different
character currently under investigation which may have some useful
applications.

     The development of in-situ treatments, particularly those
designed to remove contaminants or render them harmless, is to be
encouraged. The potential benefits would justify the effort required
to overcome the technical difficulties.

4.4   Covering Systems

Chapter 5 discusses covering systems in terms of the many, often conflicting functions, that they have to meet when used in conjunction with other isolation measures. The performance requirements of a particular covering system are largely determined by the nature of the contamination, the local geology and hydrogeology, and the projected use of the site.

It is possible to design a covering system, consisting of a single layer or a succession of layers of selected materials, to achieve the desired physical and chemical properties of the finished system. However, economics will often dictate the use of locally available materials to their best advantage so as to optimise their 'performance' rather than the use of ideal materials.

In the absence of practical experience covering systems will frequently have to be based on application of knowledge of the physical and chemical properties of the component parts.

The performance of covering systems is particularly time-dependent. The component parts of the system and contamina-tion may change with time and the environmental stress on the system may also increase. Vegetation growth can be both beneficial and detrimental; increased cover will reduce erosion but root growth may lead to penetration of synthetic barriers and to the uptake of toxic elements. The properties of all synthetic materials are likely to deteriorate with time.

Further research on covering systems is required through laboratory studies, evaluation of current field applications and retrospective studies.

4.5   In-ground Barriers/Hydraulic Measures

Vertical barriers to control the movement of groundwater and contamination can be installed using well established engineering procedures including slurry trenches, diaphragm piling and grout curtains. Such barriers will always permit passage of some water or other fluids: either because the permeability although very low is finite or because of imperfections in installation. There are, however, doubts concerning their long-term effectiveness as a means of controlling contamination owing to possible adverse interaction with contaminants and the possibility of breaches either induced by nature (eg tree roots) or by man (eg subsequent excavation).

Horizontal barriers can be installed by a number of ground injection/grouting techniques but these are not well established nor proven. They too are susceptible to adverse interactions with contaminants.

4.6 Groundwater Management and Treatment

Chapter 8 addresses the situation where groundwater may be or has been adversely affected by contaminants and the methods available to (a) enhance the quality of the degraded groundwater by in-situ treatment, (b) modify groundwater regimes, (c) treat the groundwater after extraction.

A wide range of methods are available for the treatment of contaminated water after extraction. Their economic and technical viability is restricted, however, by increasing difficulty in treating ever more dilute contaminants. After treatment, the extracted water can be reinjected or disposed of to surface waters.

In-situ treatment of groundwater suffers many of the same constraints as in-situ treatment of ground (see Section 4.3 above) including difficulty in contacting reactive agents and contaminants, uncertainty about permanence of treatment and difficulty establishing how effective it is at the time it is carried out. Nevertheless it has been successfully employed on a number of occasions and such techniques should be further developed.

4.7 Mathematical Modelling

Groundwater modelling (including the modelling of pollution movement) involves the use of field data in conjunction with certain established mathematical procedures to predict direction, quantity and rate of flow. Judicious use of a 'site-model' can assist in the placement of monitoring and treatment wells, prediction of contaminant movement and prediction of the consequences of installation of hydraulic control measures such as cut-off barriers or draw-off wells. The prediction of contaminant movement can provide a timescale (eg time for the contamination to reach a critical drinking water extraction well) against which remedial actions can be designed and installed.

It is stressed in Chapter 9 that the model results depend on the quality and quantity of the hydrogeologic data used. However, allowing for the uncertainties in the data, groundwater modelling is an important aid to rational development of a programme of remedial measures for contaminated sites.

4.8 Toxic and Flammable Gases

Toxic and flammable gases can, if released from contaminated sites, cause health problems, for example for workers at the site. Possible sources of such emissions include disposal operations, storage facilities, and treatment units. The control and correction of problems associated with toxic and flammable gases is best regarded as part of remedial action or the operation of a hazardous waste facility. Information may be required on the following:

- sources, composition and properties of gases,
- sampling and analysis,
- an evalutation of the health aspects,
- protective and safety equipment,
- construction and equipment requirements,
- remedial action measures required.

Because gas problems are site-specific, a thorough assessment of each individual site is necessary in order to identify the problem and then to develop control systems. Even though a variety of control systems are available for preventing movement of gases, buildings should not be constructed on sites likely to release gases until gas generation has stopped or the source of gases has been removed or corrected.

The use of emission models and comprehensive site surveys will aid investigators in determining the extent of any problem. The purpose and method of sample collection should always be reported, as different methods will yield different results – the data can only be properly evaluated if this information is provided. There is a need for research to improve and refine air emission models and sampling and analytical techniques. Control techniques should be monitored for their effectiveness.

4.9 Rapid Methods of On-site Analysis

Chapter 11 identifies and critically evaluates current analytical technology applicable to the on-site assessment of contaminated land. The study focused on methodology that is currently employed in the field by the United States Environmental Protection Agency and its agents in support of clean-up activities at hazardous materials spills and uncontrolled hazardous waste sites. The methods evaluated vary in complexity, ranging from non-specific essentially screening techniques to compound specific, quantitative, multi-parametric analytical protocols.

The choice of on-site techniques for evaluating contaminants will be determined by many factors including:

(i)   the type and size of site and associated environmental
        factors,
(ii)   the types and abundances of contaminants anticipated at the
        site,
(iii)   the level of investigatory effort required at the site.

There is a continuing need for the development of new techniques
and refinement of others that can provide on-site analyses for the
large array of contaminants which are encountered.

4.10 Steel Industry Sites

Chapter 12 on steel industry sites is based on a report prepared
under the CCMS Fellowship Scheme. It is included in this final
report of the Study Group because it is considered to provide a good
illustration of the benefits of such industry-based audits and the
common nature of the problems that such sites present, no matter
where they are.

The steel industry was selected because of the rapid reduction
in the size of the industry brought about by improved productivity,
the effects of the present world recession and increased pressure
from imports from new producer countries. In addition much
information relevant to coking plants is already available in the
audits made of the problems presented by coal carbonisation and
similar sites. The size of the potential problem can be seen in the
contrast in 1980 between production and capacity in the EEC
countries: 128 vs 202 M tonnes. In the UK at least six major plants
have been closed in recent years.

Generalisations about iron and steel making operations are
difficult because of the many technical changes that have occurred in
the industry. Thus location, age and integral complexity all have to
be considered. However, they all have in common, albeit to varying
degrees, a number of potential problems in terms of contamination or
engineering factors affecting redevelopment.

Principal problems include chemical contamination in the
coking plant area, pickling plant, ore and waste disposal areas such
as lagoons and slag heaps, large deposits of old physically and
chemically unstable slag, massive foundations, and the presence of
underground workings and mineshafts when plants have been located in
mining areas to make use of local materials.

5   RECOMMENDATIONS TO CCMS

The Study Group in its Summary Report to CCMS (see Appendix J)
made a number of recommendations to CCMS drawing attention to the

technical conclusions of this report and the need for appropriate policies not only for dealing with existing contamination problems but to limit their occurrence in the future.

The Committee was invited to:

(i)   note that the present Pilot Study has shown the value of international collaboration in a fast developing area of technical activity,

(ii) commend the Report of the contaminated land Pilot Study Group on contaminated land to member governments drawing attention to the general conclusions and those relating to individual projects,

(iii) encourage member governments to consider the adoption of policies:

(a) that will minimise the occurrence of contaminated land problems in the future

(b) to abate the adverse environmental impacts from contaminated land

(c) for the safe reuse of contaminated land

(iv) to continue to provide funds under the CCMS Fellowship Scheme for studies within the subject area.

6   REFERENCES

1   Slurry Trench Construction for Pollution Migration Control, US Environmental Protection Agency, Cincinnati (1984). Report EPA-540/2-84-001

2   Handbook Bodensaneringstechnieken (Handbook of soil reconstruction techniques), Ministry of Public Health and the Environment, Directorate General for the Environment (Netherlands), Staatsuitgeveriji, The Hague (1983)

3   N Grondige Aanpak (A thorough approach – what you must know about soil reconstruction). Ministry of Housing and the Environment (Netherlands), The Hague (1983)

4   Handbook – Remedial Action of Waste Disposal Sites, US Environmental Protection Agency, Cincinnati (1982), Report EPA 625/6-82-006

5   Remedial Response at Hazardous Waste Sites, US Environmental Protection Agency, Cincinnati (1984)

(a) Summary Report   – EPA 540/2/84-002a
(b) Case Studies 1-23 – EPA 540/2/84-002b

APPENDIX A: Names and addresses of Study Group members
               (National representatives and Project leaders)

Pilot Study Director:
M A Smith
                                  Department of the Environment
                                  Building Research Establishment
                                  Watford WD2 7JR
                                  United Kingdom

NATIONAL REPRESENTATIVES:

J W Assink
                                  TNO
                                  Laan van Westenek 501
                                  Postbus 342
                                  7300 AH Apeldoorn
                                  The Netherlands

M J Beckett
                                  Department of the Environment
                                  Central Directorate on
                                  Environmental Pollution
                                  Room A3.29
                                  Romney House
                                  43 Marsham Street
                                  London SW1P 3PY
                                  United Kingdom

K A Childs
                                  Senior Advisor Landfill Site
                                    Remediation
                                  Waste Management Branch
                                  Environmental Protection
                                    Service
                                  Environment Canada
                                  Ottawa
                                  Ontario, K1A 1C8
                                  Canada

P Godin
                                  Ministere de L'Environnement
                                  Direction de la'Prevention des
                                  Pollutions
                                  Service des Dechets
                                  14 Boulevard de General-Leclerc
                                  92524 Neuilly sur Seine
                                  Cedex
                                  France

A Hoschuetzky
                                  Bundesministerium des Innern
                                  Graurheindorfer Strasse 198
                                  D-5300 Bonn-1
                                  Federal Republic of Germany

J W van Lidth de Jeude   Ministry of Public Housing,
                                  Physical Planning and
                                     Environment
                                  Soil Protection Division
                                  Postbus 450
                                  2260 MB Leidschendam
                                  The Netherlands

R A Page   Environmental Protection Unit
                                  Welsh Office
                                  Cathays Park
                                  Cardiff
                                  CF1 3NQ
                                  United Kingdom

D E Sanning   USEPA Municipal Environment
                                     Research Laboratory
                                  Solid and Hazardous Waste
                                     Research Division
                                  26 West Street, Clair Street
                                  Cincinnati, Ohio 45268
                                  USA

K Stief   Umweltbundesamt
                                  Bismarckplatz 1
                                  D 1000 Berlin 33
                                  Federal Republic of Germany

Ms K Warnoe   Miljostyrelsen
                                  Standgade 29
                                  DK-1401 Kopenhagen
                                  Denmark

PROJECT LEADERS:

In-situ treatment   D E Sanning (see above)

On-site treatment   J W Assink (see above)

Covering systems   Dr G D R Parry
                                  Environmental Advisory Unit
                                  Department of Botany
                                  Liverpool University
                                  PO Box 147
                                  Liverpool L69 3BX
                                  United Kingdom

In-ground barriers/   K A Childs (see above)
hydraulic measures

Groundwater management and   K A Childs (see above)
treatment and mathematical
modelling

Rapid on-site methods of   M Gruenfeld
analysis   US Environmental Protection Agency
                                       Municipal Environmental Research
                                         Laboratory - Ci
                                       Edison
                                       New Jersey 08837
                                       USA

Toxic and flammable gases   S C James
                                       US Environmental Protection Agency
                                       Municipal Environment Research
                                         Laboratory
                                       26 West Street, Clair Street
                                       Cincinnati, Ohio 45268
                                       USA

Long-term effectiveness of
remedial actions   K Stief (see above)

Note: In addition to those listed above, and the CCMS Fellows
        listed in Appendix B, a number of other people attended
        one or more of the Study Group meetings, making valuable
        contributions to the discussion. These included:

        F J Colon   TNO, Netherlands
        V E Niemela   Environment Canada
        W H Rulkens   TNO, Netherlands

APPENDIX B: Names and addresses of recipients of CCMS Fellowships who contributed to Pilot Study

1982 FELLOWSHIPS:

J W Assink
TNO
Postbus 342
7300 AH Apeldoorn
The Netherlands

D L Barry
Atkins Research & Development
Woodcote Grove
Ashley Road
Epsom KT18 5BW
United Kingdom

Dr R M Bell
Environmental Advisory Unit
Department of Botany
University of Liverpool
PO Box 147
Liverpool L69 3BX
United Kingdom

1983 FELLOWSHIPS:

Dr E M Bridges
Department of Geography
University College of Swansea
Singleton Park
Swansea
SA2 8PP
West Glamorgan
United Kingdom

Dr T Cairney
Department of Building and
  Civil Engineering
Liverpool Polytechnic
Clarence Street
Liverpool L3 5UG
United Kingdom

Dr W G Coldewey
Westfalische Berggewerkschaftskasse
Institut fur Angewandte Geologie
Herner Strasse 45
4630 Bochum
Federal Republic of Germany

Dr U Schoettler
Institut fur Wasserforschung GmbH
  Dortmund
Zum Kellerback 52
D5480 Schwerte 1-Geisecke
Federal Republic of Germany

Dr P Geldner
Ingenieurbuero Dr-Ing Bjornsen
Kurfurstenstr 80
5400 Koblenz
Federal Republic of Germany

H Hatayama
California Department of
  Health
Hazardous Waste Management
  Branch
2151 Berkley Way
Room 119
Berkley
California 94704
USA

T Dahl
National Enforcement
  Investigations Center
US Environmental Protection
  Agency
Building 53, Box 25227
Denver
Colorado 80225
USA

APPENDIX C: Study Group programme

Meeting	Dates	Location	Hosts
Inaugural	Nov 81	Watford, UK	Department of the Environment Building Research Establishment
Second	April 82	The Hague, Netherlands	TNO
Third	2-3 Dec 82	Washington DC USA	US Environmental Protection Agency
Fourth	30 May – 2 June 83	Hamburg FRG	Umweltbundesamt & City of Hamburg
Final	April 84	Cardiff UK	Welsh Office

In addition workshops were held in The Hague on 7-9 September and Cincinnati, USA on 26-27

APPENDIX D: Subjects identified for study by the CCMS Study Group
But which were not pursued for one reason or another

E   Organic chemicals and plants

To study the tolerance to, and the uptake of, organic chemicals by
plants and their implications for site assessment, human health and
site reclamation. It was agreed that, if taken up the study should
concentrate on a few substances considered to be of general interest,
eg creosols and chlorinated hydrocarbons. Although strong support
was given to the proposal by most participants at the first meeting,
and interest was confirmed at the second meeting, no offers of
leadership were made.

I   Social impact of contaminated sites

The direct involvement of the public in such incidents as Love Canal,
Lekkerkerk and Shipham has had a significant impact on individuals
and communities. The response of these communities has however
differed and this appears to be related not only to the nature of the
situation in which they find themselves but also to the way in which
the 'problem' has been handled by the authorities. Government
institutions may not always have the technical or financial resources
or legal authority to deal with 'problem' sites. Lack of information
or dissemination of erroneous information may have detrimental
effects that exceed those of the toxic substances themselves. It was
thought that benefits could accrue from sharing information on
administrative arrangements and people's responses to 'problem'
sites.

Although considered of interest by the individuals participating in
the study it was subsequently decided that the subject was not
appropriate for detailed attention by the study group. The United
States provided some relevant information to participants.

J   Soil quality criteria

Collection and dissemination of information on national and other
guide-lines relating levels of contamination to acceptability of land
for particular end uses including site specific examples, was
proposed. It was not intended that any attempt should be made to
propose common criteria but it was thought knowledge of what had been
produced elsewhere could be of assistance to those charged with
producing such guidelines. A limited amount of information was
exchanged.

M   Contamination and specific industries

The background to this Project is explained in the main text.
Although the Study Group as a whole could not pursue the Project, a
contribution on steel industry sites was produced by a CCMS Fellow
(see Chapter 12).

0   Site identification in urban areas

Methods for identifying contaminated sites in urban areas as opposed
to identification of specific site types, eg hazardous waste
'problem' sites on a national or regional basis. The United Kingdom
provided some information to other participants.

Note: Projects A,B,C,D,F,G and H were actively pursued as part of the
Study. Projects E,I,J,M and 0 are commented on above. Projects K,L
and N formed part of a structured exchange of information on
important case studies and research in progress.

APPENDIX E: Work done by recipients of CCMS Fellowships
             participating in work of Study Group

J W Assink   Co-author of report on on-site
                                    processing of contaminated
                                    soil (see Chapter 3)

D L Barry   Report on redevelopment of
                                    former iron and steel making
                                    plants (see Summary in
                                    Chapter 12)

Dr R M Bell   Co-author of report on covering
                                    systems for contaminated land
                                    (see Chapter 5)

Dr E M Bridges   Fellowship used to meet part
                                    of costs of study on the
                                    application of multi-spectral
                                    remote sensing to the identi-
                                    fication and assessment of
                                    contaminated land including
                                    visits overseas[4]

Dr T Cairney   Fellowship used for travel in
                                    Europe in connection with study
                                    of the factors governing
                                    migration of soil moisture due
                                    to capilliary forces and
                                    development of an appropriate
                                    model

Dr W G Coldewey   Reports on experience with the
                                    covering of contaminated land
                                    in the Ruhr region of the
                                    Federal Republic of Germany[1]
                                    (to be published in
                                    Mitteilungen der Westfaelischen
                                    Berggewerkschaftskasse) and on
                                    the water permeability of
                                    cohesive soils.[2]

T Dahl   Preparation of a detailed
                                          report describing all aspects
                                          of remedial action on two major
                                          US problem sites; Occidental
                                          Chemical, Lathrop, California
                                          and Stringfellow Acid Pits,
                                          Glen Avon, California.
                                          Fellowship also used for travel
                                          in Europe

Dr P Geldner   Preparation of report on
                                          remedial measures for large
                                          scale aquifer contamination

H Hatayama   Preparation of a report with
                                          particular emphasis on the
                                          emission of volatile organic
                                          compounds from contaminated
                                          sites

Dr U Schoettler   Report on treatment of
                                          contaminated groundwater from
                                          contaminated sites (to be
                                          published by Institut fur
                                          Wasserforschung GmbH,
                                          Dortmund).[3]

C Tarus*   Study of proposal to desalinate
                                          soils in Western Turkey

* Cengiz Tarus
  Research Director
  TOPRAKSU General Directorate
  Turkey

REFERENCES

1   W G Coldewey 'Experience with the covering of contaminated land
     in the Ruhr region,' Westfaelische Berggewerkschaftskasse,
     Instutute for Applied Geology (May 1983)

2   W G Coldewey (Edit) 'Untersuchungen zur Wasserdurchlaessigkeit
     bindiger Boeden,' Mitteilungen der Westfaelishen
     Berggewerkschaftskasse, No 43 (August 1984). Translation
     available: Investigation of the water permeability of cohesive
     soils,' Reports of the Westphalian Mining Operations Fund,
     No 43)

3   U Shoettler. 'Treatment of contaminated groundwater from
     contaminated sites.' Institut fur Wasserforschung GmbH,
     Dortmund (1984)

4   M G Coulson and E M Bridges. 'The remote sensing of
     contaminated land'. International Journal of the Remote
     Sensing, Vol 4 No 5 pp 659-669(1984)

APPENDIX F: Detailed description of three on-site treatment methods
              developed in the Netherlands*

by W H Rulkens, J W Assink, W J Th van Gemert

F.1 Removal of organic bromine compounds (laboratory and pilot plant scale investigations

A site in the municipality of Wierden (the Netherlands) is contaminated with several aliphatic bromine compounds, such as tetrabromo ethanes, dibromo butanes and dibromo decane. This site, containing about 30 000 tonnes of soil and with a concentration of contaminants varying between ten and several thousands of mg bromine per kg of soil, is a threat to groundwater used for the production of drinking water.

By order of the Ministry of Public Health and Environmental Protection, TNO (PO Box 214, 2600 AE Delft, The Netherlands) and HBG (Hollandsche Beton Groep NV, PO Box 81, 2280 AB Rijswijk, The Netherlands) investigated a number of cleaning-up techniques.

Experiments on a laboratory scale were carried out to assess the possibilities of cleaning-up techniques such as:

(a) Extraction with 1,1,1 trichloro ethane or aqueous solutions of:

   –   Soft soap

   – Na_2CO_3

   –   NaOH

   –   NaClO

(b) Electrolysis of soil in a mixture of water and ethanol

(c) Flotation

(d) Thermal treatment methods:

   –   Vacuum distillation (temperatures up to $200^{\circ}C$)

   –   Steam stripping

   –   Evaporation (temperatures up to $600^{\circ}C$)

---

* Appendix to Chapter 3

An evaluation of the results of laboratory experiments showed
that extraction with an aqueous solution of NaOH is the most
promising technique for development to a practical scale. Important
criteria for choosing the extraction technique are: efficiency,
technical feasibility, costs, amount and composition of waste
streams. Extraction with NaOH-solution is successful because of its
dispersing properties. The organic bromine compounds that are
largely adsorbed to the humus-like substances in the soil are removed
along with these substances during extraction, resulting in
decontamination of the treated soil.

Experiments were conducted on a bench scale and a pilot-plant
scale to collect all relevant information for scaling-up purposes.
Attention was paid to the conditions during soil extraction,
apparatus suitable for the extraction process and cleaning of the
polluted extracting agent. Data collected from these experiments
made it possible to design an on-site treatment installation. Figure
F.1 shows a process scheme of the installation. The following
process-steps can be distinguished:

(i)   Soil pre-treatment to separate large objects (eg stones) and
        reduce the size of large clods of soil (crushing and wet
        sieving).

(ii)   Intimate mixing of soil with extracting agent (approx 0.2%
        NaOH solution); the soil-to-water ratio is about 3 to 1 on a
        weight basis.

(iii)   Extraction and washing of the soil with clean extracting
        agent in counter current flow in two modified screw
        classifiers in line.

(iv)   Dewatering of soil before redeposition. The remaining
        alkalinity of the soil will be largely neutralised by
        absorption of CO_2 from the ambient air.

(v)   The overflow of the first modified screw classifier is led
        through a settling tank for fine mineral particles dragged
        out from the screw classifier by the extracting liquid. The
        particles that settle, with diameters between approx 35 um
        and 60 um (approx 1% of the total soil), are collected from
        time to time and washed separately by mixing them with NaOH-
        solution.

(vi)   Sludge forming by adding lime as coagulant and polyelectro-
        lyte as flocculant. The sludge formed can be separated in a
        tiltable plate separator.

(vii)   Dewatering of the sludge in a solid bowl centrifuge with
        scroll discharge.

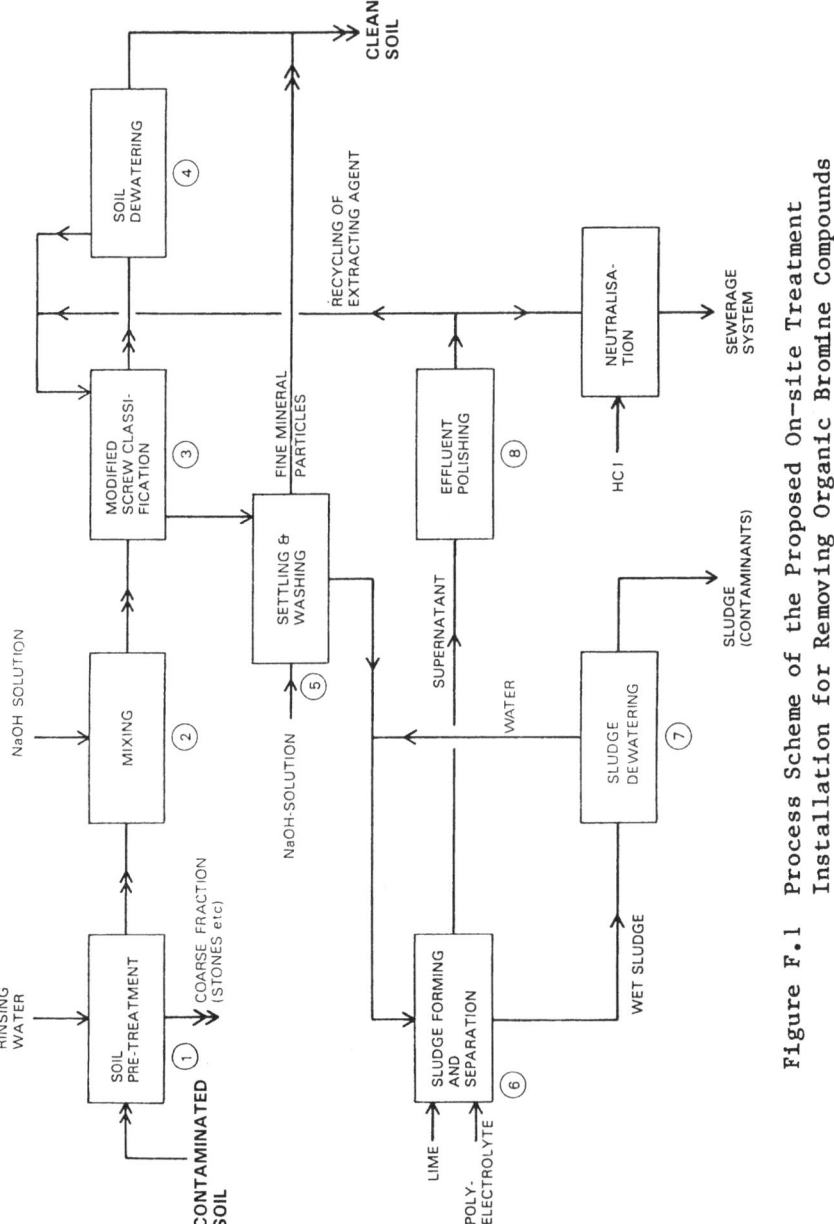

Figure F.1 Process Scheme of the Proposed On-site Treatment Installation for Removing Organic Bromine Compounds

(viii)   Effluent polishing by deep bed filtration, activated-
         carbon adsorption and finally anion exchange to remove
         any bromides formed by hydrolysis. The cleaned extracting
         agent can be recycled to the extraction process in the screw
         classifiers.

     Experiments showed that it is possible to remove the bromine
compounds from the soil down to a level of 1 mg Br/kg or less. The
cleaned extracting agent contains less than approx 0.6 mg Br/kg, the
main part of which is bromide.

     The waste sludge produced contains almost all the humus-like
substances, very fine mineral particles (40 um) and a high concentration of bromine compounds. The amount of sludge produced is about
5% of the total amount of contaminated soil owing to the high water
content of the dewatered sludge (approx 75%). The effluent polishing
step produces small amounts of spent activated carbon (approx 1 litre/
tonne of soil) and some regeneration liquid of the anion exchanger
(approx 13 litre/tonne of soil).

F.2   Removal of cyanides (laboratory and pilot-plant scale investigations)

     Abandoned gas works sites dating from the 19th and the first
half of the 20th century are often contaminated with cyanides.
Cyanides mainly occur as complexes with iron (eg ferri-ferrocyanide) which are quite immobile under normal conditions. However,
in several cases cleaning up is considered necessary from an environmental point of view. Such a site was found in The Hague (Netherlands)
in 1981.

     HBG and TNO started a study to investigate and assess different
techniques for cleaning up the sandy soil in 1982. The techniques
investigated were:

-   Scrubbing to reduce the size of ferri-ferrocyanide particles,
    followed by washing out with water (classifi-cation);

-   Flotation;

-   Extraction with aqueous alkali in order to solubilise the
    cyanides;

-   Chemical treatment with NaClO and $Ca(ClO)_2$ to form less
    harmful cyanides.

     Thermal treatment was not included, because it was thought that
complete destruction of complex iron cyanides probably needs tempera-

tures over 1400°C. Tests on a laboratory scale showed that chemical treatment is not successful and that flotation is sensitive to changes in composition of the soil.

Evaluation of the results of the experiments based on scrubbing and extraction showed that extraction is the most promising technique for development to a practical scale.

In order to assess all relevant parameters for scaling up of the extraction technique, experiments were carried out on a bench scale and a pilot-plant scale. Data collected from these experiments made it possible to design an on-site treatment installation. Figure F.2 shows a process scheme of the proposed installation. The following steps can be distinguished:

  (i)   Soil pre-treatment to separate large objects (eg wood, stones);

 (ii)   Extraction with aqueous alkali in a mixing device (eg scrubber); the pH is approx 11, and the soil-to-water ratio is about 2 to 1 on a weight basis;

(iii)   Separation of coarse sand (eg settler), dewatering (dewatering screen) and, finally, neutralisation;

 (iv)   Separation of fine sand (hydrocyclones or flocculation) followed by about four counter current extraction steps with aqueous alkali to remove cyanide to a sufficiently low level. Finally, the fine mineral fraction is dewatered in a solid-bowl centrifuge;

  (v)   Precipitation of dissolved cyanides by pH-adjustment and addition of iron salts. The precipitate is separated (eg tiltable separator) and dewatered (eg solid-bowl centrifuge).

The most troublesome step in the process is the separation between the fine mineral fraction (approx 30 to 65 um) and the extracting agent. The best results were obtained with flocculation and filtration.

Experiments showed that it is possible to remove over 99% of the cyanides present, down to levels of approximately 10 mg CN^-/kg or less. The purified water (after precipitation) contained less than 1 mg CN/kg.

The waste sludge produced contains high concentrations of cyanides and furthermore humus-like substances, clay and silt, up to particle sizes of approx 40-50 um. The amount of waste sludge is about 2% of the total amount of soil to be cleaned.

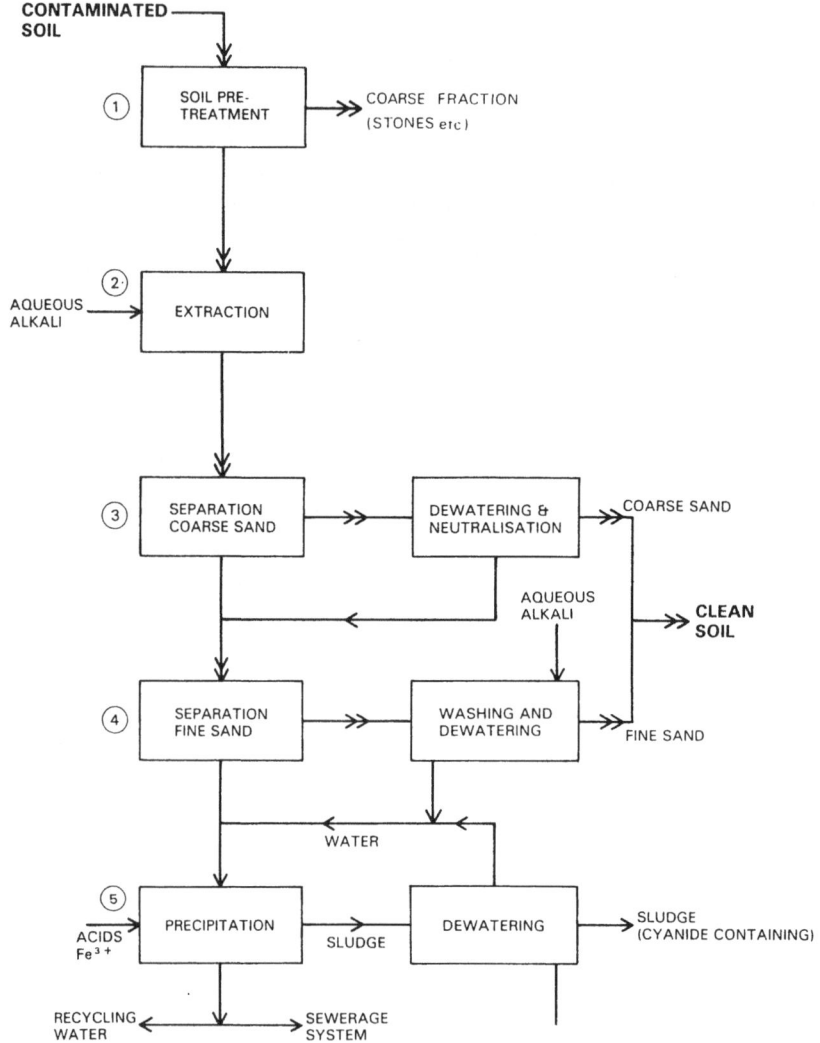

Figure F.2   Process Scheme of the Proposed Treatment
             Installation for Cyanide Removal

F.3 Removal of volatile hydrocarbons

    In the spring of 1982, approx 5000 tonnes soil containing
hydrocarbons from a former gas works site at Den Helder (Netherlands)
was cleaned up. The contaminants present in the soil comprises a
large variety of volatile hydrocarbons (boiling points less than
300°C), among which were alcohols, phenols, benzene, toluene,
naphthalene, petrol, kerosine. These volatile compounds can be
stripped in a thermal treatment installation.

    Ecotechniek (PO Box 39, 3454 ZG De Meern, Netherlands) has
developed a full scale treatment installation in which soil is heated
to $200-300^{\circ}$C and the released vapours are burned at about 800°C.
The following gives a brief description of the treatment installation
and the results obtained. The following process steps can be
distinguished in the installation (see Figure F.3):

  (i)   Soil pre-treatment to separate large objects and regulate the
        dosing rate of soil;

  (ii) Heating; two-stage rotating drum evaporator. In the first
        stage, soil is preheated indirectly by hot gases from the
        afterburner(iv). The soil is transported to the second drum
        where the soil is heated directly to $200-300^{\circ}$C by a flame;

  (iii) Cooling; the hot soil is cooled with water in a mixing
        device;

  (iv) Afterburning; the exhaust gases from the evaporator(ii) are
        burned with excess air at temperatures over 800°C with a
        residence time of approx 2 seconds;

  (v)   Heat exchanging; further process integration and energy
        recovery is achieved by preheating fresh air for the burners
        in the drum evaporator and afterburner.

  (vi) Gas washing: gas washer to separate the fine mineral particles
        from the gas expelled from the rotating drum;

  (vii) Settling; a settler is added to separate the fine mineral
        particles from the wash water. The settled particles can be
        redeposited with the cleaned soil.

    According to Ecotechniek, the general characteristics of the
process are:

-    removal of hydrocarbons with boiling points below 300°C;

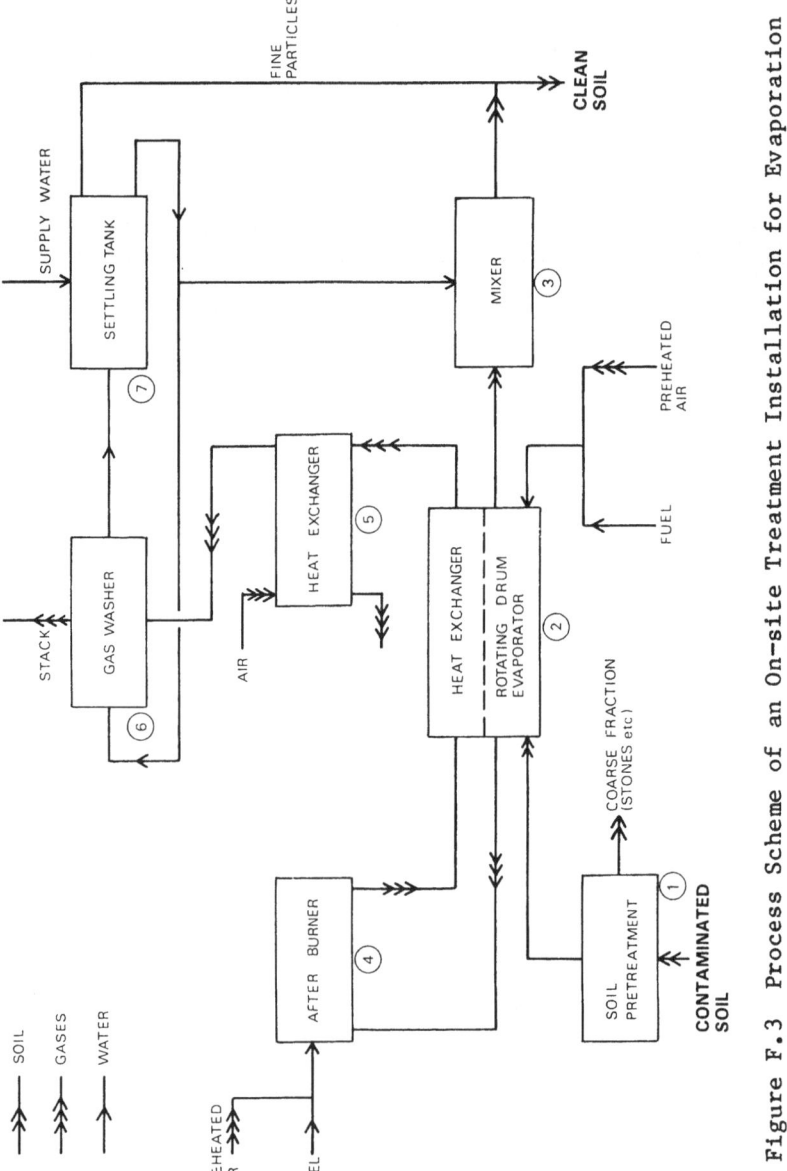

Figure F.3   Process Scheme of an On-site Treatment Installation for Evaporation
of Hydrocarbons and Destruction in an After-burner

-   hydrocarbons with sufficiently high vapour pressures at 300°C
    are removed to a large extent (approx 80-95%). Examples are
    lindane and some polycyclic aromatics;

-   not all organic chemicals decompose completely to harmless
    products in an afterburner at 800°C within a few seconds;

-   the treated soil is not severely affected in its structure.

Regular analysis of cleaned soil from the Den Helder gas works
site showed the amounts of benzene, toulene, xylene, styrene and
ethyl-benzene to be always below the detection limit (0.05 mg/kg).
Naphthalene however, being less volatile, was found in some soil
samples to be present in concentrations of up to about 1 mg/kg and in
one case 3.3 mg/kg but these values are below the target concentration (5 mg/kg). Stack gases contained approx 10 ppm CO, 8 ppm $C_x H_y$,
83 ppm SO_2, 51 ppm NO_x and 76 mg/m^3 dust.

APPENDIX G: Concentration and treatment technologies
              for contaminated groundwater *

              by K A Childs

Biological Treatment

Biological processes are the most commonly used systems for
treating aqueous waste streams containing organic constituents and
have been applied with varying levels of success to a wide range of
industrial waste streams and sanitary landfill leachates. There are
no known full-scale hazardous waste treatment facilities using
biological systems. The potential release of volatile organic
compounds to the atmosphere as a result of aeration is one of the
major concerns. If a leachate contains organic compounds which are
not readily biodegradable it may be necessary to acclimate a
biological system to the waste. The presence of compounds in the
waste stream that are refractory and/or toxic to biological systems
might even preclude use of biological treatment or may necessitate
use of another treatment process in conjunction with biological
treatment. The principal operational requirements are near neutral
pH, nutrient demands (carbon, nitrogen, and phosphorus as well as
trace elements) must be satisfied and sudden changes in organic and
hydraulic loading must be avoided. Biological treatment, processes
include the activated sludge process, which appears to have the
greatest potential for leachate treatment and anaerobic filtration or
anaerobic lagoons which are easy to operate, have minimal sludge
production, and are energy efficient. Biological treatment is a
proven technology and invariably should be considered as part of a
system (or train) for treatment of hazardous waste leachate
containing organic constituents.

Carbon Adsorption

Activated carbon adsorption is a well developed technology
having a wide range of waste treatment applications including the
removal of mixed organic contaminants from aqueous wastes. Treatment
of a variety of industrial wastewaters, cleanup of spilled hazardous
materials and the treatment of leachate and ground and surface waters
are examples of applications at working scale. No serious environmental impacts are associated with carbon systems employing
regeneration. If carbon regeneration is not employed the disposal of
carbon contaminated with hazardous materials could present a

---

* Appendix to Chapter 8

significant problem. Unit costs for carbon adsorption can vary
widely depending upon the waste to be treated. It has been shown to
be an economical approach in numerous instances, even though energy
demands are substantial, particularly those which use granular
activated carbon. Granular activated carbon is the most well
developed approach and may be used to provide complete treatment,
pretreatment or effluent polishing. Carbon adsorption processes for
complete treatment, pre-treatment or effluent polishing must be
considered as a viable candidate in almost all cases. Process trains
using biological-carbon adsorption components appear most promising
for treating aqueous waste containing hazardous constituents.

Catalysis

Potential applications of catalysis as a waste treatment process
have been identified but commercial viability remains to be
demonstrated. Catalysts are generally very selective in terms of
what reactions they will promote and do not have the broad spectrum
applicability which is preferable in complex waste streams.

Chemical Oxidation

Charcoal oxidation achieves relatively poor removal of most
organics. Chemical transformation may occur which could facilitate
treatment by other processes. Inorganics can often be transferred to
a valence state which is less toxic. Most chemical oxidation
technologies (including ozone) are fairly well developed and have
been demonstrated either at full scale or laboratory scale. To date
utilisation has been limited to treatment of dilute waste streams.
Ozonation, as a pre-treatment process prior to biological treatment,
in concert with UV irradiation as the primary treatment process or in
combination with granular activated carbon, appears to have a good
potential as a treatment process for liquids contaminated by
hazardous wastes. Advantages of oxidation with ozone or hydrogen
peroxide include the fact that no harmful residues are generated and
no chlorinated organics are formed as can be the case when alkaline
chlorination is used. The nature of intermediate products must be
assessed individually. Off-gases containing residual ozone should be
passed through activated carbon to decompose the ozone.

Chemical Reduction

Chemical reduction has little potential for treating organic
wastes. However, it is an effective means of removing inorganic
compounds or lowering their toxicity. The precipitated sludge
may require careful management because it may contain harmful
compounds. Introduction of foreign ions into the waste can be

detrimental to many of the reducing agents. The reduction of
hexavalent chromium to its trivalent form using sulphur dioxide,
sulphite salts, or ferrous sulphate as the reducing agent is a major
application of chemical reduction. Precipitation of $Cr(OH)_3$ with
lime or sodium carbonate usually follows.

Chemical Precipitation

Precipitation is a proven process and can be applied to almost
any liquid waste stream containing hazardous constitutents amenable
to precipitation. Advantages include: equipment is commercially
available, costs are relatively low, it can be applied to relatively
large volumes of liquid wastes and the energy consumption is
reasonable. The wet sludge precipitate may be considered hazardous
and may require further processing prior to ultimate disposal.
Treatability studies should be carried out prior to applying the
process to a waste stream to determine the chemicals, the removal
efficiency and the dosage. Precipitation can be considered for the
removal of metals (arsenic, cadmium, chromimum, copper, fluoride,
lead, manganese, mercury, nickel) and certain anionic species
(phosphates, sulphates, fluorides) from aqueous hazardous wastes.

Crystallisation

Crystallisation does not respond to changing waste water
characteristics and the complex nature of the process is the main
reason why this process has not found favour. Efforts to treat a
variety of industrial waste waters and sludges have had limited
success and the process appears to have little potential.

Density Separation

Density separation includes sedimentation and flotation both of
which are commonly used techniques. Sedimentation can be applied to
almost any liquid waste stream containing settleable material and has
high potential in leachate treatment systems. Sedimentation is
normally used in conjunction with other processes such as chemical
precipitation or as a pre-treatment technique. Flotation is also a
proven solids/liquids separation technique for certain industrial
applications but has higher operating costs and requires more
management than that required for gravity sedimentation. This
technique has most potential when the waste stream contains high
concentrations of oil and/or grease.

Dialysis/Electrodialysis

Both dialysis and electrodialysis appear to have limited

potential because neither process is well suited to treating mixed
constituent waste streams. Problems include membrane plugging and
deterioration and production of two output streams neither of which
can be discharged directly to the environment.

Distillation

Distillation appears to have limited potential because of its
high cost and energy requirements. If materials recovery is
objective and practical the process might be applicable especially if
the contaminants are organic solvents or halogenated organics.

Evaporation

Evaporation is not expected to have broad application to the
treatment of aqueous waste streams containing moderately volatile
organic constitutents because clean separation of these organics is
not possible without further treatment of the condensate. High
capital and operating costs, and high energy requirements, also make
this process unattractive. This process is not likely to be used
extensively for the treatment of waste waters.

Filtration

A wide spectrum of proven granular and flexible media filtration
systems are commercially available. The main advantages are: the
economics of filtration are attractive, energy requirements are
relatively low and operational parameters are well defined. It is
not a primary treatment process but usually a supplementary process
for either polishing or as a dewatering process.

Flocculation

Flocculation, often preceded by precipitation, is carried out in
conjunction with a solid/liquid separation process. It is a
relatively simple process and can be applied to almost any aqueous
waste stream containing suspended material or components which can be
precipitated. Equipment is readily available and operating costs are
low. Flocculation followed by sedimentation has potential when the
objective is the removal of suspended solids and/or heavy metals. It
may also be used in conjunction with sedimentation prior to activated
carbon adsorption treatment. The applicability of the technique, the
chemicals and the chemical dosage can be determined by experience or
by laboratory treatability tests.

Ion Exchange

Ion exchange will probably never become a treatment process of
primary importance. Use of the process will result in the removal of
dissolved salts, primarily inorganics, from aqueous solutions. If
materials recovery is possible ion exchange can be a relatively
economical process. Ion exchange has a narrower application compared
with alternative more economical processes - precipitation, flocculation, and sedimentation. There are limitations to the use of
synthetic resins, which can be damaged by oxidising agents and heat.
The presence of suspended matter or other materials will permanently
foul the resin. If organic compounds become irreversibly adsorbed by
the resins, there will be a decrease of capacity. The use of ion
exchange will probably be limited to its use as a polishing step.
Therefore, while ion exchange is believed to have some potential it
is not a process which should normally receive primary
consideration.

Resin Adsorption

Experiments using resin adsorption have shown that a broad range
of organic compounds can be adsorbed and, in certain cases, the
process has been more effective than activated carbon systems. Resin
adsorption could be employed to remove colour for solute recovery,
for selective adsorption, for 'low leakages', and when the water to
be treated contains high levels of dissolved inorganics. Polymeric
adsorbents are nonpolar with an affinity for nonpolar solutes in
polar solvents or of intermediate polarity capable of sorbing
nonpolar solutes from polar solvents and polar solutes from nonpolar
solvents. Carbonaceous resins have a chemical composition which is
intermediate between polymeric adsorbents and activated carbon and
are available in a range of surface polarities. As with activated
carbon, the only major environmental impacts relate to the
regeneration process. If not reused, spent regenerant requires
disposal, frequently by incineration or land disposal. Resin
sorption appears to be potentially viable for treatment of liquids
containing hazardous wastes.

Reverse Osmosis

Reverse osmosis is a relatively new process and can be used
for the removal of inorganic salts from rinse waters. Energy
requirements for commercially available systems are approximately
76-95 J/m^3 of product water (8-10 kWh/1000 gal). The process is
relatively costly but is capable of producing high purity effluent -
particularly when applied to dilute solutions of inorganic and some
organic solutes. Problems associated with reverse osmosis include
concentration polarisation (decreased water production with time per

unit area of membrane), the need for pre-treatment to remove solids (colloidal and suspended), the need for dechlorination when using polyamide membranes, and membrane fouling due to precipitation of insoluble salts. pH control is important. To ascertain feasibility bench and pilot scale testing prior to almost any potential application is essential. Its use will probably be limited to polishing operations subsequent to passage through some other more conventional processes or to concentrating pollutants which would be processed further.

Solvent Extraction

Solvent extraction is judged to have minimal potential compared with activated carbon. Unless a valuable product can be recovered solvent extraction will not be economically competitive.

Stripping

When ammonia removal is desired air stripping can be considered and then only when the concentration of other volatile compounds is low. Air stripping, with associated air pollution control features, does have functional appeal as a pre-treatment prior to another process such as adsorption, for it tends to extend the life of the sorbent by removing sorbable organic constituents. Air stripping of volatile components will occur during the course of any treatment process and must be accommodated in the design. Stream stripping has merit for wastes containing high concentrations of highly volatile compounds. Before treating waste streams containing multiple organic compounds, laboratory and bench scale investigations are essential. Energy requirements and other operating costs are high and not offset by product recovery. Steam stripping as a pre-treatment step to reduce the load of volatile compounds, is considered to have good potential. The overhead condensate stream will require further treatment, possibly by wet oxidation, to remove concentrated organics.

Ultrafiltration

Ultrafiltration has several industrial applications generally involving product recovery or production of high purity solvent. High capital and operating costs are normal with membrane replacement being a major factor. Ultrafiltration is judged to have limited potential for treating high volume waste streams. Treatability testing is essential.

Wet Oxidation

Laboratory studies indicate that the process may have potential for treatment of waste streams containing high concentrations of toxic organics. At pilot level and at full scale the process has been shown to be effective in treating numerous sludges and non-hazardous wastes and, at laboratory scale, in the destruction of several organic priority pollutants. The process should be considered as potentially suitable for treating concentrated organic streams generated by processes such as steam stripping. Extensive site specific treatability tests are essential to determine efficiencies, to develop design criteria, and to allow comparison with alternative technologies.

APPENDIX H: Sampling design patterns and determination
              of adequate sample numbers*

              by R E Montgomery, D P Remeta and M Gruenfeld

1 SAMPLING DESIGN PATTERNS

1.1 Non-systematic sampling patterns

1.1.1 'W' or 'X' patterns

These are simple patterns in which the individual sampling
points are located on linear traverses of the site, laid out in the
form of an imaginary letter 'W' or 'X' drawn across the site. They do
not depend on prior knowledge of the distribution of contamination
and are therefore appropriate to sites having a heterogeneous
distribution. Because relatively few sampling points are used and
the linear traverses do not cover all parts of the site, these
patterns will not define the degree and extent of contamination
adequately on contaminated sites. Their principal use has been in
sampling of agricultural land and geological formations, for such
purposes as soil surveys and mapping. When used in these
applications, the samples collected from the individual sampling
points are composited prior to analysis, so that the data obtained
represents average values of concentration. This is a further
disadvantage to their use for investigating contaminated sites.

1.1.2 Simple random sampling

The sampling points are located at random within the area of the
site. Although free of bias and capable of yielding a valid estimate
of sampling error, the technique tends to result in uneven coverage
of the site with consequent difficulty of interpolation of findings
between sampling points. It can also be difficult to re-locate the
actual sampling points used on a particular site, if it subsequently
proves necessary to collect further samples from them.

1.2 Systematic sampling patterns (Regular Grids)

Sampling points are located on a regular square or rectangular
grid, formed by the intersection of two sets of imaginary lines drawn
parallel to the sides of the site. The spacing of each set of grid
lines can be chosen to suit the dimensions of the site: intervals of
10 m, 25 m, 50 m or 100 m have commonly been used depending on the

---

*   Based on information supplied by Environmental Advisory Unit,
    Liverpool University, UK (1983)

size of the site. Intervals of more than 100 m on large sites are
likely to be of little practical value since it is then difficult to
interpolate the analytical results between adjacent sampling points.
On some sites it may be more appropriate to align the grid such that
the two sets of lines are not at right angles; the resulting pattern
is then described as a 'diamond' grid.

Regular grid sampling patterns enable the whole area of the
site to be covered, and allow the individual sampling points to be
readily relocated if further stages of sampling are necessary. They
may however introduce a risk of bias where there is periodicity in
the source data, and do not permit an entirely valid estimate of
sampling error to be derived. Despite these disadvantages, grid
sampling is most effective on many types of contaminated sites.

1.3 Other sampling patterns

1.3.1 Stratified random sampling

The site is divided into a number of 'strata' or cells of equal
size, and a given number of sampling points chosen at random within
each cell. This method combines the advantages of simple random and
systematic sampling strategies; it affords more even site coverage
and permits a valid estimate of sampling error to be derived.

1.3.2 Unequal sampling

In many types of contaminated sites it may be possible to
predict where the areas of greatest contamination are likely to be
found. When this is possible, unequal sampling can improve the
efficiency of site investigations. The technique is similar in
principle to stratified random sampling, but the size of the
individual strata or cells and the number of sampling points within
each need not be equal. Thus, where the variability of contamination
is expected to be greatest, the number of sampling points in that
area can be increased. The main disadvantage of the technique is
that accurate location of individual sampling points may be
difficult.

2   DETERMINATION OF ADEQUATE SAMPLE NUMBERS

If a grid sampling pattern is used, it is important to be aware
of the limitations of this technique. Particularly the relationship
between the number of sampling points used, the proportion of the
site which may be contaminated and the chance of the sampling points
intercepting and revealing a particular type of contamination. The

information gained by sampling will also depend on the continuity of spatial distribution of materials of concern across the site. A site which is essentially uncontaminated apart from a small pit filled with toxic waste would be described as having a discontinuous spatial distribution of contaminants. A sewage farm serving an industrial area would typically have a more continuous spatial distribution of contaminants. In the simplest case of a site uniformly contaminated with a single material or chemical species, a single sample if correctly analysed will reveal the presence of the contaminant. Such an ideal case will rarely if ever occur in practice.

The resolution of the sampling grid should be significantly smaller than the areas of contamination on the site under investigation. For a typical contaminated site (eg a gas works when there might be a dozen different contaminating species each occupying only 5% of the site areas), this sort of resolution would require an impossibly large number of sampling points. Information in this sort of detail is unlikely to be needed to make basic decisions about the development or restoration of the site.

An estimate of the proportion of a site which contains material of concern may be made from any sampling strategy and confidence limits can be attached to this estimate. For a simple random sample the estimate of proportion is simply

$$\frac{\text{number of samples containing material of concern}}{\text{total number of samples taken}}$$

The confidence limits can be calculated or read from tables of the binomial distribution. When the sample number is low, then confidence limits are unacceptably wide (see Table H.1 below) and any estimate of proportion must be treated with extreme caution.

Table H.1 Estimated Proportion and 95% Confidence Limits for a Simple Random Sample

Sample Number	Number containing material of concern	Estimated proportion	95% confidence limits
10	2	20%	3% – 56%
30	6	20%	8% – 39%
100	20	20%	13% – 29%

1   INTRODUCTION

    The countries contributing to the NATO/CCMS (Committee on
Challenges of Modern Society) Pilot Study on Contaminated Land have
each arrived at their interest in the problems encompassed by the
Study by different routes.   Nevertheless they have identified two
primary needs:

  (i) To be able to deal with sites using existing technologies in
       order to contain an existing environmental problem or to bring
       the land back into beneficial use in the short-term
       recognising that there may be some doubts about the long-term
       effectiveness of such measures.
 (ii) To identify and develop methods of dealing with problems
       presented by a particular contaminated site that will be
       effective in the long-term – ideally in perpetuity but
       certainly for a 100 years or thereabouts.

    All have also recognised that scientific and technical judge-
ments will need to be made under a variety of social, political and
economic pressures.

    The Pilot Study is based on the definition of 'contaminated
land' as 'land that contains substances, that when present in
sufficient quantities or concentrations, are likely to cause harm
directly or indirectly to man, to the environment, or on occasions,
to other targets'.

    This broad definition of 'contaminated land' encompasses the
special problems posed by 'uncontrolled hazardous waste sites'(US
Environmental Protection Agency), hazardous waste 'problem' sites
(OECD) and 'Les points noirs' (France) as well as a much wider range
of sources of contamination.

    Any land previously used for industrial or commercial purposes
or for the disposal of wastes, including domestic refuse and sewage
sludge, may be contaminated.   In practice much contamination on
former industrial sites, has arisen from the disposal of solid and
liquid wastes within the curtilage.

    Two important points should be noted about this definition:

_____

* Cross referenced to Chapters, Tables and Appendices in the
main report

(i)   the emphasis on the <u>presence</u> of <u>potentially</u> harmful
       substances rather than on the past use of the land.
(ii)   the implication that a 'problem' is only defined after site
       investigation, and evaluation of information specific to the
       site and taking into account its intended use.

Although the problem of 'contaminated land' has taken on a new
importance during the past five years or so, it is not, in reality, a
new problem. The construction industries in all countries have long
been aware of the need to select the materials used in construction
to withstand potentially aggressive conditions, for example sulphate
and acid attack on concrete. Throughout history, towns and cities
have expanded over the accumulated wastes deposited outside the
former boundaries.

Incidents such as Love Canal in the USA and Lekkerkerk in the
Netherlands have however engendered a determination in a number of
countries to try to avoid the unwitting creation of such problems in
the future as well as to develop the means of tackling them when they
are found. Such contaminated sites present a complex technical and
scientific challenge both in evaluation and in treatment, that has to
be met by bringing together a wide range of specialists. Although
the background to action differs from country to country the
technical problems are similar. The CCMS Pilot Study on Contaminated
Land was an attempt to enhance the resources available in any one
country through a programme of mutual co-operation.

2   THE STRUCTURE OF THE STUDY

The Pilot Study started in November 1981 with a planning
meeting in the United Kingdom. Further meetings were held in the
Netherlands (April 1982), the United States of America (December
1982), Federal Republic of Germany (May 1983) and the United
Kingdom (April 1984). A number of meetings of experts were also
held.

The Pilot Study consisted of seven projects (see Table 1.1),
each led by a Project Leader (see Appendix A) assigned by one of the
participant countries, of which five were directly concerned with
remedial action. These seven projects were supported by a structured
exchange of information co-ordinated by the United Kingdom as lead
country for the Pilot Study as a whole. A number of other topics
were identified by the Study Group as meriting study but for which
the time and resources were not available for them to be actively
pursued (these are listed in Appendix D). Some contributions to
these other topics, were however made by some of the CCMS Fellows who
particpated in the Study.

All participants at the first meeting in November 1981 identified remedial measures as the subject about which, despite active
research programmes in progress in a number of countries
much remained to be learnt, especially about long-term
effectiveness. It was considered that benefits could be achieved
by learning from each other's experience.

3   OPTIONS FOR REMEDIAL ACTION

Actions taken in response to the identification of a contamination problem may in the first instance be designed to provide a
temporary amelioration of an immediate hazard or they may be designed
to control pollution of the environment on a longer-term basis.
However, in all but a minority of cases, the aim will be to either
eliminate the identified risk or at least reduce it to an 'acceptable'
level.

The most direct and obvious solution in many cases may appear
to be to remove the offending materials, deposits and contaminated
ground, from the site for disposal or treatment elsewhere. However,
this is not always practicable, nor is it always desirable, nor is
it necessarily a permanent solution in the broadest sense. Thus
there are strong environmental, economic and social reasons for
dealing with contamination where it is found. This was the main
theme of the CCMS Pilot Study on Contaminated Land. There are three
ways of achieving this aim, each of which was addressed in the
study:

   (i)   On-site treatment – defined as being methods of decontami-
          nating or otherwise reducing the environmental impact of the
          bulk of contaminated materials on a site.
   (ii)   In-situ treatment – defined as the treatment without
          excavation of the bulk material on a contaminated site.
  (iii)   Macro-encapsulation – in which a site is contained within a
          'box' usually consisting of a cover, a peripheral barrier and
          sometimes a base.

The liquid phase within and on contaminated sites is of great
importance. It is the usual means by which the contamination moves
out into the wider environment and the treatment option should not
be adopted without first considering how to deal with the liquid
phase.

Methods for the treatment of heavily contaminated leachate,
surface waters, etc, are well established and continuously being
improved upon. Such methods were therefore not included in the Pilot
Study. However, the study looked at the liquid phase in three
respects:

  (i)   The use of hydraulic measures as an adjunct to other treatment
         options.
 (ii)   The treatment of contaminated groundwater after extraction
         or by in-situ means.
(iii)   The modelling of groundwater movement.

     Other topics considered were (i) the occurrence of toxic and
flammable gases on contaminated sites, the problems they present and
how these can be overcome and (ii) analytical methods for use in the
investigation of sites.

4   SUMMARY OF INDIVIDUAL PROJECTS

4.1   Long-term Effectiveness of Remedial Measures

     This Section is based on the briefs agreed for each of the
Projects at the start of the Study.

     This Project, led by the Federal Republic of Germany, dealt
with the philosophical and technical problems surrounding the design
of long-term effective remedial measures and also with methods for
evaluating the effectiveness of remedial actions (Chapter 2).

     In the majority of reclamation or remedial schemes it is
necessary to use methods that should be effective on the basis of
current knowledge and professional judgement but which cannot be
demonstrated to have been so. Nor is it usually possible to
evaluate experimental techniques before they are employed because
of the difficulties of simulating field conditions or providing
satisfactory accelerated testing regimes. Thus is often necessary
to predict behaviour over a long time period (which might range
from 20 years to an indeterminate time span) on the basis of very
little practical experience. In these circumstances information
on already restored sites that subsequent monitoring has shown to
have been successful to date would be of great value.

     The potential of the various options for remedial action dealt
with in the other Projects was to provide either 'once-and-for-all-
time' or long-term effective solutions was considered. Methods for
evaluating long-term effectiveness were reviewed.

4.2   On-site Processing of Contaminated Soils

     This Project (Chapter 3), led by the Netherlands, was concerned
with methods of decontaminating or otherwise reducing the environmental impact of the bulk of contaminated material on a site by:
excavation of the soil; treatment to destroy, render harmless,

neutralise, stabilise or fixate the contaminants; and usually redeposition of the treated soil on site.

Excavating then processing the material avoids three of the main difficulties encountered with in-situ techniques:

  (i)   difficulty in getting 'reactants' in contact with
         contamination;
 (ii)   difficulty in establishing that treatment has worked;
(iii)   danger of groundwater contamination by treatment agents.

On-site processing may be as simple as scraping off surface layers, mixing them and then respreading, perhaps with prior removal of and disposal of the most contaminated material. More typically it involves some form of physical, chemical, microbial or heat treatment to separate, destroy, render harmless, neutralise, fix or stabilise the contaminants.

Most of these treatments are derived from experience with the treatment of hazardous wastes and from mineral processing. Some are well established; others are the subject of research. The long-term effectiveness of others require further investigation ie those involving stabilisation rather than degradation or removal of the contaminants. Although having much in common with the problems of treating any hazardous waste the material excavated from a contaminated site may present additional difficulties due to its chemical complexity, heterogeneity and bulk.

4.3   In-situ Treatment of Contaminated Sites

This Project (see Chapter 4) led by the United States, dealt with methods of treating, without excavation, the bulk material on a contaminated site by detoxifying, neutralising, degrading, immobilising or otherwise rendering harmless contaminants where they are found.

Current methods for dealing with contaminated or 'problem' hazardous waste sites typically involve excavation and removal of the offending material for disposal elsewhere or permanent destruction, or macro-encapsulation ie covering and/or surrounding the site with barriers in order to contain the contamination. The effectiveness of the latter cannot be guaranteed whilst the former presents a number of technical and other problems. The excavation and transport of the contaminated material may be hazardous and it may be difficult to find a new safe disposal site. Redeposition may also only move the problem elsewhere to be rediscovered by later generations. Destruction by incineration or other means may be equally difficult to arrange particularly in the case of large sites. Hence means of treating the mass of material where it lies are attractive.

Possibilities include: grouting; liquid extraction using water or some other solvent; microbial degradation; chemical neutralisation, fixation or degradation; and electro-osmosis. Most of these treatment systems have in common the need to inject fluids into the ground in one way or another, although less complex treatments such as deep ploughing to invert strata where contamination goes only to shallow depths is also a form of in situ treatment. Some systems have the added potential advantage of improving some other property of the ground: for example grouting may improve engineering properties.

4.4   Covering and Barrier Systems

This Project (Chapters 5-9), led jointly by the United Kingdom and Canada, dealt with the design of systems to isolate potential targets from contamination by preventing the migration of contaminants out of contaminated sites and by preventing ingress of surface or groundwater into contaminated sites.

Many contaminated sites are at present dealt with by total or partial isolation or macro-encapsulation of the site. In the extreme this may involve trying to contain totally the contamination within an impermeable barrier that will both prevent or control egress of contamination, and ingress of water. Often however treatment is restricted to the superimposition of cover, which may be a simple soil layer or combination of several layers including an 'impermeable' barrier which may be of clay or a synthetic membrane. Such covering systems were the subject of the study prepared by the United Kingdom (see Chapter 5).

Covering systems are frequently required to perform a number of possibly conflicting functions and the major task of the designer is to identify the functions required and the possible conflicts. As with many remedial actions there is little past evidence on which to predict long-term behaviour. Due allowance must be made for the influence of external environmental factors such as climate including extremes in rainfall from drought to flood conditions, wind and water erosion, plant growth and the activities of burrowing animals. Allowance must also be made or interaction between contaminants and materials, changes with time in the materials used in construction and any potential for movement of contaminants in liquid or gaseous form.

In-ground vertical and horizontal barriers to the movement of contaminants and groundwater (Chapter 7) may be required either as a primary remedial measure or as an essential supplement to some other form of remedial action, particularly for example in-situ treatment. These were the subject of a study led by Canada.

The construction of cut-offs by means of grout injection or installation of slurry trenches or diaphragm walls are well established procedures required for a variety of engineering purposes, many of which however are of a short-term nature. Interactions with contaminants and the affects of environmental factors such as plant-root penetration have to be taken into account. In-ground horizontal barriers may be required to complete the isolation of the contamination from surrounding ground. In contrast to the construction of vertical barriers this requires relatively untested engineering procedures.

4.5 Control and Treatment of Groundwater

This Project, led by Canada, was concerned mainly with these operations designed to control or treat the liquid phase on contaminated sites including design of cut-off systems, hydrogeological modelling and groundwater treatment.

Pollution of surface and groundwater has been a major feature of 'problem hazardous waste sites' in a number of countries. The OECD Experts Seminar in 1980 paid particular attention to these problems. The control of and chemical and microbial treatment of leachate or surface water arising on sites may be an essential feature of the clean-up programme. Systems have also been proposed for the draw-off of contaminated groundwater for purification prior to reinjection or disposal (dealt with in Chapter 8). Such processes may be more effective if the volume of groundwater to be treated must be limited by insertion of appropriate cut-off barriers eg by grout curtains or clay fill trenches (as discussed in Chapter 7). Such studies, and the evaluation of alternative treatment options for specific sites, can be assisted by the use of mathematical modelling (see Chapter 9).

4.6 Toxic and Flammable Gases

This Project (Chapter 10), led by the United States, was concerned with; volatile organic emissions from contaminated sites, production, migration and control of gases from land disposal sites including typical landfill gases such as methane and carbon dioxide.

Volatile organic compounds (VOC's) present on contaminated sites including uncontrolled hazardous waste sites can enter the atmosphere causing local hazardous or unpleasant (due to odour etc) conditions and causing general environmental pollution that may be of significance at large distances from the source. Such problems and occurrences are also relevant to the creation of new treatment, storage and disposal facilities and to the co-disposal of hazardous waste with domestic and inert industrial wastes to landfill.

The emission of toxic and odourous VOC's from landfills may
be aided by the presence of landfill gases. Remedial actions taken
to control or utilise landfill gas must take into account the
presence of VOC's. An understanding of the factors governing
landfill gas production can be important to a proper appraisal and
subsequent design of remedial actions on sites where volatile organic
compounds are present.

If buildings are to be built on, or in close proximity to, sites
that are producing toxic (eg carbon monoxide) or flammable gases (eg
methane) it is necessary to have reliable data on the amount and type
of gas being produced so that any protective measures can be properly
designed. Whilst much has been published on the assessment of
landfill sites as a source of fuel gas, particularly in the United
States, and much on general prescriptions for protective measures,
the latter very rarely contain any guidance relating design to
information on quantitative production of gas. Published descriptions
of buildings incorporating counter-gas measures rarely contain any
numerical data on the design of the system or evidence on the
successful working of the installed system. Such information would
be of great value to those engaged in the design, of what must
usually be a site-specific solution.

4.7 Rapid Methods of On-site Analysis

This Project (Chapter 11) led by the United States differed from
the other Projects in not being primarily concerned with remedial
measures. It was concerned with identifying methods of analysis
allowing determinations to be made on 'soil', water and gas samples
on-site so as to speed-up and reduce site investigation and reduce
costs.

Rapid on-site methods of analysis are potentially useful in the
initial appraisal of sites and in the identification of highly
contaminated areas of sites. They are not normally expected to
substitute for proper laboratory-based analyses but can help in the
identification of the analytical parameters to be looked at
subsequently in the laboratory. Great precision or accuracy are
usually not required and simple yes/no answers of whether a certain
concentration is exceeded are often sufficient (a common feature of
systems designed to monitor personnel safety in respect of atmospheric pollution in the work-place). Any lack of accuracy may be
partly offset by the ability to take a much greater number of
readings. Systems are needed for both inorganic and organic
contaminants. The US has an active programme devoted to developing
methods for use in specially equipped mobile laboratories and a
description of these laboratories was included in the Project
report.

4.8 Steel Industry Sites

Although generally, and for the purposes of the Pilot Study, contaminated land should be viewed in terms of problems arising from the presence of particular substances rather than from previous land use there is nevertheless a link between past land use and the nature of contamination that is likely to be found. It is possible that although the site types are common to different countries, the perception of potential problems may be different. It is also possible that one country has already tried remedial procedures that could be of use elsewhere.

There has been a sharp decline in the steel industry in a number of NATO member countries resulting in large areas of land falling vacant offering the opportunity for redevelopment. The Study Group was unable to pursue this subject area itself but it was chosen as a topic for study by a CCMS Fellow and a summary of the resulting report is included in the full report of the Study Group (see Chapter 12).

The Study included a historical and technical review of iron and steelmaking processes and the principal sources of associated ground contamination. The hazards that can arise from the contamination and the range of engineering problems associated with the development of such sites were also reviewed. Some examples of actual and proposed site redevelopment in the UK were described briefly.

5   CCMS FELLOWSHIP SCHEME

The success of the Pilot Study was greatly assisted by the work of ten recipients of CCMS Fellowships. The Fellowships enabled the recipients to attend meetings of the Study Group, to contribute directly to the various Project reports and to produce reports of their own (see Appendices B and E). Assistance was also received from the Study Tour Scheme.

6   GENERAL CONCLUSIONS

The definition of 'contaminated land' used for the Pilot Study deliberately embraces a spectrum of sites which differ widely in size and in the degree and nature of contamination: they range from small sites with a shallow superficial deposit of easily removed solid waste to large dumps of unidentified chemicals that are already causing environmental damage. In reality the majority of sites lie somewhere between the two extremes. The problems are compounded if the site has already been built upon.

Some options for the treatment of contaminated land provide
a once and for all time solution to the problem; for example 'the
contamination is removed or destroyed'. Many other options,
however, leave contaminants in the ground either separated from
potential targets by a complete or partial containment system or more
intimately, chemically or physically 'locked up'. These
containment/stabilisation processes must be viewed with care in
relation to long-term effectiveness.

Very few of the technologies described have been sufficiently
proven in applications specific to the treatment of contaminated land
although they may have been tried for other purposes. This is
equally true of those intended to provide an immediate and permanent
remedy and those providing containment/ stabilisation. Consequently,
it is important that long-term evaluation studies are established.
Properly planned and monitored demonstration projects have a valuable
role to play in this regard.

The Study Group on Contaminated Land came to the following
general conclusions:

(i)   Retrospective studies of already reclaimed sites should be
       encouraged in order to obtain information on the performance
       of the treatment strategies adopted.
(ii)   The development of on-site and in-situ processes resulting in
       the removal or destruction of contaminants is to be
       encouraged as providing once and for all solutions.
(iii) The long-term nature of research into remedial measures
       designed for containment or stabilisation must be recognised.
       Some projects may need several years for completion and it
       may take longer to accumulate sufficient confirmatory
       evidence. Funding arrangements may need to reflect these
       limitations.
(iv)   New methods of treatment can not be proved under laboratory
       conditions alone. There is a need for properly designed
       field trials and demonstration projects.
(v)   Monitoring of remedial actions should be part of the scheme.
       Such monitoring is an essential part of many civil
       engineering projects and should not be viewed as casting
       doubt on the effectiveness of the selected solution.
       Maintenance is likely to be required whenever a non-permanent
       treatment option is adopted. Funding should take into
       account the need for monitoring, maintenance and sometimes
       for a phased approach to remedial action.
(vi)   Records of contaminated sites and of the treatment carried
       out should be kept unless the contamination is removed or
       destroyed. Such records are important since the use of the
       site might change, the remedial measures deteriorate, or
       thereby changing knowledge of the contaminants may change the
       perception of the risks. It follows that the future

development of such sites should be subject to control by
appropriate authorities. The development of contaminated
land for residential use, recreational or other uses likely
to bring people into close contact with it requires special
care and the responsible authorities will wish to satisfy
themselves that controls are adquate.

(vii) Policies for the disposal of wastes to land should take
after use of the land into account.

(viii) Industries should avoid contaminating their sites. The
potential for contamination that might inhibit after-use
should be a material consideration when development takes
place.

(ix) Remedial actions require a multi-disciplinary approach.
Scientific and professional judgement assists in the
selection of remedial actions.

7   CONCLUSIONS OF INDIVIDUAL PROJECTS

7.1 Long-term Effectiveness

    With regard to the individual Projects the Pilot Study Group the
came to following conclusions:

    This Project has in large measure provided the basis of the
general conclusions above. However it is stressed that because of
the dynamic nature of contamination, remedial systems and the
environment in which the remedial system has to operate, remedial
measures that do not result in elimination of the contamination are
likely to have limited life spans.

    It is difficult to predict with any degree of certainty the
long-term effectiveness when planning a remedial action because of
the uniqueness of each case and the lack of data on long-term
performance of the separate unit operations that will make up the
overall remedial system. It is therefore necessary to include
monitoring of effectiveness as an integral part of the remedial
action, for as long as the contaminants remain and may be released to
the environment due to the decreasing effectiveness of the remedial
scheme. The effective life span of a remedial action may be
lengthened by appropriate monitoring and maintenance. Research may
help to reveal why closely similar schemes have different long-term
effectiveness and enable the development of better means for
prediction of long-term effectiveness.

7.2 On-site Treatment

    In principle there are several different procedures which might
be used for on-site treatment. These include extraction, thermal

treatment, chemical treatment, mechanical and physical separation, steam-stripping, microbiological treatment, stabilisation and flotation. In practice only thermal treatment, certain stabilisation methods sludge farming, and a number of methods involving treatment with a liquid phase (extraction, flotation, classification) are sufficiently developed for application but others are under active investigation particu-larly in the Netherlands. Further intensive studies of such processes should be encouraged.

7.3   In-situ Treatment

There are a number of attractive options for the in-situ treatment of contaminated land but with the exception of a few special cases where they have already been used, they are generally unproved. The major technical difficulties are:

(a)   how to ensure intimate contact between treatment agent and
        contaminants,
(b)   physical and chemical heterogeneity of many sites,
(c)   possible unwanted interactions between agents and contaminants,
(d)   how to check treatment has been effective,
(e)   limitations on treatment depths,
(f)   disposal and treatment of waste stream.

The thermal treatment of soil by direct application of electrical energy to cause melting is an option of a different character currently under investigation which have some useful applications.

The development of in-situ treatments, particularly those designed to remove contaminants or render them harmless should be encouraged. The potential benefits would justify the effort required to overcome the technical difficulties.

7.4   Covering Systems

Covering systems may be employed on their own or in conjunction with other remedial measures. The performance requirements of a particular system are largely determined by the nature of contamination, the local geology and hydrogeology, and the projected use of the site. Economics will often dictate the use of locally available materials which have to be used to their best advantage to optimise performance rather than the use of 'ideal' materials. In the absence of practical experience covering systems will frequently have to be based on applying the physical and chemical properties of the component parts. The long-term effectiveness of covering systems is particularly time dependent as the component parts of the system, contamination and environmental stress are subject to change.

Further research is required through laboratory studies, evaluation of current field applications and retrospective studies.

7.5 In Ground Barriers Hydraulic Measures

Vertical barriers to control the movement of groundwater and contamination can be installed using well established engineering procedures. Doubts arise however concerning their effectiveness as a means of controlling contamination owing to possible adverse interaction with contaminants and the possibility of breaches either induced by nature (eg tree roots) or by man (eg subsequent excavations).

Horizontal barriers can be installed by of a number of ground injection/grouting techniques but these are not well established nor proven. They too are susceptible to adverse interactions with contaminants.

The use of hydraulic measures, eg the use of pumping to control groundwater levels, with or without treatment of the extracted water (see 7.6 and Chapter 8) will frequently be an essential adjunct to the use of in-ground barriers and other remedial actions.

7.6 Groundwater Management and Treatment

The quality of contaminated groundwater may be improved by in-situ treatment or by treatment after extraction. A wide range of methods are available for treatment in the latter case but the economic and technical viability is restricted by increasing difficulty in treating ever more dilute contaminants. After treatment the extracted water may be re-injected or disposed of to surface waters. In-situ treatment of groundwater suffers from many of the constraints suffered by in-situ treatment of soil (above) including difficulty of contacting reactive agent and contaminants, establishing how effective treatment has been at the time it is carried out and uncertainty about permanence. Nevertheless it has been successfully used on a number of occasions and such techniques should be further developed.

7.7 Mathematical Modelling

Groundwater modelling, including modelling of pollution move-ment, involves the use of field data with certain established mathematical procedures to predict direction, quantity and rate of flow. Use of a 'site model' can assist in the placement of monitoring and treatment wells, prediction of contaminants movement and prediction of the consequences of installation of hydraulic

control measures. The prediction of contaminant movement can provide
a timescale (eg time for the contaminants to reach to critical
drinking water extraction well) against which remedial measures can
be designed and installed.

7.8 Toxic and Flammable Gases

Toxic and flammable gases may be generated at both remedial
action sites and operating hazardous waste facilities. The primary
concern relates to health and safety both on-site and off-site. Air
emission modelling, monitoring and sample analyses are currently
being investigated. Control systems are being designed and tested
for protection against gas migration and for protection of existing
and planned structures. In order to avoid serious problems that
could occur at operating facilities or remedial action activities,
a management plan for the control of these gases should be an
integral part of any overall operating or remedial action plan.
Research is in progress to define and document toxic and flammable
gas problems.

7.9 Rapid Methods of On-site Analysis

The study concentrated on methods currently in use within the
US Environmental Protection Agency and its agents. A number of
methods varying in complexity were identified. There is a continuing
need for the development of new techniques and refinement of existing
techniques which can provide on-site analyses for the large array of
contaminants that are encountered.

7.10 Steel Industry Sites

A summary of the report of a CCMS Fellowship Study was included
to illustrate the value of such industry audits for predicting the
potential problems that may arise on sites. The steel industry is
considered particularly relevant subject for such an audit because of
the wide range of contamination and engineering problems that can be
encountered. Many of the sites cover a large area and the current
reduction scale of the steel industry in many NATO countries brought
about by improvements in productivity and current economic
difficulties, has resulted in a considerable number of sites being
defunct, particularly in the UK and USA.

One of the most significant sources of contamination associated
with iron and steel making is the coke making processes. These give
rise to a range of organic chemicals (eg coal tars) and gas
purification wastes. The coking plant may be located on the works
(UK) or on the coal field (Germany). Other significant contaminants

are flue dusts (heavy metals), and cold finishing wastes (eg pickle liquors). These are frequently found on sites mixed with the major solid wastes, namely blastfurnace and steelmaking slags. Plans for the development of former iron and steel making sites must take into account not only the extensive chemical contaminanation that can be present but also the engineering problems associated with massive foundations and physically unstable steel making slags. The chemical contamination in the ground can give rise to different types of hazards to a range of 'targets'; investigation and construction workers, site users, building materials and plant life can be adversely affected. A number of major reclamation schemes are in progress in the United Kingdom.

8   RECOMMENDATIONS TO NATO CCMS

      The Committee is invited to:

  (i)   note that the present Pilot Study has shown the value of
         internation collaboration in a fast developing area of
         technical activity,
  (ii)   to commend the report of the Pilot Study Group on Contaminated
         Land to member governments drawing attention to the general
         conclusions listed above and to the conclusions relating to
         individual projects,
  (iii) encourage member governments to consider the adoption of
         policies:

         (a)   that will minimise the occurrence of contaminated land
               problems in the future;
         (b)   to abate the adverse environmental impacts from
               contaminated land;
         (c)   for the safe reuse of contaminated land.
  (iv)   to continue to provide funds under the CCMS Fellowship scheme
         for studies within the subject area.

1   INTRODUCTION

   This Appendix lists some of the more common chemicals found at
uncontrolled hazardous waste sites that have resulted in toxic and
flammable gas problems with some of the relevant chemical properties
such as molecular weight, boiling point, vapour pressure, flash
point, and solubility. Established limits are reported in the form
of Provisional Limits in Air (PLA), the Threshold Limiting Value
(TLV), and the Short Term Exposure Limit (STEL). The concentrations
that have been found at specific uncontrolled hazardous waste sites
are also listed. These are air concentration ranges measured at the
site unless indicated otherwise in the notes.

   The concentration ranges given in the Table that follows are
taken from current literature. These sources do not generally state
the conditions under which the measurements were made. Sampling and
analytical techniques must be included with reported data if the
information is to be useful for health protection and other
purposes.

   The references in the following text and Table refer to
Chapter 10.

2   DEFINITIONS

   The recommended Provisional Limits in Air represent the maximum
constituent concentration considered safe in terms of continuous
exposure in the air outside the physical boundaries of any processing
facilities.[82] These limits are based on TLVs for man, the Maximum
Allowable Concentrations (MAC) for man, the Median Tolerance Limits
(TL_m) for fish, and Lethal Dose (LD_{50}) values for various forms
of animal life, plant reactions following exposure, and exposure
symptoms in man. This limit is equal in value to one-hundreth of
the established TLV.

   The TLV is the concentration of an airborne constituent to which
workers may be exposed repeatedly, day after day without adverse
effect.[47,82] Those reported in this table are based on time
weighted average for an 8-hour work day or 40-hour work week.

   The STEL is the maximal concentration to which workers are
exposed for a period up to 15 minutes continuously without suffering
from (i) irritation, (ii) chronic or irreversible tissue change, or
(iii) narcosis of sufficient degree to increase accident proneness,
impair self rescue, or materially reduce work efficiency, provided

407

that no more than four excursions per day are permitted, with at least 60 minutes between exposure periods, and provided that the daily TLV also is not exceeded.[47] These are tentative values used by the American Conference of Governmental Industrial Hygienists (AGGIH) and are based on one or more of the following criteria: (i) Adopted TLVs, (ii) Pennsylvania Short-Term Limits for Exposure to Airborne Contaminants, (iii) OSHA Occupational Health and Safety Standards, and (iv) NIOSH criteria for recommended standards for occupational exposure to specific substances.[47]

The Threshold Odour Concentration (TOC) given is the concentration at which 50% of an odour panel recognised the odour as being representative of the odourant being studied.[49]

Odour characteristics of each chemical are given to provide an idea of what the chemical may smell like.

The volatility is noted for each chemical. This is based on the boiling point of each compound. In general, the lower the boiling point, the more volatile the chemical.

Health effects are reported for each chemical. These are reported as symptoms, or organs or external areas of the body that may be affected. For some chemicals, concentrations are given at which these effects may result.

3   NOTES ON TABLES

T   = Trace
ND   = Not Detected
NF   = Non-flammable
MW   = Molecular weight
BP   = Boiling Point ($^{\circ}$C)
VP   = Vapour pressure expressed in mm Hg at 20°C
FP   = Flash point ($^{\circ}$F) (Closed Cup = CC: Open Cup = OC;
                                 Tagged Closed Cup = TCC)
Sol   = Solubility expressed in mg/1 at 20°C
PLA   = Provisional limit for air for man in mg/m^3
TLV   = Threshold limit value in mg/m^3
STEL   = Short Term Exposure Limit in mg/m^3
DO   = Distinct Odour in mg/m^3
TOC   = Threshold Odour Concentration (50%) in mg/m^3
a   = represents range over 9 landfills (mean values)
b   = BKK concentration range in landfill gas
c   = Ambient air concentration range at stations around BKK
          site
d   = off site concentrations
e   = represents range over 3 landfill sites
f   = vapour pressure in mm Hg at 25°C
g   = solubility in mg/1 at 25°C
h   = peak concentration; head space analysis results
i   = peak concentration
j   = vapour pressure in atm. at 20°C
k   = vapour pressure in mm Hg at 22°C

References refer to Chapter 10

CHARACTERISTICS AND EFFECTS

Type	MW	BP	VP	FP	Sol	PLA	TLV	STEL	DO	TOC	Odour
Methylene chloride (Dichloromethane, methane dichloride, methylene dichloride, solaesthin)	84.9	40	349	None	20 000	17.4	1740	870	790	90-550	Strongly odorous, sweetish; not pleasant
Carbon tetrachloride (tetrachloromethane, perchloromethane, necatorina, benzin-o-form)	153.82	76.8	90	NF	800	0.65	64	130	175 ppm	130-1260	Strongly odorous; sweet; pungent; ether-like
Chloroform (Trichloromethane)	119.4	61.5	160	NF	80,000	1.2	120	225	-	1000	Sweet
Benzene (Benzol, 78.1 cyclohexatriene)		80.1	76	12(TCC)	1780	0.8	80	75	310	1-180	Strongly odorous sweet; aromatic
Chlorobenzene (Monochlorobenzene, benzene chloride, chlorobenzol, chlorobenzene, phenyl chloride)	112.6	132	8.8	85(CC)	500	3.5	350	-	-	1	Chlorinated moth balls; aromatic; faint; pleasant
Dichlorobenzene (o,m,p) (1,2-dichlorobenzene, dichlorobenzol, dichloricide, paramoth)	147.0	173(m) 174(p) 180(o)	1	151(o)(CC) 153(p)(OC) 163(m)	100	3.0(o) 4.5(p)	300(o) 450(p)	675(p)	100 ppm(o) 15--0(p)	300(o)	Strong, irritating; aromatic; nauseating

EFFECTS etc

OCCURRENCES ON CONTAMINATED SITES

	Volatility	Health Effects	Conc. Ranges	Site	Ref	Other Refs
Methylene chloride (Dichloromethane, methane dichloride, methylene dichloride, solaesthin)	Very	Narcosis; does not cause organic injury; mildly irritating to skin on repeated contact if free to evaporate; painful to eyes, but no permanent damage may be expected	T- 1000 ng/m^3 0.7-11.6 ug/m^3 95.0 ppb 0.03-4.89 ppb	Kin Buc Love Canal Sylvester Midco I	17 26 11 35	8,46,48,49,50 58,82,83
Carbon tetrachloride (tetrachloromethane, perchloromethane, necatorina, benzin-O-form)	Volatile	Acutely and chronically toxic; enters body by inhalation, ingestion and absorption through the skin; single or repeated exposures can cause serious or fatal injury; reports of illness from breathing air containing 25 ppm	111-13687 ng/m^3 5.0 ug/m^3	Kin Buc Love Canal	17 26	8,46,48,49, 50,51,82,83
Chloroform (Trichlorormethane)	Very	Dizziness; dullness, nausea, headache, fatigue, anethesia; liver and kidney damage may result from repeated exposures to low concs, in air; eye injury, skin burns, heart irregularities	T-6389 ng/m^3 0.5-24.0 ug/m^3 5.1 ppm 0.2-1.0 ppb 24 ug/m^3 10.0 ppb	Kin Buc Love Canal BKK[b] BKK[c] Love Canal Sylvester	17 26 18 18 1 11	8,46,48,49, 50,51,82,83, 84
Benzene (Benzol, cyclohexatriene)	Volatile	Irritation of eyes, nose, and respiratory system; headache, nausea, anoremia, fatigue; dermatitis, bone marrow changes, blood abnormalities; carcinogenic; regarded as a cumulative poison due to the slow build up in the body tissues and fluids because of its low solubility in the circulating blood.	522.7 ug/m^3 10-2000 ppm 3.0-4.8 ppb 270 ug/m^3 2066 ppm 19.2 ppb T-29.5 ppm 0.03-3.56 ppb	Love Canal BKK[b] BKK[c] Love Canal Ohio River Pk[h] Chem-dyne Lees Lane Midco I	26 18 18 1 28 11 87 35	8, 58,82,83,84
Chlorobenzene (Monochlororbenzene, benzene chloride, chlorobenzol, chlorobenzene, phenyl chloride)	Immediate	Eye and nasal irritation at about 200 ppm, at which concentration the odour is unpleasant; dangers of chronic exposure; avoid skin contact; mildly narcotic, causes drowsiness and some lack of co-ordination; may cause depression of central nervous system.	T-1127 ng/m^3 0.1-172 ug/m^3 ND-500 ppm 240 ug/m^3	Kin Buc Love Canal BKK[b] Love Canal	17 26 18 1	8,48,51,82,83
Dichlorobenzene (o,m,p) (1,2-dichlorobenzene, dichlorobenzol, dichloricide, paramoth)	Intermediate	Affects liver primarily, kidney secondarily; short exposures to high concentrations may result in depression of the central nervous system (o). Painful to eyes and nose at concentrations of 80-160 ppm; less toxic than O-isomer; skin contacted will burn and become red (p).	T-33783 ng/m^3 0.3-100.5 ug/m^3	Kin Buc Love Canal	17 26	8,46,48,49, 51,82,83

CHARACTERISTICS AND EFFECTS

Type	MW	BP	VP	FP	Sol	PLA	TLV	STEL	DO	TOC	Odour
Ethylbenzene (Phenylethane; ethyl benzol)	106.2	136.2	7	64(TCC)	152	4.35	435	545	-	-	Strongly odorous
1,2-Dichloroethylene (1,2-dichloroethene, acetylene dichloride, sym-dichlorothylene, dioform)	96.9	48 (trans) 61 (cis)	200f (cis)	39(CC)	600 (trans) 800 (cis)	-	790	1000	-	-	Ethereal; slightly acrid
1,1 Dichloroethane (Ethylidene chloride, ethylidene dichloride)	98.9	57.3	180	-	5500	-	400	1010	-	-	Chloroformlike; distinctive; irritating
1,2 -Dichloroethane (EDC, ethylene dichloride, ethene dichloride, Ethylenechloride, Glycoldichloride, Dutch liquid, Brocide)	99	83.5	61	56(CC)	8690	2.0	200	60	750 200ppm	25-450	Sweet; chloroformlike; unpleasant to neutral; aromatic
Trichloroethanes- (1,1,1-trichloroethane, methyl chloroform, chloroethane, (1,1,2-trichlorothane)	133.4	74	100	None	4400	19	1900	2450	3900	2100	Strongly odorous; sweet (1,1,1); chloroformlike
chloroethane) (1,1,2-trichlorothane)	133.4	13.7	19	-	4500	-	-	90	-	-	

EFFECTS etc			OCCURRENCES ON CONTAMINATED SITES			
	Volatility	Health Effects	Conc Ranges	Site	Ref	Other Refs
Ethylbenzene (Phenylethane; ethyl benzol)	Intermediate	Eye irritation (men exposed to 1000 ppm); absorption is chiefly by inhalation; tolerance; tolerance develops to irritation; higher concentrations may cause dizziness; acute skin contact causes burning sensation; chronic effects – repeated exposure to concentrations greater than 200 ppm produces fatigue, nervousness, sleeplessness, loss of weight & appetite; symptoms disappear when worker removed from exposure; chronic skin contact may defat and irritate skin	8.6-16.6 ppm 0.03-0.43 ppb	Lees Lane Midco I	87 35	8,48,49,51, 58,82,83
1,2-Dichloroethylene (1,2-dichloroethane, acetylene dichloride, sym-dichlorothylene, dioform)	Very	Eye irritant; toxic to breathing, swallowing, skin contact; repeated & prolonged exposure is hazardous; but does not have a high toxicity in low concentrations	T-5263 ng/m^3 334 ug/m^3	Kin Buc Love Canal	17 26	46,48,49,51
1,1 Dichloroethane (ethyli dene chloride, ethylidene dichloride)	Very	One of the most toxic of the commonly used chlorinated hydrocarbons; chronic toxicity somewhat less than carbon tetrachloride; affects liver	364-470 ng/m^3	Kin Buc	17	46,48,49,51
1,2-Dichloroethane (EDC, ethylene dichloride, ethane dichloride, ethylenechloride, glycoldichloride, Dutch liquid, Brocide	Volatile	Significant acute & chronic effects; can become adapted to odour at low concentrations; therefore, odour not reliable warning; toxic by inhalation, contact with skin or mucous membranes; vapour in air irritates eyes, nose, throat; may cause nausea, vomiting, drowsiness	T-2173 ng/m^3 ND-5000 ppm 0.4-3.0 ppb	Kin Buc BKK[b] BKK[c]	17 18 18	46,48,49,51 82,83
Trichloroethanes- (1,1,1-trichloroethane, methyl chloroform, chloroethene) (1,1,2-trichloroethane)	Volatile	Eye and mucous membranes irritation on contact as liquid or vapour; vapour may cause anesthetic effects headache at 500 ppm, 180 min; exposure to 2-3% volume for more than a few minutes will cause complete inco-ordination, unconsciousness, and possible death (1,1,1). Main effect of acute exposure is depression of central nervous system (response proportional to degree of exposure); no effects when concentration in air is less than 350 ppm by volume; mild eye irritation, temporary impairment of co-ordination, though minimal, is quickly indicated at 900 to 1,000 ppm by volume in air	73 ug/m^3 T (1,1,1) 294-357 ng/m^3 (1,1,2)	Love Canal Kin Buc Kin Buc	1 17 17	8,48,49,51, 58,82,83

CHARACTERISTICS AND EFFECTS

Type	MW	BP	VP	FP	Sol	PLA	TLV	STEL	DO	TOC	Odour
Tetrachloroethane (1,1,2,2-tetrachloroethane, Acetylene tetrachloride, cellon, Bonoform, sym-Tetrachloroethane)	167.9	146.2	5	NF	2900	0.35	35	70-skin	5 ppm	3 ppm	Sweet; pleasant, chloroformlike
Trichloroethylene (TCE, Trichloroethane, Ethylene Trichloride, Ethinyl Trichloride, Triclene, Trielene, Trilene, Trichloran, Trichloran, Algylen, Trimar, Trilline, Tri, Threthylen, Trethylene, Westrosol, Chlorylen, Gemalgene, Germalgene)	131.4	86.7	60	NF	1.1 g	5.35	520	800	580	110	Strongly odorous; soft; solventy; ethereal; chloroformlike
Tetrachloroethylene (Perc, Tetrochloroethene, Nema, Ethylene Tetrachloride, Tetracap, Tetropil, Perclene, Ankilostein, Didakene)	165.8	121.2	14	NF	150 g	6.7	670	1000-skin	480	50 ppm	Etheric; Chloroformlike
Vinyl Chloride (Chloroethene, Chloroethylene)	62.5	-13.9	2660f	-108(OC)	1.1 g	7.7	3	-	-	25000	(Slight odour at 4,000 ppm) Mild; sweetish; faintly pleasant at high concentrations
Toluene (Methylbenzene, Toluol, Phenylmethane, Methacide)	92.1	110.8	22	40(TCC)	515	3.75	750	560-skin	260	1-140	Strongly odourous; burnt; unpleasant to neutral

EFFECTS etc

OCCURRENCES ON CONTAMINATED SITES

	Volatility	Health Effects	Conc Ranges	Site	Ref	Other Refs
Tetrachloro-ethane (1,1,2,2-tetrachlo-ethane, Acetylene tetrachloride, cellon, Bonoform, sym-Tetrachloro-ethane)	Intermediate	Highly toxic by inhalation, repeated skin and mucous membrane contact; severe irritation to eyes on contact with liquid or vapor; burning & swelling of eyes occur	$1140 \ ug/m^3$ $32-54 \ ng/m^3$	Love Canal Kin Buc	1 17	48,49,51 82,83
Trichloroethylene (TCE, Trichloro-ethane, Ethylene Trichloride, Ethinyl Trichloride, Triclene, Trichloran, Trilene, Trichloran, Trichloran, Algylen, Trimar, Trilline, Tri, Threthylen, Threthylene, Westrosol, Chlorylen, Gemalgene, Germalgene)	Volatile	Depresses the central nervous system, parti-cularly from acute exposures; causes visual disturbance and mental confusion accompanied by inco-ordination; harmful by inhalation, by repeated and prolonged contact with skin and mucous membranes or when swallowed	$T-10052 \ ng/m^3$ $73 \ ug/m^3$ ND-1000 ppb 0.2-1.8 ppb 1.8 ppb 1.54 ppb 0.05-0.58 ppb	Kin Buc Love Canal BKK[b] BKK[c] Chem-Dyne[i] Sylvester[d] Midco I	17 26 18 18 11 11 35	8,48,49,51, 58,82,83
Tetrachloro-ethylene (Perc, Tetrochloro-ethene, Nema, Ethylene Tetrachloride, Tetracap, Tetropil, Perclene, Ankilostein, Didakene)	Volatile	Depresses the central nervous system at high concentrations; toxic by inhalation, prolonged or repeated contact with skin or mucous membranes; affects lungs when volume in air exceed maximum allowable concentration; eyes tear, nose and throat irritated, full-ness of head; mental con-fusion evidenced	$T-2896 \ ng/m^3$ 0.2-52 ug/m^3 ND-1500 ppm 1.4-3.7 ppb 0.62 ppb 0.04-0.24 ppb	Kin Buc Love Canal BKK[b] BKK[c] Chem-dyne[i] Midco I	17 26 18 18 11 35	17,46,48,49 51,82,83
Vinyl Chloride (Chloroethene, Chloroethylene)	Highly volatile	Dizziness; disorienta-tion at 25,000 ppm for 3 minutes; narcotic effects appear with exposure to concentra-tions greater than 1000 ppm; - drowsiness, blurring of vision, inability to concentrate, numbness of feet and/or hands, staggers when trying to walk	$15,000- 48,750 \ ng/m^3$ 83-12,800 ppm 2-7.3 ppb 17.9-122.6 ppm 0.07-1.10 ppm	Kin Buc BKK[b] BKK[c] Lees Lane	17 18 18 87 88	8,46,48,50, 51,82,83
Toluene (Methyl-benzene, Toluol, Phenylmethane, Methacide)	Volatile	Powerful narcotic; produces mild fatigue, headache, giddiness, weakness, confusion, and parethesia of the skin (at 200 ppm for 8 hrs); dangerous to inhale; irritating to skin & eyes. Acute systematic effects for 8 hrs exposure[169]: 50-100 ppm - no serious effects; slight drowsiness and headache in new worker. 200 ppm - new worker fatigued, muscular weakness, skin sensations of burning, nausea; fatigue and restless sleep might persist. 300-400 ppm produces varying degrees of fatigue, headache, mental confusion, muscular weakness, skin sensations of burning, nausea; fatigue & restless sleep might persist. 600 ppm - mental confusion, lack of co-ordination, staggering; even after 3 hrs at this concentration - marked fatigue, mental con-fusion, exhilaration, headache, dizziness; fatigue and weakness may last several hours with nausea, nervousness, headache. 800 ppm - nausea, pronounced confusion, marked lack of co-ordination, staggering; may last several days with marked sleeppessness	$0.1-6.2 \ mg/m^3$ $570 \ ug/m^3$ 85 ppm T-23.6 ppm	Love Canal Love Canal Ohio River Pk[h] Lees Lane	26 1 28 87	8,46,48, 49,51,58, 82,83

CHARACTERISTICS AND EFFECTS

Type	MW	BP	VP	FP	Sol	PLA	TLV	STEL	DO	TOC	Odour
Chlorotoluene (Benzylchloride, alphy-chloro-toluene, Chloro-toluol)	126.6	179	1	153	-	0.05	5	-	-	0.25	Lacromator; aromatic; pungent; irritating
Vinylidene Chloride (1,1-Dichloro-ethane, 1,1-Dichloro-ethylene)	96.9	31.6	500	14(OC)	-	-	99	80	1000ppm	500ppm	Sweet; chloro-formlike
Xylenes (o-xylene: 1,2-dimethylbenzene m-xylene: 1,3-dimethyl-benzene p-xylene: 1,4-dimethylbenzene, xylol dimethyl-benzene)	106.2	139.1(m) 138.3(p) 144.4(o)	6 (m) 6.5(p) 5 (o)	77(m&p)(TCC) 63(o)(TCC)		198 (p) 175 (o)		4.35	435 650- skin	170(o)	Strongly odourous; sweet; aromatic
PCBs (polychlori-nated biphenyls)	-	-		312	Insoluble in water	-	-	-	2-skin	-	-

EFFECTS etc			OCCURRENCES ON CONTAMINATED SITES			
	Volatility	Health Effects	Conc Ranges	Site	Ref	Other Refs
Chlorotoluene (Benzylchloride, alphy-chloro-toluene, Chloro-toluol)	Intermediate	Moderate health threat; eye and skin irritant; can cause serious eye burns; vapors can injure respiratory tract, but worker unlikely to stay in area long enough for injury due to irritating odor	$0.008-7650$ ug/m^3 $6,700$ g/m^3	Love Canal Love Canal	26 1	48,51,83
Vinylidene Chloride (1,1-Dichloro-ethane, 1,1-Dichloro-ethylene)	Very volatile	Irritating to skin and eyes & may cause burns; over exposure to vapors may cause anesthesia; may be injurious to liver and kidneys; severe lung irritation may produce conjestion	$454-555$ ng/m^3 ND-1200 ppm $0.3-1.3$ ppb	Kin Buc BKK[b] BKK[c]	17 18 18	8,46,48,49, 51,82
Xylenes (o-xylene: 1,2-dimethylbenzene m-xylene: 1,3-dimethyl-benzene, p-xylene: 1,4-dimethylbenzene, xylol dimethyl-benzene)	Intermediate	Absorption by breathing vapors primarily; nose, throat, eyes, skin irritant; acts as an anesthetic in high con-centrations; acute toxicity (inhalation): mild exposure may cause dizziness, weakness, feelings of well being, headache, nausea, vomiting, sensation of tightness in the chest, lack of co-ordination. More severe exposure may cause blurring and vision, tremors, shallow and rapid breathing, irregular heart rate, paralysis, unconsciousness, convulsions	91 ppm ND-10.7 ppm 140 ug/m^3 (m&p) 73 ug/m^3 (o)	Ohio River Pk[h] Lees Lane Love Canal Love Canal	28 87 1 1	8,46,48,49, 51,82,83
PCBs		Irritation of eyes, nose, throat, chloracne, liver damage, still birth	$0.05-3$ ug/m^3 (W) $246-300$ ug/m^3 (S) 0.0075 ug/m^3 $0.41-1.5$ ug/m^3 (S) 0.021 ug/m^3 (W)	Caputo Caputo Lehigh Elect New Bedford New Bedford	1 1,3 11 2 2	46,84,87

* (W) Winter
  (S) Summer

Reactive Group	With	Results*
(1) Acids, mineral, non-oxidising	Cyanides	GT, GF
	dithiocarbamates	GF, H, F
	fluorides, inorganic	GT
	halogenated organics	GT
	mercaptans and other organic sulphides	GT, GF
	metals, alkali and alkaline earth, elemental	GF, H, F
	metals, other elemental and alloys as powders, vapours, or sponges	GF, H, F
	metals, other elemental and alloys as sheets, rods, drops, moldings	GF, H, F
	Nitrides	GF, H, F
	Nitriles	GT, GF, H
	Organophosphates phosphothiotes, phosphodithiotes	H, GT
	Sulphides	GT, GF
	oxidising agents, strong	GT, H
	reducing agents, strong	GF, H
	Explosives	H, E
(2) Acids, mineral, oxidising	Amides	H, GT
	Amines, aliphatic and aromatic	H, GT
	Azo compounds, diazo compounds and hydrazines	H, GT
	carbamates	GT
	cyanides	GT, GF
	diothiocarbamates	H, GF, F
	fluorides, inorganic	GT
	halogenated organics	H, F, GT
	isocyanates	H, F, GT
	mercaptans and other organic sulphides	H, F, GT
	metals, alkali and alkaline earth, elemental	GF, H, F
	metals, other elemental and alloys as powders, vapours or sponges	GF, H, F

* See end of table for key

(2)	Acids, mineral, oxidising (Contd)	metals, other elemental and alloys as sheets, rods, drops, moldings	GF, H, F
		nitrides	F, E, H
		nitriles	H, F, GT
		nitro compounds, organic	H, F, GT
		hydrocarbons, aliphatic, unsaturated	H, F
		hydrocarbons, aliphatic, saturated	H, F
		peroxides and hydro- peroxides, organic	H, E
		phenols and cresols	H, F
		organophosphates, phos- phothioates, phosphodi thioates	H, GT
		sulphides, inorganic	H, F, GT
		combustible and flammable materials, misc	H, F, GT
		explosives	H, E
		reducing agents	H, F, GT
(3)	Acids, organic	cyanides	GT, GF
		dithiocarbamates	H, GT, GF
		fluorides, inorganic	GT
		metals, alkali and alkaline earth, elemental	GF, H, F
		metals, other elemental and alloys as powders, vapours, or sponges	GF
		nitrides	H, GF
		sulphides, inorganic	GT
		explosives	H, E
		oxidising agents, strong	H, GT
		reducing agents, strong	H, GF
(4)	Alcohol and glycols	metals, alkali and alkaline earth, elemental	GF, H, F
		nitrides	GF, H, E
		peroxides and hydro- peroxides, organic	H, F
		oxidising agents, strong	H, F
		reducing agents, strong	H, GF, F
(5)	Aldehydes	diothiocarbamates	GF, GT
		metals, alkali and alkaline earth, elemental	GF, H, F
		nitrides	GF, H
		oxidising agents, strong	H, F
		reducing agents, strong	GF, H, F

(6) Amides	metals, alkali and alkaline earth, elemental	GF, H
	oxidising agents, strong	H, F, GT
	reducing agents, strong	H, GF
(7) Amines, aliphatic and aromatic	halogenated organics	H, GT
	metals, alkali and alkaline earth, elemental	GF, H
	peroxides and hydro- peroxides, organic	H, GT
	oxidising agents, strong	H, F, GT
	reducing agents, strong	H, GF

(8) Azo compounds,
     diazo compounds
     and hydrazines
	metals, alkali and
	  alkaline earth, elemental
		GF, H

	metals, other elemental
	  and alloys as powders,
	  vapours, or sponges
		H, F, GT

	metals, other elemental
	  and alloys as sheets,
	  rods, drops, moldings,
	  etc
		H, F, G

	peroxides and hydro- peroxides, organic	H, F, E
	explosives	H, E
	oxidising agents, strong	H, E
(9) Carbamates	metals, alkali and alkaline earth, elemental	GF, H
	peroxides and hydro- peroxides, organic	H, F, GT
	oxidising agents, strong	H, F, GT
(10) Caustics	halogenated organics	H, GF
	metals, alkali and alkaline earth, elemental	GF, H

	metals, other elemental
	  and alloys as powders,
	  vapours, or sponges
		GF, H

	nitro compounds, organic	H, E

(10) Caustics (Contd)	organophosphates, phos-
	  phothioates, phosphodi-
	  thioates
		H, E

	explosives	H, E
(11) Cyanides	metals, alkali and alkaline earth, elemental	GF, H
	nitrides	GF, H
	peroxides and hydro-	

		peroxides organic	GT, H, E
		oxidising agents	GT, H, E
(12)	Dithiocarbamates	metals, alkali and alkaline earth, elemental	GT, GF, H
		nitrides	GF, H
		peroxides and hydro- peroxides, organic	H, F, GT
		oxidising agents, strong	H, GT, F
		reducing agents, strong	H, GT
(13)	Esters	metals, alkali and alkaline earth, elemental	GF, H
		nitrides	GF, H
		explosives	H, E
		oxidising agents, strong	H, F
		reducing agents, strong	H, F
(14)	Ethers	oxidising agents, strong	H, F
(15)	Hydrocarbons, aromatic	oxidising agents, strong	H, F

(16)	Halogenated
	  organics
		metals, alkali and
		  alkaline earth, elemental
			H, E

		metals, other elemental and
		  alloys as powders, vapors,
		  or sponges
			H, E

		metals, other elemental
		  and alloys as sheets,
		  rods, drops, moldings
			H, F

		nitrides	GF, H
		peroxides and hydro- peroxides, organic	H, E
		oxidising agents, strong	H, GT
		reducing agents, strong	H, E
(17)	Isocyanates	metals, alkali and alkaline earth, elemental	GF, H

		metals, other elemental
		  and alloys as powders,
		  vapors, or sponges
			GF, H

		oxidising agents, strong	H, F, GT
		reducing agents, strong	H, GF
(18)	Ketones	metals, alkali and alkaline earth, elemental	GF, H
		nitrides	GF, H
		peroxides and hydro- peroxides, organic	E

	oxidising agents, strong	H, F
	reducing agents, strong	GF, H
(19) Mercaptans and other organic sulfides	metals, alkali and alkaline earth, elemental	GF, H
	metals, other elemental and alloy as powders, vapors, or sponges	GF, H, F
	nitrides	GF, H
	peroxides and hydro- peroxides, organic	H, F, GT
	oxidising agents, strong	H, F, GT
	reducing agents, strong	GF, H
(20) Metals, alkali and alkaline earth, elemental	nitrides	E
	nitro compounds, org.	H, GF, E
	peroxides and hydro- peroxides, organic	H, E
	phenols and creosols	GF, H
	combustible and flammable materials, Misc.	H, G, F
	explosives	H, E
	oxidising agents, strong	H, F, E
	water and mixture containing water	H, GF
(21) Metals, other elemental and alloys as powders, vapors, or sponges	hydrocarbons, aliphatic, unsaturated	H, E
	explosives	H, E
	oxidising agents, strong	H, F, E
	water and mixtures containing water	GF, H
(22) Metals, other elemental and alloys as sheets, rods, drops, moldings, etc	explosives	H, E
	oxidising agents, strong	H, F
(23) Metals and metal compounds, toxic	explosives	E
(24) Nitrides	nitriles	GF, H
	nitro compounds, org.	H, GF, E
	peroxides and hydro- peroxides, organic	H, GF, E
	phenols, cresols	GF, H
	combustible and flammable materials, misc.	H, GF, F
	explosives	E

		oxidising agents, strong	H, F, E
		water and mixtures	
		containing water	GF, H
(25)	Nitriles	peroxides and hydro-	
		peroxides, organic	H, GT
		oxidising agents, strong	H, F, GT
		reducing agents, strong	H, GF
(26)	Nitro compounds,	oxidising agents, strong	H, E
	organic	reducing agents, strong	H, E
(27)	Hydrocarbons,	oxidising agents, strong	H, F
	aliphatic,		
	unsaturated		
(28)	Hydrocarbons,	oxidising agents, strong	H, F
	aliphatic,		
	saturated		
(29)	Peroxides and	sulphides, inorganic	H, GT
	hydroperoxides,	combustible and flammable	
	organic	materials, misc	H, F, GT
		explosives	H, E
		reducing agents, strong	H, E
(30)	Phenols and	explosives	H, E
	cresols	oxidising agents, strong	H, F
		reducing agents, strong	GF, H
(31)	Organophosphates,	oxidising agents, strong	H, F, GT
	phosphothioates,	reducing agents, strong	GT, GF, H
	phosphodithioates		
(32)	Sulphides,	explosives	H, E
	inorganic	oxidising agents, strong	H, F, GT
		water and mixtures	
		containing water	GT, GF
(33)	Epoxides	explosives	H, E
		oxidising agents, strong	H, F, G
(34)	Combustible and	explosives	H, E
	flammable	oxidising agents, strong	H, F, G
	materials,	reducing agents, strong	GF, H
	miscellaneous		
(35)	Explosives	polymerisation compounds	H, E
		oxidising agents, strong	H, E
		reducing agents, strong	H, E

(36)	Polymerisation compounds	oxidising agents, strong	H, F, GT
		reducing agents, strong	H, GF
(37)	Oxidising agents, strong	reducing agents, strong	H, F, E
(38)	Reducing agents, strong	water and mixtures containing water	GT, GF

---

* From H K Hatayama et al, 'A method for determining the compatibility of hazardous waste', USEPA (1980). This report is no longer available from EPA or NTIS. An updated version entitled 'A proposed guide for estimating the incompatibility of selected hazardous wastes based on binary chemical reactions,' is to be published by the American Society for Testing and Materials (ASTM) D34 Committee on Waste Disposal.

Key:

F	Fire
GT	Toxic gas generation
GF	Flammable gas generation
H	Heat generation
E	Explosion
G	Innocuous and non-flammable gas generation

427